Hillslope Processes

Edited by
A.D. Abrahams

Department of Geography, State University of New York at Buffalo

Boston
ALLEN & UNWIN
London   Sydney

Allen & Unwin, Inc.,
8 Winchester Place, Winchester, Mass. 01890, USA

Allen & Unwin (Publishers) Ltd,
40 Museum Street, London WC1A 1LU, UK

Allen & Unwin (Publishers) Ltd,
Park Lane, Hemel Hempstead, Herts HP2 4TE, UK

Allen & Unwin (Australia) Ltd,
8 Napier Street, North Sydney, NSW 2060, Australia

First published in 1986

Library of Congress Cataloging-in-Publication Data
Main entry under title:

Hillslope processes.
    Papers from the 16th Annual Geomorphology Symposium,
held at the State University of New York at Buffalo,
Amherst Campus, Sept. 28–29, 1985.
    Includes bibliographies and index.
    1. Slopes (Physical geography)—Congresses.
2. Slopes (Physical geography)—Rocky Mountains—
Congresses.   3. Slopes (Physical geography)—West (U.S.)
—Congresses.   I. Abrahams, Athol D., 1946–
II. Geomorphology Symposium (16th: 1985: University
of New York at Buffalo)
GB448.H55   1986   551.4'36   85–22973
ISBN 0–04–551102–0 (alk. paper)

British Library Cataloguing in Publication Data

Abrahams, Athol D.
    Hillslope processes.—(The 'Binghampton symposia
in Geomorphology, ISSN 0261–3174; v. 16)
1. Slopes (Physical geography)   2. Weathering
I. Title   II. Series
551.4'36   GB448
ISBN 0–04–551102–0

Set in 10 on 12 point Times by Mathematical Composition Setters Ltd.,
Salisbury, UK
and printed in Great Britain by Mackays of Chatham

To my mother,
for her sacrifices and encouragement

Preface

Hillslopes occupy most of the land surface of the Earth. In areas of erosional topography the entire landscape, except the valley floors, consists of hillslopes. Consequently, hillslopes are a major focus of research in geomorphology. The study of hillslopes is essential not only in order to understand better the natural landscape but also for numerous practical reasons, such as controlling soil erosion and sedimentation on agricultural lands and mitigating the hazard posed by landslides. Geomorphologists have studied hillslopes since the latter part of the 19th century, but until the late 1950s their research focused almost exclusively on morphology. During the past 30 years, however, there has been a growing interest in hillslope processes, and today research on processes dominates this branch of geomorphology.

Hillslope processes are also studied in a variety of disciplines other than geomorphology. Moreover, studies within geomorphology owe a considerable debt to and are often closely linked with these other disciplines, which include hydrology, pedology, agricultural engineering, civil engineering, and engineering geology. Thus the study of hillslope processes is truly an interdisciplinary science. At the heart of this science is geomorphology: insofar as it is concerned with both hydraulic and gravitational processes, it alone seeks to integrate the more specialized knowledge generated by the other disciplines.

Because the scientists conducting research on hillslope processes work either within or on the margins of several disciplines, they bring to the subject different sets of skills, experiences, and approaches. For these very reasons it is highly desirable for them to interact and exchange information and ideas. However, precisely because they are associated with disparate disciplines, they have few opportunities to meet and discuss their research. The 16th Annual Geomorphology Symposium on 'Hillslope Processes' was therefore convened with the objective of bringing together representatives of the various disciplines concerned with hillslope processes for the first international meeting on the subject. The goals of the symposium were to provide a forum for the presentation of current research on hillslope processes, to facilitate the exchange of ideas and the dissemination of new developments across disciplinary boundaries, and to generate new initiatives for future research and cooperation.

The proceedings of the symposium are contained in this volume. Following the opening chapter of general interest by Young and Saunders on the comparative rates of denudation by the various slope processes, there are nine papers on hydraulic processes and eight on gravitational processes.

The first chapter on hydraulic processes is by Dunne and Aubry and demonstrates by means of model building and field experiments that rill development depends on a balance between sediment transport by sheetwash and rainsplash. Sheetwash on sufficiently steep slopes incises rills, whereas rainsplash diffuses sediment from protuberances and fills rills, thereby smoothing the surface. The two chapters by Meyer and by Morgan and his colleagues are primarily concerned with the effects of soil aggregates and vegetation on runoff and soil erosion and reflect the current research emphasis in agricultural engineering. The next three chapters all deal with hillslope processes in alpine environments. Gardner's study reveals the important role of intermittent fluvial activity in bedrock gullies as an erosional agent on steep mountain slopes in the Canadian Rocky Mountains. In the Colorado Rocky Mountains, on the other hand, Caine reports that although there is considerable downslope movement of sediment, especially by mass transfers within the cliff–talus system, virtually none of this sediment reaches the streams; most of the silts and clays transported by the streams are eolian in origin. Dixon's research was carried out in the same area as Caine's, but whereas Caine was concerned with sediment, Dixon focused on the downslope movement of solutes. There then follow two chapters that utilize mathematical and computer techniques to develop hillslope hydrology models. Anderson and Howes' model is based on spatial variations in infiltration, whereas Beven's model takes into account hillslope form and spatial variability in soil characteristics. The final chapter dealing with hydraulic processes is by Kirkby and describes a computer simulation model for the evolution of a valley network on a surface subjected to wash and creep/splash. The conceptual convergence of this chapter and that by Dunne and Aubry is striking, specifically with regard to the mechanism controlling the process of incision by sheetwash.

In the first chapter on gravitational processes, Moon uses techniques developed in rock mechanics to show that the gradients of many bare rock slopes in the fold mountains of South Africa are determined by the rock mass strength, which is largely a function of intact rock strength and the spacing of partings in the rock. Statham and Francis review the processes that control the formation of scree slopes, paying particular attention to the factors influencing the input of debris from the headwall, scree processes, the morphology and sedimentology of scree slopes, and the effect of weathering on the subsequent evolution of such slopes. Of late there has been a surge of interest in debris flows, and the nature of this research can been seen in Pierson's innovative study of the depth, composition, velocity, and flow characteristics of ten channelized debris flows on Mount St. Helens. The application of soil mechanics to the analysis of landslides is exemplified by the next four chapters. In a chapter of broad significance, Iverson develops a new mathematical theory of unsteady, nonuniform landslide motion and applies this theory to Minor Creek landslide, a large, slow-moving landslide complex in northwestern California. The theory employs a constitutive model that can represent landslide deformation styles ranging from dilatant viscoplastic flow to rigid–plastic frictional slip. Bovis also examines the mechanics of large, slow-moving landslides in southwestern British Columbia, and reports that they show only slight internal deformation and move largely as a result of displacement along boundary shear zones. Chandler proposes a new classification of processes

causing landslides and goes on to discuss these processes and the problems of
assessing field shear strength in the laboratory. Dietrich, Wilson, and Reneau
describe the geometry of colluvium-filled bedrock hollows, develop a model
for the filling of these hollows with colluvium, and analyze the processes that
cause them to empty by landsliding. In the last chapter on gravitational
processes, Nilsen summarizes the types of investigations conducted on landslides as part of the well-known San Francisco Bay Region Environment and
Resource Study, and the use of these investigations in the preparation of slope-
stability maps for land-use planning in the Bay region.

During the past 30 years the study of hillslope processes has become increasingly technical and complex. Allegorically, this field of study has grown from
childhood to adolescence. Although it has come a long way in three decades,
it is still far from maturity, and much basic work remains to be accomplished.
For instance, the basic hydraulics and mechanics of most hillslope processes
remain poorly understood, and there is an urgent need for field experiments
on the scale of the entire hillslope rather than the small plot. Finally, obvious
problems exist with extrapolating the results of process studies over space and
through time in order better to understand hillslope morphology. Clearly,
many challenges lie ahead, and it is hoped that this volume will be of assistance
in meeting those challenges.

Athol D. Abrahams
State University of New York at Buffalo
April, 1985

Acknowledgments

The high quality of the chapters in this volume is due at least in part to the diligent efforts of the following individuals who gave freely of their time to review the manuscripts: M. G. Anderson, Lawrence E. Band, Michael J. Bovis, Rorke B. Bryan, T. Nelson Caine, M. A. Carson, R. J. Chandler, John E. Costa, Donald R. Coates, George R. Foster, R. Allan Freeze, James S. Gardner, Alan D. Howard, Richard M. Iverson, Arvid M. Johnson, Peter G. Johnson, R. H. King, Laurence A. Lewis, Brian H. Luckman, David M. Mark, Robert H. Meade, Jr., Daniel R. Muhs, Tor H. Nilsen, A. J. Parsons, M. J. Selby, Terrence J. Toy, and Ming-ko Woo. To these people I express my sincere appreciation for their help and generosity.

The 16th Annual Geomorphology Symposium, the proceedings of which are contained in this volume, was supported by grants from the National Science Foundation (SES-8508459), the SUNY-wide program Conversations in the Disciplines, and the SUNY-Buffalo program Conferences in the Disciplines. These grants permitted participants to travel from three continents to the symposium and contributed greatly to its ultimate success.

The publishers and I thank John Wiley & Sons Ltd. for permission to reproduce Figure 4.1 and the Editor, *South African Geographical Journal*, for permission to reproduce Figures 11.1, 11.3, 11.4, 11.7, and 11.8.

Contents

List of tables

Contributors

M. G. Anderson
Department of Geography, University of Bristol, Bristol BS8 1SS, England

Brian F. Aubry
Department of Geological Sciences, University of Washington, Seattle, Washington 98195, U.S.A.

Keith Beven
Institute of Hydrology, Wallingford OX10 8BB, England

Michael J. Bovis
Department of Geography, University of British Columbia, Vancouver, British Columbia V6T 1W5, Canada

T. Nelson Caine
Department of Geography, University of Colorado, Boulder, Colorado 80309, U.S.A.

R. J. Chandler
Department of Civil Engineering, Imperial College of Science and Technology, London SW7 2BU, England

William E. Dietrich
Department of Geology and Geophysics, University of California, Berkeley, California 94720, U.S.A.

John C. Dixon
Department of Geography, University of Arkansas, Fayetteville, Arkansas 72701, U.S.A.

Thomas Dunne
Department of Geological Sciences, University of Washington, Seattle, Washington 98195, U.S.A.

S. C. Francis
Department of Geology, Chelsea College, University of London, 552 Kings Road, London SW10 0UA, England

H. J. Finney
Department of Agricultural Engineering, Silsoe College, Silsoe, Bedford MK45 4DT, England

James S. Gardner
Department of Geography, University of Waterloo, Waterloo, Ontario N2L 3G1, Canada

S. Howes
Department of Geography, University of Bristol, Bristol BS8 1SS, England

Richard M. Iverson
U.S. Geological Survey, Cascades Volcano Observatory, 5400 MacArthur Boulevard, Vancouver, Washington 98661, U.S.A.

M. J. Kirkby
School of Geography, University of Leeds, Leeds LS2 9JT, England

H. Lavee
Department of Geography, Bar-Ilan University, 52 1000 Ramat Gan, Israel

L. D. Meyer
USDA Sedimentation Laboratory, P.O. Box 1157, Oxford, Mississippi 38655, U.S.A.

Elaine Merritt
Department of Agricultural Engineering, Silsoe College, Silsoe, Bedford MK45 4DT, England

B. P. Moon
Department of Geography, University of the Witwatersrand, Johannesburg 2001, South Africa

R. P. C. Morgan
Department of Agricultural Engineering, Silsoe College, Silsoe, Bedford MK45 4DT, England

Tor H. Nilsen
RPI Pacific, Inc., 507 Seaport Court, Suite 101, Redwood City, California 94063, U.S.A. (formerly with U.S. Geological Survey, 345 Middlefield Road, Menlo Park, California 94025, U.S.A.

Christine A. Noble
Department of Agricultural Engineering, Silsoe College, Silsoe, Bedford MK45 4DT, England

Thomas C. Pierson
U.S. Geological Survey, Cascades Volcano Observatory, 5400 MacArthur Boulevard, Vancouver, Washington 98661, U.S.A.

Steven L. Reneau
Department of Geology and Geophysics, University of California, Berkeley, California 94720, U.S.A.

Ian Saunders
Department of Geography, Simon Fraser University, Burnaby, British Columbia V5A 1S6, Canada

I. Statham
Ove Arup & Partners, Cambrian Buildings, Mountstuart Square, Cardiff CF1 6QP, England

Cathy J. Wilson
Department of Geology and Geophysics, University of California, Berkeley, California 94720, U.S.A.

Anthony Young
International Council for Research in Agroforestry, P.O. Box 30677, Nairobi, Kenya

Part I:

General

1
Rates of surface processes and denudation

Anthony Young and Ian Saunders

Abstract

After an initial introduction on the development of studies of rates of slope processes, the present state of knowledge is summarized for rates of soil creep, solifluction, surface wash, solution, landslides, cliff and slope retreat, and total denudation. Other slope processes are noted, with special reference to the activities of animals. Representative rates for denudation as a whole are 50 B (predominant range 10–100 B) for gentle relief and 500 B (predominant range 100–1000 B) for steep relief. Denudation rates reach a maximum in the semiarid and tropical subhumid (savanna) climatic zones, and in montane climates with steep slopes. Problems in the interpretation of the results of process rate studies include sediment transfer paths, the spatial and temporal sampling frames, and relict landforms. Questions of the control of slope retreat, equilibrium and the steady state, time, space, and causality are discussed in the light of process rates. The influence of Man can accelerate denudation by 2, 10, or over 20 times, depending on the type of land use. Recommendations for future process studies are to employ a spatial sampling design; short-term instrumentation and measurement could usefully be supplemented by other independent means of checking the validity of the data, such as estimates of rates of landscape change over geologic time.

Introduction

The aim of this chapter is to provoke discussion on the problems that arise in the interpretation of data on rates of slope processes and slope retreat—that is, change in form as well as processes. Also, on the grounds that most landscapes are made up largely of slopes, we cover rates of total ground lowering, here referred to for convenience as denudation. Particular attention is given to those problems encountered when attempting to adopt a straightforward measure-and-extrapolate procedure.

The chapter begins with some reflections on the historical development of process measurements. This is followed by a summary of the major results achieved to date, based upon a previous review (Saunders and Young, 1983). The remainder of the chapter is directed towards a wider-ranging discussion of the interpretation of the results than was previously attempted, including placing them within the context of some classical questions in the geomorphology of slopes.

An historical perspective

Recognition of the fact that landscapes are not immutable dates back to Hutton and Playfair, although 19th-century geologists had to work hard to combat the alternative hypotheses of Creation or Noachian catastrophism. Lyell (1841, p. 161) had seen that old marine cliffs, no longer undercut, were "reduced to gentle slope." Scrope clearly set out how surface wash operates: "The direct fall of rain removes particles from the ground surface and carries them away to the lowest accessible levels. The general surface is more or less lowered" (Scrope, 1866, p. 193). Landscape evolution on a massive scale was apparent to the great American geologists who described the West in the closing decades of the 19th-century.

Until the 1950s, however, little was known about the rates of geomorphic processes. A few isolated individuals possessed of natural curiosity had spotted "something that moved" and measured it, mostly in montane periglacial environments in Europe (Schmid, 1925; Morawetz, 1932, Krumme, 1935); but, for the most part, geomorphological discussion of processes was based upon reasoning derived from form: Davis (1892), Fenneman (1908), Gilbert (1909), Lawson (1932), Baulig (1940), and Birot (1949) formed the undergraduate reading of the senior author in the early 1950s. Thus, wash increases in volume downslope and so can carry the same sediment load over a gentler slope and thus forms concavities; but, of course the existence of concavities shows that wash must be predominant! Then there was the indisputable reasoning which led to Horton's (1945) "belt of no erosion"— indisputable, that is, until Yair (1972) set up wash traps right in the middle of it.

The problem of "the everlasting hills" was more severe than in most other branches of geomorphology. Rivers obviously flow, carry sediment, and from time to time their banks collapse; marine cliffs in soft rocks recede rapidly, whereas glaciers move perceptibly even within the timespan of an expedition. But in most parts of the world, especially the smooth, soil-covered landscapes of western Europe, change on slopes is not apparent in a human lifetime. Both authors experienced the revelation, on their first visits to the Rocky Mountains, that landscapes really are alive. For a time, geomorphology was imbalanced, with a modern, process-oriented textbook on rivers (Leopold et al., 1964) but nothing comparable in other branches.

Michaud began to survey movements of painted stones in the Alps in 1947,

Schumm measured ground loss on badlands from 1952 onwards, and Rougerie started wash measurements in the Ivory Coast. Grove recorded the daily movement of a mudflow, and Young put wooden pegs into the ground surface below rock outcrops.

Since that time, instrumentation has been developed to measure all the processes on slopes: soil creep, solifluction, surface wash, landslide movement, and latterly throughflow and solution. The stages in obtaining rates of movement are basically similar whatever the process: (1) develop the instrumentation necessary to record very small changes; (2) measure these changes, usually over a period of 1–3 years; and (3) extrapolate the results in time and space.

Within the context of the exponential growth in geomorphology as a whole, this procedure gave rise to a steadily growing stream of published rates of slope processes, from some 10 per year in the 1960s to over 40 per year in the 1980s. Three times we have attempted to summarize and compare such reports, for total denudation (Young, 1969), and for slope processes, slope retreat, and total denudation (Young, 1974; Saunders and Young, 1983). On the last occasion our report covered 419 publications and was by no means complete. Further updatings of this nature will probably have to be made for individual processes or climatic environments.

The present state of knowledge

Design of Procrustean bed

Figures 1.1 and 1.2 present a highly generalized summary of available data on rates of surface processes on slopes, slope retreat, and total denudation. These are derived from the corresponding figures in Saunders and Young (1983), to which reference can be made for details.

The environments from which these data originate are extremely diverse, including variations in climate, vegetation, rock, soil type, and slope angle. In an attempt to derive consistencies from this diversity, some gross simplifications have been made. Vegetation cover is assumed to correspond to climatic type, thus eliminating vegetation as an independent variable. Soil type is also excluded for lack of sufficient data; it would certainly be worthwhile for someone to attempt a summary of process rates in relation to the FAO or other soil classification, for certain relations would clearly emerge, such as faster creep on expanding clays, rapid surface wash on impermeable clays. Rock type is grouped into limestones (for chemical solution records only), unconsolidated rocks, and all other rocks.

Having thus for the most part eliminated three environmental factors, the main ways in which the data are classified are by climatic type and slope. Climates are grouped into eight broad classes as shown in Table 1.1. Tropical subhumid (sometimes called tropical "wet-and-dry" corresponds to savanna vegetation (cerrado in South America) and tropical humid to rain forest. We

Table 1.1 The eight broad climatic classes.

Climatic class	Abbreviation in Figures 1.1 and 1.2	Approximate Köppen equivalent
polar/montane	P/M	E
temperate maritime	Tm	Cfb, Cfc
temperate continental	Tc	Df
Mediterranean	Med	Cs
semiarid	S-A	BS
arid	Arid	Bw
subtropical humid	ST	Cfa
tropical subhumid	TrS	Aw
tropical humid	TrH	Af, Am

have been taken to task for grouping polar with montane climates on grounds of climatic differences of importance to processes, such as the high diurnal temperature range of mountains. However, the mountains in which periglacial geomorphologists often work are in high latitudes. Through lack of data, arid climates are omitted from all except one figure.

Finally, slopes are divided into "gentle" and "steep." Steep covers individual slopes over $25°$ and mountainous or steeply dissected areas in which such slopes are frequent. Gentle is short for 'gentle to moderate' slopes, and also covers river basins or other area which possess a wide range of slope angles. It is realized that the above are generalizations which have been made because the aim of the present chapter is to identify broad trends. Readers who seek more specific detail may refer initially to Tables I–X in Saunders and Young (1983) and from there back to the 419 primary sources.

For measurements of ground loss and slope retreat the Bubnoff unit B is employed, where $1 B = 1.0$ mm per 1000 years, equivalent to $1.0 \text{ m}^3/\text{km}^2 \text{ yr}$ and, in terms of rock (not soil) mass, 0.026 t/ha yr.

The way in which the ranges have been derived from the original data is explained in Figure 1.1. We should like to stress that there is no element of *a priori* reasoning in these diagrams—that is, no adjustments or interpolation for what the results ought to be. They are wholly based on the observed records, with ranges omitted where data for particular climate/slope combinations is insufficient.

Processes

Soil creep   Measurements of soil creep have shown that movement near the surface in temperate maritime climates is typically 0.5–2.0 mm/yr (Fig. 1.1A). In temperate continental climates, probably because of the more severe ground freezing in winter, it ranges up to 15 mm/yr. More data are needed from the deep soils of the tropical humid zone. No doubt clays creep faster than sandy

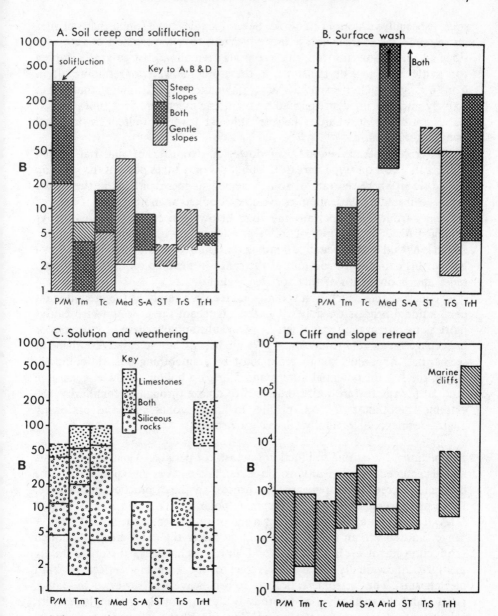

Figure 1.1 Generalized ranges of observed rates of surface processes and slope retreat grouped according to climate (see text), slope angle, and (for 1C only) rock type: (A) soil creep and solifluction, (B) surface wash, (C) solution, including weathering, and (D) cliff and slope retreat. Ranges are generalized from data in Saunders and Young (1983, Figs. 1–4). A continuous line at the upper or lower limit of a range indicates that three or more records existed, of which the single most extreme value (high or low) has been excluded from the range. A broken line indicates that the boundary is based on only one record.

soils, especially if montmorillonitic. Several studies have found that wet sites move much faster than dry, a fact which destroys the relation of rate with slope angle. This means, too, that creep may remain as fast on gently sloping concavities as higher up the slope, as these become waterlogged more often. Considering the relative spatial scales, Slaymaker's (1972) finding that slopes actively undercut by a stream had a volumetric creep rate four times that of slopes not undercut is hard to believe, although it was of quite high statistical significance.

Soil creep operates entirely by downslope transfer of material, which reduces its effect on slope retreat. Ground loss only takes place at the very top of the slope where the rate of soil movement is increasing, as in the classic Davis–Gilbert inference about convexities. Looked at another way, if 1 m³ of soil exits from the slope into the river below at a creep rate of 4 mm/yr averaged for 250 mm depth of soil over a period of 1000 years, then a slope 100 m long will only retreat by 10 mm in that time, or a rate of 10 B. As shown below, this comes at the bottom of the range of denudation rates for normal relief and is one tenth of that for steep relief.

The inescapable conclusion is that soil creep can rarely be the predominant denudational process on slopes, by which is meant the process which causes more slope retreat than any other. Only on substantially curved convexities or in seepage zones (cf. Anderson, 1977) might this be the case. Elsewhere, it seems that the major role of creep must be a smoothing one, differentially taking the soil off potential humps and filling up hollows. How else can one account for the feature which only fails to amaze through its familiarity: the extreme smoothness of slope profiles in humid zones, either as uniformly angled segments or regularly curved convexities?

Solifluction Measuring solifluction is a task for periglacial geomorphologists and also students on expeditions, since results are easier and quicker to come by than soil creep measurements. The process can be of practical importance. Rates of movement appear in column 1, polar/montane climates, of Figure 1.1A. They span a wide range, but a majority cluster in the 10–100 mm/yr range, and so are an order of magnitude faster than soil creep. Although solifluction involves both heave and flow mechanisms, these are rarely measured individually, and rates refer to the combined effect of both mechanisms. The instrumentation to record heave was reviewed by James (1971), and the methods used to study soil creep can also be employed with solifluction. The technological capability for microscale measurements is therefore available, but a major proportion of solifluction studies still rely on the relatively crude method of monitoring surface pegs and stones.

Of all the factors involved, soil moisture content has proved repeatedly to be of primary importance in promoting downslope regolith movement, being a fundamental ingredient for both flow and frost heave. Slope angle, soil texture, and vegetation tend to play secondary roles, though other things being equal, solifluction is faster on steeper slopes. Silty soils are favored over clays, which are too cohesive, and sands, which drain too easily. Vegetation may help solifluction by decreasing runoff and so increasing infiltration, or it may hinder it with a restrictive root system.

The tendency for measurements to be made where such obvious evidence as lobes or terraces are apparent means that average rates are overestimated. Solifluction can occur over the whole of a slope, and so measurements on lobes only represent the maximum rates of soil movement. These are obviously unrepresentative of slopes where the spatial extent of lobes is small in comparison to the remainder of the slope. Solifluction on slopes without microrelief may approach rates comparable with those of soil creep.

There is also an indication that contemporary rates of solifluction are somewhat faster than those of past millenia. Buried organic layers within lobes may be radiocarbon assayed to yield long term estimates of movement rates. In a review of such data, Benedict (1976) reported average rates of 0.6–3.5 mm/yr for time periods of up to 5000 years, or one to two orders of magnitude lower than present-day rates. This illustrates well the potential problems involved in temporal extrapolation of contemporary process measurements.

Although this process also operates by downslope transfer, the reservations given above concerning the effect on slope retreat do not apply with such force, as we are dealing with a faster process. Measured solifluction rates, put into a process–response model, could account for something like half of total slope retreat, although in comparative studies it does not come out on top (Rapp, 1960; Iveronova, 1969; Söderman, 1980). The smoothing effects can also be noted in relict form in some contemporary temperate areas and is discussed in a later section.

Surface wash   Surface wash includes the subprocesses of raindrop impact and surface flow or runoff, which partly but not entirely correspond to the achievement of soil detachment and downslope transport. A most elegant field experimental design is that of suspending fine netting just above the soil surface, thereby eliminating raindrop impact without altering runoff; in the subhumid tropics this can greatly reduce erosion. Most studies attempt to convert downslope sediment transport into ground loss, often by artificially enclosing their catchment area (as in standard soil erosion plots). The few who have tried pairs of wash traps down the slope have found that ground loss in the short term is irregular, with parts of the slope aggrading—perhaps another smoothing action (Townshend, 1970).

However much one tries to generalize the data, the inescapable fact is that rates of wash vary widely, from 1 B (the lower limit of measurement accuracy due to ground disturbance) in some temperate deciduous woodlands up to 1000–10 000 B on badlands (Fig. 1.1B). Leaving these two situations aside, most records cover two orders of magnitude, the range being 2–200 B. Reasons for this diversity are not hard to find: variations in rainfall intensity, vegetation cover, soil erodibility, slope angle, and slope length—that is, the variables in the various models for estimating agricultural soil erosion. In a British or European woodland there is virtually complete cover by the canopy, leaf litter and the humus horizon, and it tends to drizzle all day long. Badlands occur where a weak, erodible geological formation exists where rainfalls commonly exceed 250 mm/hr, causing such rapid erosion that plant seedlings are washed away before they can take root.

There can be little doubt that surface wash is the predominant denudational process in semi-arid and tropical subhumid climates, and probably in deserts too (McGee, 1897); contrary to early supposition, it is substantial under rain forest as well.

Solution Anyone who reads a textbook on soil formation cannot fail to be convinced that chemical weathering, coupled with the removal of dissolved substances in solution, must be a process of importance in all climates (bar the extremely arid) and on almost all rocks. What happens in the conversion of rock to soil? As every soil science student learns, first the soluble salts are removed, then the exchangable cations, and finally part of the combined silica is lost. True, some of these substances recombine as clay minerals, but a good proportion are removed in the groundwater, by lateral throughflow or deeper groundwater movement, to emerge as dissolved load in the rivers where geomorphologists measure the concentration. This process, which is here called solution, has long been recognized as predominant for limestones, even in pre-experimental days. In classical geomorphology, however, it was largely ignored as a process on other kinds of rock. We wrote our essays on "Which is the most important, soil creep or wash?," little suspecting that the answer might turn out to be "Neither."

Figure 1.1C presents the evidence, mostly derived from the conversion of catchment data to average ground loss. Rates are never very high, not often above 100 B, but neither are they very low except in some records from cool or cold climates. Limestones do indeed have a predominant range higher than that for siliceous rocks, but with substantial overlap: 10–100 B compared with 2–60 B. Measurement of rock weathering by loss from weighed tablets or the microerosion meter, fall within the same order of magnitude as data derived from dissolved load of rivers.

In 1956 the senior author worked out the first process–response models for slopes, unaware that Ahnert had simultaneously hit upon the same idea and was computerizing it (Young, 1963; Ahnert, 1966). The first results were devastating: an exceedingly low rate of solution produced more slope retreat than the fastest reasonable assumptions about creep and wash. The reason is that creep and wash operate by downslope transport, solution by what may be called direct removal. True, if we are thinking of lateral eluviation, then the groundwater can become chemically saturated and there will be no further weathering. But so long as leaching is downwards into the deeper ground-water, the weathered chemicals are taken directly away from all parts of the slope and the effect is the same as if they had simply evaporated. Solution at a rate of 10 B means slope retreat at 10 B, and this for a slope of any length and any angle, including erosion surfaces, a disturbing conclusion for the geomorphology of the 1950s.

How ironic that Walter Penck's big mistake, that of assuming that once regolith became "mobile" it was lost from the slope with infinite speed, should turn out to be true—for a process which he did not apparently consider!

Landslides When discussing landslides we are on less firm ground. Of the many published case studies, very few contain the necessary data to estimate average rates of downslope transport or slope retreat, volume moved, area covered, and frequency of occurrence. Of the 18 rate values assembled by Saunders and Young (1983, Table VI), taking each end of a range as a value, 11 lie between 100–1000 B and 5 between 1000–10 000 B. The latter range is retreat of an order of magnitude reached in geomorphology only by badlands, river bluffs, and marine till cliffs.

The conclusion must be that landsliding far outweighs all other processes, but only at the places and times at which it occurs. The conditions conducive to active landsliding are well known: shallow slides affecting the regolith only on slopes above a certain limiting angle ($\sim 25°$) on most kinds of rock. Deeper movements can occur on gentler angles in a variety of special situations (e.g., seepage zones). Elsewhere, the argument that the once-in-a-millenium landslide might move more tonne-meters than eons of continuously acting processes is contradicted by slope form: if it were true, our hillsides would be all bumps and hollows. Ungraded slopes, in the Davisian sense, retreat fast and mostly by mass movements; but on achievement of a graded slope—that is, a continuous soil and vegetation cover—slow and continuously acting processes take over.

Cliff and slope retreat There have been a limited number of attempts to estimate slope retreat directly, as contrasted with measuring a process. Most of these refer to cliffs (free faces, bare rock), and where this is not the case, the slopes concerned are usually steep. In Figure 1.1D the records for cliffs and slopes are therefore not separated.

The largest number of records are from the polar/montane zone, in which all but three records lie in the range 20–1000 B, and the mean value (logarithmic) is close to 100 B. Values from other climates give the appearance of being faster, mostly 100–10 000 B, but this may be a false comparison caused by differences in rock types. In periglacial regions many of the records refer to cliffs in hard rocks left by glacial erosion, whereas studies in other climates are often on softer rocks. The world record seems to be 250 MB, or 250 m/yr, from a bluff of the Mississippi (D. Brunsden, personal communication).

Marine cliffs demonstrate the enormous range in rates of retreat caused by lithology. Where the sea chooses to erode, till cliffs retreat at about 1 m/yr, fast enough to remove a Roman village and allow children to collect the human bones falling from the eroding graveyard at Dunwich. Rocks of medium hardness yield to the sea at 4000 B and upwards. Conversely, there are hard-rock cliffs in the Western Isles of Scotland, most certainly a high-energy environment, which appear to have remained totally unaffected by erosion at the present sea level, maintaining near-verticality in situations ranging from highly exposed to sheltered inlets.

Other processes. Lest it be thought that the above discussion is comprehensive, let us run briefly through some of the fringe processes, for most of which there are few or no rate measurements.

One may consider as creep-related the extremely slow processes of *rock creep*, *deep creep*, *snow creep* (the probable cause of tree curvature in snowy places), and *tree uprooting*. This last refers to the large amount of soil that is pulled up attached to the root system when a tree blows over, a cause of substantial downslope soil transfer in rain forest.

Splash creep is a half-way process, whereby rainsplash without any runoff can effect a net transfer of soil particles down the slope. Wash-related processes include *litter wash*, whereby mineral soil particles are washed away attached to leaf litter (van Zon, 1980), and the supposed movement of *suspended clay particles in throughflow*, a virtually unmeasurable process. *Snowpatch erosion* is mostly by wash after snow has lain long enough to prevent establishment of vegetation. Then there are processes transitional between slope retreat and river erosion: successively *percolines*, *piping*, *spring sapping*, and *gullying*.

One might class as solution-related the uptake of minerals by *accumulator plants*, which no doubt then find their way downslope. These minerals include silica (McKeague and Cline, 1963), whereas an element otherwise very stable in soils, aluminum, is taken up in large amounts by the tea bush. *Deflation* is *sui generis*. And beware! You may be struck by a *ploughing block*, a *sliding stone*, or the distinctly more dangerous-sounding process of *mountain sculpture by rolling debris* (Blackwelder, 1942).

Then there is the diverse and sometimes devastating change to slopes wrought by the *animal kingdom*. Two examples are familiar, termites and earthworms. They contribute to the classic mechanism of soil creep by tunnel formation, mound building, and ingestion: disturb a soil particle and it will preferentially fall back downhill. Terracette formation, a process on the margin between soil creep and shallow landsliding, may be assisted by farm livestock.

Furthermore, there is the undoubted substantial soil disturbance achieved by moles (Jonca, 1972), rabbits, pocket gophers (Mielke, 1977; Thorn, 1978), and burrowing animals in general (Imeson, 1976). In the Negev, Israel, procupines can play a decisive role along with representatives from the insect world, isopods (Yair, 1974). In part of the Rajasthan desert, India, the desert gerbil opens up 60 000 new burrows per square kilometer *per day* (Sharma and

Joshi, 1975). Last but not least, we owe to Best (1972) our knowledge of the mound-building and degrading activities of the banner-tailed kangaroo rat!

Denudation

Denudation refers to the general lowering of the ground surface by removal of sediment (solid and dissolved) from within a drainage basin. No specific slope processes need be identified, and indeed rarely are. Denudation has received more interest than any individual slope process. Its attraction must lie partly in the relative ease by which data may be obtained. Also, the fact that the spatial extent of measurement usually covers a much greater area than that for other processes, which rely heavily on point-source information, has the apparent advantage of better representation of the landscape as a whole. This avoids some of the spatial variability problems of surface processes by integrating the sediment yield over the entire drainage basin.

Recorded rates of denudation are summarized by Saunders and Young (1983) and in Figure 1.2. The apparent wide scatter of data can be rationalized by close examination of individual climates and relief. Typical ranges are given in Table 1.2. However, these generalizations are cautiously presented, given the limited amount of attention that many environments have seen. Denudation rates reach a maximum in the semiarid and possibly the savanna environments. High rates are further recorded in montane, subtropical, and rain forest regions. The temperate zone has received abundant interest, and it

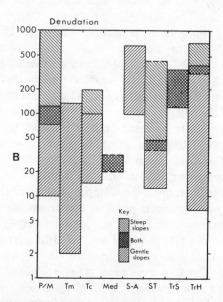

Figure 1.2 Generalized ranges of observed rates of denudation, grouped according to climate (see text) and angle of dominant slopes. Ranges are generalized from Saunders and Young (1983, Fig. 5) using the same procedures as for Figure 1.1.

Table 1.2 Typical ranges of climate and relief.

Climate	Relief	Typical range (B)
glacial	gentle (ice sheets)	50–200
	steep (valley glaciers)	1000–5000
polar/montane	mostly steep	10–1000
temperate maritime	mostly gentle	5–100
temperate continental	gentle	10–100
	steep	100–200+
Mediterranean	—	10–?
semiarid	gentle	100–1000
arid	—	10–?
subtropical humid	—	10–1000?
tropical subhumid	—	100–500
tropical humid	gentle	10–100
	steep	100–1000
any climate	badlands	$1000–10^6$

is clear that denudation is generally slower here, with the humid and heavily vegetated zone showing the slowest global rates. The erosive capability of glacier ice is obvious for valley glaciers, but ice sheets are not necessarily so active.

It is commonly observed that sediment yield per unit area decreases from small to large basins. This is usually interpreted as being due to the more intense erosion in upland headwater areas and to alluvial storages having increasing residence times with distance downstream. The reverse situation (increasing sediment yield per unit area with increasing basin size) may also be encountered. Trimble (1977) suggested that to assume an equilibrium state in the fluvial system, which is necessary to extend contemporary denudation estimates over time, is invalid in these instances. Only when stream sediment yields balance primary sediment production can equilibrium, and consequently confidence in denudation measurements, be presumed. Under nonequilibrium conditions, an understanding of sediment transfer paths and delivery ratios is needed. Since this problem is applicable to other processes besides fluvial transport, it is discussed in the following section.

At least two previous studies have compared rates of denudation with the time available (between orogenic periods) for peneplanation to occur. Both concluded that sufficient time did exist, Schumm (1963) calculating a need for 15–110 Myr, Ahnert (1970) for 11 Myr, or 18.5 Myr if isostatic rebound is accounted for. The orders of magnitude in Figure 1.2 support these conclusions. Assume that 1000 m of relief is to be removed by erosion. Initially this will be fast, the rate probably declining as a negative exponential with time. Let it be supposed for simplicity that half is removed at the rate typical for mountainous areas, 500 B, and half at that typical for gentle slopes, 50 B. Planation will then be achieved in some 11 Myr. Moreover, as it is the slow later stages that take up most of the time, it makes little difference whether uplift is 2000 or 3000 m, raising the time needed only to 12–13 Myr. This is well within the interval between major orogenic episodes. In areas of active uplift denudation is generally an order of magnitude less than the rate of uplift (Saunders and Young, 1983, Table XI), and only very vigorous erosion in mountainous catchments seems able to keep up. This adds further support to the broad validation of the cyclic concept.

Many denudation studies, especially the earlier ones, partly or completely ignored potential sources of error in the data and their geomorphological consequences. Of the problems, those of sediment sources and transfer paths relate to some extent to the slope/river interface, whilst that of temporal extrapolation is common to all process studies. These are therefore discussed for convenience in the following section. One problem that is specific to studies based on river sediment is that of bed load estimation. Dissolved and suspended loads can readily be sampled, but there is great difficulty in obtaining reliable estimates of bed load. As a result, most such studies have simply ignored it.

Problems

Some problems that are specific to individual processes have been noted in the preceding section. Here we draw together some more general problems in the interpretation of slope processes. We discuss the implications of process studies in the context of some of the fundamental concepts of geomorphology, both early and modern, including questions of time, space, and the relations between processes and landform evolution. Finally, having for the most part intentionally sought to assess rates in the natural landscape, brief attention is given to the impact on process rates of the activities of Man.

Sediment sources and transfer paths

In denudation studies, sediment sources are usually assumed to be from within the drainage basin. For solid sediment this is generally safe, though an extreme example is provided by the East Twin catchment in the Mendip Hills, England, where the atmospheric input of suspended solids actually exceeded the amount exported by the stream (Finlayson, 1978). Atmospheric inputs can make highly significant contributions to the solute budget and can be elucidated by careful analysis of precipitation and subsequently subtracted from the natural sediment budget. Dissolved and particulate organic matter is also a further cause of nondenudational sources of error. Janda (1971) suggested that previous calculations of denudation based upon measured dissolved loads were overestimated by 1.4–2.4 times, and Meade (1969) lists further potential sources of error. A geochemical inventory of solutes is thus required in order to obtain unadulterated process rates (Walling and Webb, 1978).

What goes in does not necessarily come out, at least not in the short term.

In the progress of sediment down a single slope there may be periods of storage, short-term as between rainstorms, and somewhat longer in colluvial deposits. For river basins as a whole, the problems of sediment storage are much greater. Take the case of river bank erosion, which in some areas appears to provide the greater part of the suspended load. How is such material to be interpreted in terms of denudation? It could be argued that it is ultimately colluvial in origin and so from a long-term perspective acceptable for estimates of denudation rates. This, however, introduces a disparity of timescales between periods of sediment yield and sediment supply.

Some of the material taken up as river load may not be of colluvial origin at all. Widespread cases occur in the reworking of glacial sediments, lake deposits, or alluvium deposited in response to changes of sea level. An exceptional but graphic example is the recent rapid denudation of the Mount St. Helens volcanic deposits, taking place mostly by gullying, including abundant landsliding from oversteepened valley sides (Lehre et al., 1983).

In the assessment of contemporary rates from river basins, an estimate of the sediment delivery ratio is essential. This defines that fraction of material produced by primary weathering and erosion that is outputted from the basin, and thus indicates the importance of storage. The magnitude of the ratio may vary widely: for example, 50 percent for a small mountainous catchment (Dietrich and Dunne, 1978) versus only 6 percent for upland basins in southeastern U.S.A. (Trimble, 1977). Sediment may be stored in the slopes, for example as talus, or in rivers as bars or floodplains. This problem is compounded with increasing basin size, as storage times lengthen. If sediment going into storage was balanced by the release of stored material back into the sediment transport system, then we need not be concerned. Only a thorough analysis of the sediment transport paths, however, could lead to an assessment of the magnitude and significance of sediment delivery ratios within a catchment. The situation could be elaborated into a model somewhat along the lines of nutrient cycling models in soil science, with stores and flows; at different points in time a given store might be growing, decreasing, or in a steady state.

The spatial sampling frame

In many kinds of scientific research it is considered a *sine qua non* that observations should rest on a sampling design which permits determination of statistically valid averages from the whole area of the object under study. Surprisingly, this is not the case in slope process measurements in that a valley-side slope and a river basin are such obvious spatial units. This applies particularly to denudation studies, where the assumption that the sediment output has been evenly derived from the whole of the basin is rarely realistic. Spatial variability is large even for what might be supposed a relatively uniform process, solution, as exemplified by the mapping of solute yields in the Exe basin, England (Walling and Webb, 1978) or the Polish Carpathians (Welc, 1978).

Only rarely does one read that an observer identified the different kinds of slope unit, surveyed their distribution, and sited the recording instruments on the stratified base thus established. Slaymaker (1972) is such an exception. So too is the study of an alpine environment in Colorado by Bovis and Thorn (1978), who in setting such an example found that 50 percent of the sediment came from 3 percent of the surface area, that occupied by late-lying snow patches. No wonder nivation hollows are formed.

The reason may lie in the difficulty of maintaining recording instruments in the natural landscape for years on end. T-bars, Gerlach troughs and Schefferville bedsteads are unlikely to survive the inquisitive attentions of cattle, moose, elephants, hikers, and nomadic pastoralists. Given that a fully rigorous instrument siting design may not often be practicable, the best that can be hoped for is an awareness of spatial variability in sediment sources and transport, and preferably inclusion of records from both relatively inactive and obviously eroding parts of the landscape.

The temporal sampling frame

"Events may have progressed both faster and slower in the past than during the brief interval which we call the present" wrote W. M. Davis (1895, p. 8–9); or, what is here today, may or may not be gone tomorrow. A high proportion of records on which this chapter is based refer to periods of three years or less. Hauswirth and Scheidegger (1976) with 21 years and Jahn (1981) with 17 years are exceptions; and Young (1978) hopes to extend his 12-year record of soil movement to 32 years by re-excavation in 1995. But even this added order of magnitude is derisory compared with the period of evolution of landforms. Church (1980) demonstrated that short-term observations of natural phenomena will not fully sample their variability. In saying "My PhD measurements show that this valley is half a million years old" we are extrapolating a line to 10^5 times its length. The magnitude/frequency concept is relevant here: as the return period of the dominant geomorphic event increases we are increasingly unlikely to sample it.

Circumstances are occasionally found that permit estimates of average rates for long periods. These have great value for providing some kind of check on the validity of extrapolation. Fortunately, they prove to be of the same order of magnitude as contemporary rates. For geological timespans, denudation rates of 12–40 B were noted in our previous review together with King's (1940) estimate of 30–49 B for the retreat of the Natal Drakensberg. Since these are averages they will encompass intrabasin storage, which could account for values being on the low side compared with present-day measurements.

Magnitude and frequency

The magnitude–frequency diagram plots total work accomplished against work accomplished during individual events of a given magnitude multiplied by the frequency of occurrence of such events (Wolman and Miller, 1960). The return frequencies of varying intensities of rainfall, which trigger most processes, is a convenient measure of the magnitude of an event. The concept has been extended to incorporate the recovery time, that needed for slow but continuous processes to counteract the effects of the extreme event (Woman and Gerson, 1978). What does the shape of the curve look like for slope processes, and at what recurrence interval does the peak lie? Expressed more simply, when does most slope retreat take place: during rainfalls such as are experienced every year, in the 2-, 5-, or 10-year storm, or the still rarer catastrophic event?

Most landslides occur infrequently, set off by intense rainfall after the ground has previously become saturated. Where their scars occupy much of the slope, the landslide is clearly the dominant event, and the recurrence interval at a given site is shorter than the recovery time required by continuous processes to smooth out the surface. On steep slopes in the tropical rain forest zone, however, landslides may be common and yet there is no marked microrelief; the recovery time is sufficient.

The effectiveness of surface wash is increased during the big storm in at least four ways: raindrops are larger, infiltration capacity is more likely to be exceeded (Hortonian runoff), the water table is more likely to rise to the surface (saturation runoff), and rate of transport increases more than linearly with volume of flow. This suggests that where wash is the major process it may be at least the middle-intensity events, with 2–10-year recurrence intervals, that cause most slope retreat. In Malaysian rain forest, Young had set up wash traps at the time of a 30-year storm; these were not merely filled with sediment, but their rims were left standing above the surface, 10–20 mm of soil having been removed in 24 hours.

Solifluction by its nature occurs annually, no doubt faster during unusually wet springs. Soil creep presumably carries on all the time, nudging forward a little faster when the regolith gets unusually wet. Solution would be faster when excess water leaches solutes away, bringing less saturated water in contact with the rock and so increasing chemical reaction rates. Much of this is old-style *a priori* reasoning, but short of maintaining observations for 100 years we must seek such convergent evidence as may become available.

Climatic change

The existence of major climatic fluctuations during the Quaternary, and the recency of the last of these, raises the question of relict landscapes: specifically, to what extent are the temperate landscapes of today inherited more or less unchanged from periglacial conditions, and (erosional) desert landforms left over from pluvial periods? The latter might reasonably be supposed to be true to a large degree, in that rates in deserts are thought (without evidence) to be very slow, whereas high rates of denudation certainly occur in semiarid conditions.

As so many geomorphologists live in former periglacial regions, a clearer view on the first question would be welcome. The abundant evidence of periglacial activity in such areas (e.g., Budel, 1937; Kotarba, 1977; Harris, 1981) conjures up a worrying vision of slopes stripped bare by solifluction and torrents surging down dry valleys, only to be choked by massive inputs of Coombe Rock. L. C. King (personal communication) believes this is why most European slopes lack the free face which he regards as normal; the slopes have been smoothed by periglacial activity, to which they are now monuments. Comparing periglacial deposits with their slope sources, Williams (1968) estimated that 3–10 m of regolith has been removed by solifluction in southern England.

The evidence from rates of processes is ambiguous. For denudation, the predominant ranges overlap, 10–1000 B for periglacial climates and 5–200 B for temperate, although most of the periglacial records are from mountainous regions, whereas the temperate ones mostly refer to gentler relief. For the latter half of the Pleistocene epoch, warm and cold periods have been more or less equal in total duration. As periglacial processes are more active in a denudational sense than are temperate ones, it could be expected that landforms and deposits from this regime will be dominant in those areas that have recently experienced cold climates. In a similar vein to the magnitude–frequency concept, the slower temperate processes might then require a "recovery time" after periglacial conditions have ended before they can begin to produce their own suite of landforms. Many temperate landscapes are apparently in such a state at present.

Control of slope retreat

The basic idea of control of slope form was set out by Gilbert (1877, p. 99), refined by Savigear (1960), and subsequently assumed by Young (1972, p. 23–24), and Carson and Kirkby (1972, p. 104–106). In control by weathering (or on weathering-limited slopes), transport processes can potentially remove more material than is produced by weathering, and the rate of slope retreat is controlled by the latter. In control by removal (or on transport-limited slopes), weathering initially produces comminuted material faster than it can be removed, and slope retreat is thus limited by the rates of net transport plus direct removal (cf. Jahn, 1968). This state of affairs is temporary, the regolith thickening until its protection slows down weathering such that the rock–regolith interface retreats at the same rate as the ground surface.

Bare rock slopes, whether the whole length from crest to base as in desert mountains or free faces within regolith-covered slopes, are clearly subject to control by weathering. Slopes on unlithified rock, like clay strata or sand dunes, are necessarily subject to control by transport. In those tropical landscapes (by no means all) where the regolith reaches 10–50 m or more in depth, it may be presumed that weathering has got ahead of the capacity for removal at some time, irrespective of possible two-cycle origins. This reasoning is strengthened by the fact that nearly the whole of these thick weathered layers consists of saprolites, and signs of movement are uncommon below 2 m (below the termites?).

Most temperate landscapes (and many tropical) have a regolith 0.3–3 m thick. Is this simply a measure of the time since periglaciation, multiplied by the weathering rate, and thus the regolith is (*pace* Man's intervention) becoming thicker? Or has a steady state been attained in which weathering would speed up if the regolith became thinner, and transport or solution removal would become faster if it became thicker? This latter state of affairs can be simulated by a computer: initially you get wild alternations between bare rock and impossibly thick weathering mantles, but persistence in modifying the control equation leads to a plausible simulation of reality (Ahnert, 1976).

Lastly, we would reiterate the view (cf. Young, 1972, p. 206–207) that the fundamental cause of the hillslope–pediment landform is that the slope is under control by weathering (albeit with some coarse regolith cover), whereas the pediment is subject to control by transport. The pediment is in a steady state in the medium term, but slowly retreating in the long term. (see below).

Equilibrium and the steady state

The concept of slopes as open systems in a steady state (Hack, 1960; Chorley, 1962) was foreshadowed by an earlier idea, that of the profile of equilibrium. Slopes with a continuous regolith cover represent a profile of equilibrium, the position of every point on which depends on all the others, wrote Baulig (1940). The shape of this profile is maintained by the moving soil layer. Interdependent factors are the rates of weathering and transport, particle size, and debris thickness. The profile is mobile, and depends partly on the rate of river erosion.

If this were true, how could we have the valley-in-valley forms that so obviously indicate rejuvenation? Presumably because that question is phrased in the wrong context. The lower parts of such slopes, turf-scarred and terracetted, have conveyed the message to their upper parts, still basking in the indolence of late maturity. It is clear that, except in badlands and other special cases, slopes do not achieve equilibrium in 10^4 years; but inasmuch as they retreat some 50 m in 10^6 years, most do not need that long. We therefore suggest that typical hard-rock slopes reach some form of equilibrium, free of abrupt changes in angle except where related to lithological differences, over a period of the order of 10^5 years.

It is these equilibrium slopes that become the steady-state landforms of a region, the form dependent on lithology, climate, vegetation, and the consequent balance of processes. Carson and Kirkby's "characteristic forms" present the same concept, derived from theoretical process–response models. Such equilibrium forms are not a single shape but a suite of slope profiles of varying steepness (where slope decline occurs) or varying proportions of their component elements (under parallel retreat). This is why each climatic and lithological environment has its own characteristic landscape: for example, compare those of the English chalklands, New Zealand feral relief, South African Karroo sediments, or Malaysian rain forest. Uplift and rapid dissection can destroy these equilibrium forms, imposing the near-uniformity of the 35° V-shaped valley.

Time, space, and causality in slope evolution

The subheading is derived from the classic statement about cyclic, graded, and steady time in geomorphology (Schumm and Lichty, 1965). This has more recently been extended in terms of thresholds and of pulsed and ramped perturbations, subsuming some of the concepts of magnitude and frequency, recovery time, and equilibrium discussed above (Brunsden and Thornes, 1979; Schumm, 1979). These powerful ideas were mainly based upon or illustrated in terms of changes in the gradient of a river channel. Let us briefly re-examine them in the light of slope evolution under the action of surface processes.

Assume that rapid river downcutting produces a steep, rocky slope, following which basal erosion becomes slow or zero. Slope retreat is initially large by rapid mass movements and in relative terms is very rapid. In the course of time, the slope approaches stability in the engineering sense; the interval between landslides increases as they come to occur only under storms of long recurrence interval, until recovery time becomes sufficient for continuously acting processes to smooth out irregularities of form. The slope acquires a continuous soil and vegetation cover, except if interrupted by a free face, and achieves the first of its equilibrium or character forms.

It now passes through a suite of such forms. If evolving by slope decline these will become progressively gentler in angle and, in the early stage at least, develop a longer convexity and concavity. If, on the other hand, they are evolving by parallel retreat, then the convexity, free face and debris slope may retain their form, angles, and relative proportions while the pediment progressively extends. Such evolution takes place within *cyclic time*, perhaps of the order of 10^5-10^6 years. Within such time intervals the whole slope profile is continuously retreating (Fig. 1.3).

Over *graded time*, 10^2 or possibly 10^3 years, the situation is different.

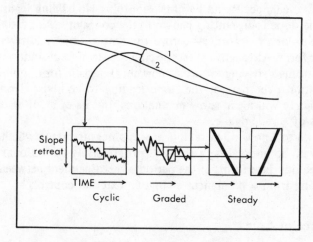

Figure 1.3 Slope retreat during cyclic, graded, and steady time. The vertical scale refers to ground loss perpendicular to the slope surface.

External conditions undergo pulsed changes, mainly drier and wetter spells of varying durations and individual storms, but also, for example, a floodplain river swinging against and away from the slope base. The slope cannot be in equilibrium under all of these conditions, hence there will be corresponding fluctuations in transport and removal of regolith. This means temporary aggradation on parts of the slope: mostly on the concavity under the action of creep or solifluction, but almost anywhere in the case of wash. A graph of the position of a point on the ground surface against time is a jagged line, with irregularities of varying lengths; but the 1000-year running mean of this line becomes the smooth and steady retreat over cyclic time.

Steady time is the period over which one set of the pulsed changes of graded time prevails: one crowded hour of glorious storm or, for the most part, a normal year. Most of our process measurements take place within steady time. Lest you should be lulled into reassurance by the word "steady," note that this could occupy any segment on the fluctuating line of graded time, with fast or slow retreat or advance. By placing wash traps at intervals down a slope and sampling monthly, it might be possible to identify by time and place the intervals of steady time which combine to produce the fluctuations over graded time.

One specific aspect that is clarified by such analysis is how a pediment at the same time can be subject to control by transport, which should imply no net ground loss, and yet in the long term be retreating. Over graded time its form has reached a profile of equilibrium conditioned by transport capacity. Over any section, what goes in comes out or very nearly so, as over cyclic time a net excess or removal of 0.1 percent or less will lead to a very slightly gentler slope. Thus, the pediment regrades.

Returning to a longer timescale, suppose now that one of the equilibrium forms is interrupted by a major, ramped change, such as rejuvenation at the base or climatic change. Rapid basal erosion plus landsliding leads to a 35° segment at the slope foot, cutting back into the concavity. As noted above, it is a long time—of cyclic order—before the whole slope has reattained its characteristic form. Alternatively, let the climate become substantially drier or wetter. This means a striving towards a new characteristic form, quite possibly with aggradation over part of the slope leading to colluvial deposits. The graded time curve will again show fluctuations, but its cyclic time trend may include areas of slope advance.

How are we to know when taking process measurements whether we are studying the cyclic interval between two, progressively transitional stages of the suite of equilibrium forms or the period of readjustment between different forms resulting from a permanent change in external controls?

The influence of Man

Throughout the above review we have intentionally sought to assess process rates under natural conditions. In large part this is a matter of scientific curiosity, coupled with the need for correct interpretation of landforms. In relation to the modern demand for applied studies, it can be justified by the need to provide a baseline of natural rates against which to judge the effects of land use by Man.

The impact of Man on slope processes ranges, like the Richter scale, from mild disturbance through serious disruption to catastrophic devastation. We have two kinds of evidence for this: the rather limited number of studies where some means of comparing pre- and post-Man rates has been found, and simple comparison between rates of natural and accelerated erosion.

The increase in atmospheric pollutants during the last century has unquestionably increased the amounts of solutes that are precipitated into terrestrial systems. This can affect more than exposed medieval stonework. Increases in solute concentrations could affect rates of weathering and so the primary production of sediment. Winkler (1970) suggests a doubling of weathering rates due to anthropogenic alteration of Earth–atmosphere hydrochemistry. This aspect of denudation remains largely unresearched, but acid rain's recent publicity boom should rectify this.

The major effect, however, is upon accelerated erosion by surface wash and its consequences for river sediment loads and total denudation. Some guesses at the orders of magnitude of different kinds of impact are as follows.

1. Clearance of natural forest and its replacement by nonintensive agricultural use, such as Neolithic to medieval forest clearance in Europe, and ecologically balanced shifting cultivation in tropical forests. On very limited evidence, acceleration of denudation is 1.5–3 times. Humid temperate pastures with matted grasses might not erode any faster than woodlands. As soon as bare soil is exposed, by hoe or plough, accelerated erosion of that area is inevitable, although the sediment may not travel far.

2. A change to intensive agriculture with adequate soil conservation. It now seems likely that once arable use becomes widespread, denudation is likely to be accelerated by 5–10 times. Gospel among soil conservationists is the "range of tolerable erosion," the target for design of conservation practices. This is typically assessed at 2.5–12.5 t/ha yr (Stocking, 1978). These values are equivalent to 100 and 500 B, respectively, the former being twice the typical natural rate on gentle slopes, the latter equal to that on steep slopes.

3. A change to agriculture with poor soil conservation, including overgrazed pastures. Soil erosion measurements reveal losses of 50–200 t/ha yr and upwards, equivalent to 2000–8000 B. On steeply sloping landforms this is an acceleration of natural rates by 4–16 times. If such rates occur on gentle slopes, they are ten times that acceleration. An individual site which has degraded into severe sheet erosion or gullies may be losing upwards of 10 mm/yr or 10 000 B.

Conclusions

We have pointed out some limitations and problems involved in the interpretation and collection of the data. It would be easy for us to suggest methodological additions and improvements that are needed, but many would succumb to the practicalities of long-term field research.

Slope retreat and denudation are the result of complex interactions between the atmosphere and the Earth, acting over many different timescales, most of them far beyond the bounds of human experience. Such a multifaceted problem deserves a correspondingly varied approached to research. A holistic approach to the systems involved would help overcome some of the more obvious limitations of any single technique. Specifically, when embarking on the standard operation of instrumentation and recording, it would be well to search for other means of either checking the data or placing them in perspective, including a recourse to older methods such as dating of deposits and denudation chronology. The current trend for examination of the problem of geomorphological rates from a variety of perspectives is certainly beneficial and a recommended path for future research to follow.

References

Ahnert, F., Zur Rolle der elektronische Rechenmaschine und des mathematischen Modells in der Geomorphologie, *Geomorphologie Zeitschrift, 54*, 118–133, 1966.

Ahnert, F., Functional relationships between denudation, relief and uplift in large mid-latitude drainage basins, *American Journal of Science, 268*, 243–263, 1970.

Ahnert, F., Brief description of a comprehensive three dimensional process–response model of landform development, *Zeitschrift fur Geomorphologie Supplementband, 25*, 29–49, 1976.

Anderson, E. W., Soil creep: an assessment of certain controlling factors with special reference to Upper Weardale, England, Ph.D. thesis, 593 pp. University of Durham, 1977.

Baulig, H., Le Profil d'équilibre des versants, *Annales de Géographie, 49*, 81–97, 1940. Reprinted with additions in *Essais de Géomorphologie*, 2nd. ed., pp. 125–147, Les Belle Lettres, Paris, 1950.

Benedict, J. B., Frost creep and gelifluction features: a review, *Quaternary Research, 6*, 55–76, 1976.

Best, T. L., Mound development by a pioneer population of the banner-tailed kangaroo rat, *Dipodomys spectabilis baileyi* Goldman, in eastern New Mexico, *American Midland Naturalist, 87*, 201–206, 1972.

Birot, P., *Essai sur quelques problèmes de morphologie générale*, 176 pp., Centro de Estudos Geographicos, Lisbon, 1949.

Blackwelder, E., The process of mountain sculpture by rolling debris, *Journal of Geomorphology, 5*, 325–328, 1942.

Bovis, M. J., and C. E. Thorn, Soil loss variations within a Colorado Alpine area, *Earth Surface Processes and Landforms, 6*, 151–163, 1981.

Brunsden, D. and J. B. Thornes, Landscape sensitivity and change, *Transactions of the Institute of British Geographers, 4*, 463–484, 1979.

Budel, J., Eiszeitliche und rezente Verwitterung und Abtragung im ehemals nicht vereisten Teil Mitteleuropas, *Petermanns Geographische Mitteilungen, Erganzungheft, 229*, 71 pp. 1937.

Carson, M. A., and M. J. Kirkby *Hillslope Form and Process*, 475 pp., Cambridge University Press, Cambridge, 1972.

Chorley, R. J., Geomorphology and general systems theory, *US Geological Survey Professional Paper*, *500-B*, 1962.

Church, M., Records of recent geomorphological events, in *Timescales in Geomorphology*, edited by R. A. Cullingford, D. A. Davidson, and J. Lewin, pp. 13–29, Wiley, Chichester, 1980.

Davis, W. M., The convex profile of badland divides, *Science*, *20*, 245, 1892.

Davis, W. M., Bearing of physiography on uniformitarianism, *Bulletin of the Geological Society of America*, *7*, 8–11, 1895.

Dietrich, W. E., and T. Dunne, Sediment budget for a small catchment in mountainous terrain, *Zeitschrift fur Geomorphologie Supplementband*, *29*, 191–206, 1978.

Fenneman, N. M., Some features of erosion by unconcentrated wash, *Journal of Geology*, *16*, 746–754, 1908.

Finlayson, B. L., Suspended soilds transport in a small experimental catchment, *Zeitschrift fur Geomorphologie*, *22*, 192–210, 1978.

Gilbert, G. K. *Report on the Geology of the Henry Mountains*, 160 pp., U.S. Geological Survey, Washington, 1877.

Gilbert, G. K., The convexity of hilltops, *Journal of Geology*, *17*, 344–350, 1909.

Grove, A. T., Account of a mudflow on Bredon Hill, Worcestershire, April 1951, *Proceedings of the Geologists Association*, *64*, 10–13, 1953.

Hack, J. T., Interpretation of erosional topography in humid temperate regions, *American Journal of Science*, *258-A*, 80–97, 1960.

Hauswirth, E. K. and A. E. Scheidegger, Geomechanisch Untersuchungen der Grosshand-bewegung Hallstatt-Plassen (Osterreich), *Rivista Italiana di Geofisica e Scienze Affini*, *3*, 85–90, 1976.

Horton, R. E., Erosional development of streams and their drainage basins; hydrophysical approach to quantitative morphology, *Bulletin of the Geological Society of America*, *56*, 275–370, 1945.

Imeson, A. C., Some effects of burrowing animals on slope processes in the Luxembourg Ardennes. Part 1: The excavation of animal mounds in experimental plots, *Geografiska Annaler*, *58A*, 115–125, 1976.

Iveronova, M. I., An attempt at quantitative analyses of contemporary denudation processes, *Izvestiya Akademiya Nauk SSSR, Seriya Geograficheskaya*, *2*, 13–24, 1969.

Jahn, A., Denudational balance of slopes, *Geographica Polonica*, *13*, 9–29, 1968.

Jahn, A., Some regularities of soil movement on the slope as exemplified by observations in the Sudety Mountains, *Transactions of the Japanese Geomorphological Union*, *2*, 321–328, 1981.

James, P. A., The measurement of soil frost-heave in the field, *British Geomorphological Research Group Technical Bulletin 9*, 43 pp., Geo Abstracts, Norwich, 1971.

Janda, R. J., An evaluation of procedures used in computing chemical denudation rates, *Bulletin of the Geological Society of America*, *82*, 67–79, 1971.

Jonca, E., Water denudation of molehills in mountainous areas, *Acta Theriologica*, *17*, 407–412, 1972.

King, L. C., Attempt at a measure of erosion in Natal, *Transactions of the Geological Society of South Africa*, *43*, 153–157, 1940.

Kotarba, A., Traits dynamique des versants de la haute montagne dans les Tatras Poloniases, *Studia Geomorphologica Carpatho-Balcanica*, *11*, 177–187, 1977.

Krumme, E., Frost und Schnee in ihrer Wirkung auf den Boden im Hochtaunus, *Rhein-Mainische Forschungen*, *13*, 1935.

Lawson, A. C., Rain-wash erosion in humid regions, *Bulletin of the Geological Society of America*, *43*, 703–724, 1932.

Lehre, A. K., B. D. Collins, and T. Dunne, Post eruption sediment budget for the North Fork Toutle River drainage, June 1980–June 1981, *Zeitschrift fur Geomorphologie Supplementband*, *46*, 143–163, 1983.

Leopold, L. B., M. G. Wolman, and J. P. Miller, *Fluvial Processes in Geomorphology*, 522 pp., Freeman, San Francisco, 1964.

Lyell, C. *Elements of Geology*, *Vol. 1*, 2nd. ed., 437 pp. Murray, London, 1841.

McGee, W. J., Sheetflood erosion, *Bulletin of the Geological Society of America*, *8*, 87–112, 1897.

Meade, R. H., Errors in using modern stream load data to estimate natural rates of denudation, *Bulletin of the Geological Society of America*, *80*, 1265–1274, 1969.

Michaud, J., Emploi de marques dans l'étude des movements du sol, *Revue de Géomorphologie Dynamique*, *1*, 180–194, 1950.

Mielke, H. W., Mound building by pocket gophers (*Geomyidae*): their impact on soils and vegetation in North America, *Journal of Biogeography*, *4*, 171–180, 1977.

Morawetz, S., Eine Art von Abtragungsvorgang, *Petermanns Geographisch Mitteilungen*, *78*, 231–233, 1932.

Rougerie, G., Etudes des modes d'erosion et due faconnement des versants en Cote d'Ivoire equatoriale, *Premier rapport de la Commission pour l'Étude des Versants, Union Géographique Internationale*, *1*, 136–141, 1956.

Saunders, I. and A. Young, Rate of surface processes on slopes, slope retreat and denudation, *Earth Surface Processes and Landforms*, *8*, 473–501, 1983.

Savigear, R. A. G., Slopes and hills in West Africa, *Zeitschrift fur Geomorphologie Supplementband*, *1*, 156–171, 1960.

Schmid, J., *Klima, Boden und Baumgestalt in beregneten Mittelebirge*, Neudam, 1925.

Schumm, S. A., The role of creep and rainwash on the retreat of badland slopes, *American Journal of Science*, *254*, 693–706, 1956.

Schumm, S. A., The disparity between present rates of denudation and orogeny, *U.S. Geological Survey Professional Paper 454-H*, 1963.

Schumm, S. A., Geomorphic thresholds: the concept and its applications, *Transactions of the Institute of British Geographers*, *4*, 485–515, 1979.

Schumm, S. A., and R. W. Lichty, Time, space, and causality in geomorphology, *American Journal of Science*, *263*, 110–119, 1965

Scrope, G. P., On the origin of valleys, *Geological Magazine*, *3*, 193–199, 1866.

Sharma, V. N., and M. C. Joshi, Soil excavation by desert gerbil *Meriones hurricane* (Jerdon) in the Shekhawati region of the Rajastan desert, *Annals of Arid Zone*, *14*, 268–273, 1975.

Slaymaker, H. O., Patterns of present sub-aerial erosion and landform in mid-Wales, *Transactions of the Institute of British Geographers*, *55*, 47–68, 1972.

Söderman, G., Slope processes in cold environments of Northern Finland, *Fennia*, *158*, 83–152, 1980.

Stocking, M., A dilemma for soil conservation, *Area*, *10*, 306–308, 1978.

Thorn, C. E., A preliminary assessment of the geomorphic role of pocket gophers in the alpine zone of the Colorado Front Range, *Geografiska Annaler*, *60A*, 181–188, 1978.

Townshend, J. R. G., Geology, slope form and process and their relation to the occurrence of laterite in the Mato Grosso, *Geographical Journal*, *136*, 392–399, 1970.

Trimble, S. W., The fallacy of stream equilibrium in contemporary denudation studies, *American Journal of Science*, *277*, 876–887, 1977.

Walling, D. E., and B. W. Webb, Mapping solute loadings in an area of Devon, *Earth Surface Processes*, *3*, 85–99, 1978.

Welc, A., Spatial differentiation of chemical denudation in the Bystrzanka Flysch catchment (the West Carpathians), *Studia Geomorphologica Carpatho-Balcanica*, *12*, 149–162, 1978.

Winkler, E. M., Errors in using modern stream load data to estimate natural rates of denudation: discussion, *Bulletin of the Geological Society of America*, *81*, 983–984, 1970.

Wolman, M. G., and R. Gerson, Relative scales of time and effectiveness of climate in watershed geomorphology, *Earth Surface Processes*, *3*, 189–208, 1978.

Wolman, M. G., and J. P. Miller, Magnitude and frequency of forces in geomorphic processes, *Journal of Geology*, *68*, 54–74, 1960.

Yair, A., Observations sur les effets d'un ruissellement dirigé selon la pente des interfluves dans une région semi-aride d'Israël, *Revue de Géographie Physique et de Géologie Dynamique*, *14*, 537–548, 1972.

Yair, A., Sources of runoff and sediment supplied by the slopes of a first order drainage basin in an arid environment (northern Negev—Israel), *Akademie der Wissenschaften, Mathematisch-Physikalische Klasse*, *29*, 403–417, 1974.

Young, A., Soil movement by denudational processes on slopes, *Nature*, *188*, 120–122, 1960.

Young, A., Deductive models of slope evolution, *Neue Beitrage zur internationalen Hangforschung*, Vandenhoeck and Ruprecht, Göttingen, p. 45–56, 1963.

Young, A., Present rate of land erosion, *Nature*, *224*, 851–852, 1969.

Young, a., *Slopes*, 288 pp. Oliver and Boyd, Edinburgh, 1972.

Young, A., The rate of slope retreat, *Institute of British Geographers Special Publication*, *7*, 65–78, 1974.

Young, A., A twelve-year record of soil movement on a slope, *Zeitschrift fur Geomorphologie*, *29*, 104–110, 1978.

van Zon, H. J. M., The transport of leaves and sediment over a forest floor, *Catena*, *7*, 97–110, 1980.

Part II:

Hydraulic processes

2
Evaluation of Horton's theory of sheetwash and rill erosion on the basis of field experiments

Thomas Dunne and Brian F. Aubry

Abstract

Robert Horton's theory of sheetwash erosion and rill incision was based on his analysis of sheetwash without rainsplash. We conducted experiments on these processes on savanna hillslopes in Kenya to obtain information on the hydraulics of sheetflow and on sediment transport. These data were used to construct a simple, two-dimensional model of flow and sediment transport over natural microtopography, and thus of the spatial pattern of erosion. The model correctly reproduced Horton's prediction and our field observation that, on a sufficiently steep slope, sheetflow would become unstable and would incise a dense network of rills. When sheetwash was accompanied by rainsplash, the rills were filled with sediment and eradicated. Larger rills, formed by large volumes of sheetflow from hillslope segments 600–900 m long, were also degraded by altering the relative roles of raindrop impact and channel flow. The results support the theory that the initiation and maintenance of rills depends upon the balance between sediment transport by sheetwash, which tends to cause channel incision, and rainsplash, which tends to diffuse sediment from protuberances and to fill channels, thereby smoothing the surface.

The Horton theory

Horton (1945) proposed a theory of sheetflow, sheetwash erosion, and rill formation which injected a new level of rigor into hillslope geomorphology. The theory stimulated agricultural engineering studies of soil erosion and conservation, and in geomorphology *sensu stricto* it formed the basis for interpreting landscape geometry (Schumm, 1956; Melton, 1958) and for mathematical modeling of hillslope profiles (Hirano, 1966, 1975; Kirkby, 1971; Smith, 1971; Smith and Bretherton, 1972).

The important components of Horton's theory are the following. Rainfall intensity exceeds the infiltration capacity of the soil, and water accumulates in depressions on the soil surface before spilling over to run downslope as an irregular sheet of turbulent flow. The sheet imposes a shear stress on the soil surface. Close to the divide, the shear stress is less than the critical value required to entrain particles, but as the flow thickens and accelerates with increasing distance downslope the "critical tractive force" of the soil is eventually exceeded and erosion occurs beyond a critical distance. Horton derived this important length scale as a function of the difference between rainfall intensity and infiltration capacity, the hydraulic roughness and critical tractive force of the surface, and the local gradient. He was not able to take advantage of the research on rainsplash erosion that Ellison (1944, 1947) was doing at the time and thus did not appreciate the significance of erosion at shorter distances from the divide.

Beyond this "belt of no erosion" (Horton, 1945) sheetwash erosion is proportional to total shear stress (that shear stress taken up by vegetation, large immobile grains, and tiny bedforms is ignored), and therefore to the local product of flow depth and water surface gradient (taken as equal to the land surface gradient). A consequence of this relationship is that deeper filaments of the flow should erode faster than neighbouring filaments and thus become even deeper, evolving into rills. Thus, an erosive sheetflow is fundamentally unstable, and "sheet erosion implies the formation of either a rilled or gullied surface" (Horton, 1945, p. 332). According to this model, the critical distance required for the sheetflow to become erosive is the same as that required for channel incision and is equal to one-half the reciprocal of drainage density.

Developments of the Horton theory

Numerical analysis

Cordova et al. (1983) modeled the consequence of two-dimensional flow and sediment transport over a hillslope, the initial surface of which was assigned a random field of microtopography and of erodibility (critical shear stress). A flow concentration was induced at the lower boundary. It is not clear why this last step was necessary, since the initial erosion could have been located by the interaction of flow and the randomized initial conditions at the boundary. The authors followed Horton in limiting erosion near the divide by specifying a critical shear stress above which erosion begins. Beyond this belt of no erosion, sediment transport was taken as a power function of local velocity, and bed elevations were computed for sequential time steps, showing the growth of a rill network. The diagrams contained in the chapter by Cordova et al. imply only headward extension of the rills, although it is not clear why this should be the sole process of channel extension given their description of the erosion physics in the model. Also, the authors, following Horton, imply that the network is finally stabilized by the belt of no erosion.

Perturbation analysis

The issue of sheetflow instability and rill incision was taken up more formally by Smith (Smith, 1971; Smith and Bretherton, 1972). He cast the problem in terms of the equation of continuity for sediment transport:

$$\nabla \cdot \mathbf{e} \frac{q_s}{\rho_b} = \frac{\partial z}{\partial t} \tag{1}$$

where $\mathbf{e}$ is the unit vector parallel but in the opposite direction to the local gradient, ρ_b is the bulk density of soil, q_s is the mass flux rate of sediment per unit width of hillside, and $z(x, y, t)$ is the elevation of the land surface, which is twice continuously differentiable in space and once in time. It was assumed that the transporting water always moves down the local gradient of the surface (i.e., there are no convective accelerations due to microtopography).

Smith went on to postulate that the sediment flux rate is some positive monotonic function of the magnitudes of the local water flux rate per unit width of hillside q and the local gradient s $(= | \nabla z |)$:

$$q_s = F(q, s) \tag{2}$$

and in particular for sheetwash:

$$q_s = k_s q^n s^m \tag{3}$$

where m and n are greater than 1.0. He then examined the consequence of instantly modifying the surface by some infinitesimal amount. If such a perturbation is removed during continued erosion, the surface is stable; the surface is unstable if the initial perturbation grows and modifies the shape of the hillslope.

The sediment transport rate increases with local water discharge. If, for example, a perturbation of the surface, occurring without altering to a significant degree the magnitude of s, caused flow to converge so that the local discharge were doubled, its capacity for sediment transport would increase by 2^n, where $n > 1$. Yet the local sediment influx would only have been doubled. The difference in sediment transport between this local influx and the capacity would be made up by scour from the surface—that is, the flow convergence would cause incision. Thus, the mechanism triggering incision is a perturbation of the surface that causes flow convergence, leading to a local downslope increase in flow and therefore in sediment transport. Smith used a perturbation analysis to demonstrate that the downslope rate of increase of water flow is proportional to the transverse curvature of the perturbed surface:

$$\frac{\partial q'}{\partial x} = \frac{x}{s} \frac{\partial^2 z'}{\partial y^2} \tag{4}$$

where the primes indicate the perturbed state. A confusing aspect of this analysis, referred to by Smith and Bretherton (1972, p. 1521) as "unrealistic," is that perturbations with a small wavelength should grow faster than larger ones because they have greater transverse curvatures, and thus they force a greater rate of flow convergence. Such a result implies incision around every particle on the surface and the rilling of the entire surface.

The arguments of Horton and Smith are fundamentally sound, as they refer to sheetwash alone. The more uncomfortable aspects of Smith's finding that the smallest scale of perturbation is the one that should grow most quickly can be countered by modifying his and Horton's assumption that flow and sediment transport always follow the local gradient. In a paper on the growth of spatial structure in a class of biological and social systems, Smith (1976) used an analysis similar to that for flow and sediment transport. In this case, individuals migrate up local density gradients. Perturbations of all frequencies grow, and the smallest frequencies of perturbation grow at the fastest rate, precluding any stable spatial structure. Smith (1976, p. 368–369) goes on to show that if the individual migrant responds not to the gradient at a point but over some area of width a, the wavelength of perturbation that grows fastest and dominates the spatial structure is a.

Our interpretation of how this effect might be relevant to rill formation lies in the field observation that the velocity of overland flow over a rough hillside is influenced not only by local gradient but by the microtopography over some upslope and lateral distance. Convective accelerations, caused for example when flow is driven over microtopography by the local pressure gradient, can complicate the relationships between bed gradient, water surface gradient, and velocity, so that the flow and sediment transport fields are influenced by the average gradient over some area. According to Smith's (1976, p. 369) derivation, the width of this area of influence should be equal to the spacing of rills (the wavelength of the spatial pattern). On many hillsides this value exceeds 1 m, which from our qualitative field observations of runoff seems too large for an area of influence for flow over microtopography. The resolution of this problem awaits an analysis of the full Navier–Stokes equations for two-dimensional flow over rough, realistic microtopography. However, the mechanism of stabilizing a rill pattern by averaging stresses over some area does not explain decreases of rill density over time. For example, Collins (1984) found that as the infiltration capacity of a tephra surface increased through time and the frequency and intensity of sheetwash declined, small rills were filled and the drainage density decreased.

Effects of rainsplash

Laboratory observations

Moss et al. (1979) observed during laboratory experiments that on low slopes raindrops impacting a thin sheetflow tended to suppress channel formation. As the slope was increased at a constant total water discharge consisting of different proportions supplied by rainfall and inflow at the upper boundary, a greater application of rainfall energy was needed to suppress channels, until on the steepest gradient (0.30) a rainfall intensity of 100 mm/hr was insufficient. The same authors (Moss et al., 1982) conducted another set of experiments on laboratory channels in sand. They observed that thin sheetflows tend to develop secondary circulations transverse to the main flow, and that scour resulted when neighboring flow cells lowered fast surface water to the bed. The resulting scour localized the cells, causing enlargement to "protochannels." However, they observed that intense bombardment by raindrops suppressed the secondary flow, stabilizing the surface. This mechanism may be important at the small perturbation scales examined in the experiments, but it cannot extend to the scale at which most rills are stabilized or eradicated on natural hillslopes (Collins, 1984) or even on laboratory slopes used by Moss et al. (1979). In these cases it has been observed that rills with depths and widths of up to several centimeters have been eradicated by intense rainfall.

Field observations

Dunne (1980) objected to the initial Horton statement that sheetwash implies the formation of rills and gullies. He pointed to examples of hillslopes in Kenya known to be eroding rapidly that showed few signs of rilling on the upper several hundred meters of the longitudinal profiles. Such hillslopes have significant microtopography. We have surveyed cross-profiles at various distances along these hillsides, measuring elevations reproducibly to within 1 mm at intervals of 0.1 m. The amplitudes and wavelengths of microtopography range from 0.01 to 0.25 m and from 1 > 10 m respectively. Over most of the hillslope length, the depressions in this microtopography are not incising to form an integrated network of rills. Flow and sediment transport converge and diverge around microtopographic highs. The surface appears to be stable against rilling until a distance of several hundred meters from the divide is reached. Above this point rills are not formed even though sheetwash dominates sediment transport. Thus, Horton's critical length of sheetwash erosion is not the same as the critical distance needed for the development of rills. The latter distance exceeds the former by an amount that is small in the case of clay badlands (Schumm, 1956) or new, fine-grained tephra (Collins et al., 1983) but in other cases is very large.

Theoretical reasoning

The fact that sheetwash tends to incise a surface yet many hillslopes are stable against such incision implies that there is another process which tends to counter the incision (Dunne, 1980). The process could be any form of sediment transport that diffuses particles from high to low areas at a rate that depends on the local gradient, and thus tends to smooth out perturbations in the surface. Rainsplash, soil creep, frost action, tree-throw, and trampling or digging by animals will each behave in this manner, and will tend to destroy rills. Rainsplash is the most intense and widespread of these processes in environments where sheetwash occurs, although frost action is known to eradicate rills in badlands (Schumm, 1956). Dunne (1980) argued that as rainsplash transport is usually taken to be roughly proportional (via k_r) to the local gradient, the continuity equation for lateral sediment transport derived from (1) is

$$-\frac{\partial z'}{\partial t} = k_r \frac{\partial^2 z'}{\partial y^2} \tag{5}$$

The local lowering rate by rainsplash is proportional to the lateral curvature of microtopography. This tendency is of the same form and opposite sign to the tendency for incision due to sheetwash alone, obtained by substituting (4) into (1).

Near the divide, sheetflow is thin. If microtopography forces this sheetflow to converge locally, there is a corresponding thinning of the sheet and emergence of microtopographic mounds. The intensity with which sediment is mobilized by raindrop impact is extremely sensitive to the depth of water covering the soil surface. For example, McCarthy (1980, p. 174–175) found that the volume of aerial transport decreased by approximately one-half as the depth of water covering the surface increased from 0 to 0.2–0.25 times the raindrop diameter. Thus, flow convergence would be associated with neighboring flow divergence and thinning with a consequent strong increase in aerial rainsplash from the mound. Even if the mound did not emerge completely, raindrop impact through a thinned sheetflow would accelerate sediment transport from the mound. As the average flow depth increases downslope, a larger proportion of mounds will be submerged by deeper sheetflow until at some critical distance (which would depend on the runoff rate, gradient, and soil erodibilities with respect to both rainsplash and sheetwash) the diffusion of sediment from mounds will not keep pace with the incision by sheetwash and the rilling tendency will dominate.

If gradient and/or runoff rate are high, the shear stress and therefore sediment transport capacity of the sheetwash are high. The tendency for incision is strongest under these conditions. Dense networks of rills on clay badlands, road cuts, and silt-rich cultivated fields reflect the dominance of this tendency. In the experiments of Moss et al. (1979), the transport capacity of sheetwash increased with gradient and successively larger applications of raindrop energy were required to cause enough splash transport into the flow concentrations to balance the sheetwash transport capacity and prevent incision. Eventually, the gradient and shear stress became sufficiently high that the rainsplash could no longer supply enought sediment to do this.

Thus, the formation of rills depends on the balance between sheetwash, which incises, and rainsplash, which diffuses or smooths the surface. If sheetwash is rare due to a high infiltration capacity, but rainsplash occurs often, then rills may be absent or may begin far downslope, where x is large and s is small in (4). However, if the infiltration capacity is very low so that sheetwash occurs in most rainstorms, and particularly if the slope is steep, the incision process could exceed the ability of rainsplash to diffuse sediment from the protuberances even close to the divide. Collins (1984) used such reasoning to explain the dramatic decrease in rill densities on Mount St. Helens tephra as the infiltration capacity increased through time.

Deductions

If the formation of rills reflects the relative intensities of rainsplash and sheetwash, the following features of the process should be observable in the field:

1. A sheetflow without raindrops should become unstable and form rills where the product of depth and gradient of the sheetflow is sufficiently high.
2. If water is then applied to the surface as rainfall, there should be a tendency to fill the rills with sediment, and if the rainfall intensity is sufficiently high the rills should be eradicated.
3. If artificial rainfall is applied only to the head of a rill at the foot of a long hillslope that sheds large volumes of runoff in natural rainstorms of the same intensity, the rill form should be degraded and possibly eradicated, because of the altered balance of rainsplash and flow transport, which may cause faster lowering of the rill divide than the channel or even filling of the channel.

Field experiments

We have conducted a set of field experiments to test the hypotheses outlined above and to obtain information on runoff and erosion processes that can be used to extend the ideas of Horton and Smith. The experiments involved subjecting plots with various gradients and vegetation covers on one soil type to artificial rainstorms, and measuring the resulting runoff characteristics and sediment transport. These values were used in the construction of models of sediment transport and erosion. We have also applied runoff without rainsplash to the upper ends of plots in order to examine and model the process to which the theoreticians were referring. This chapter focuses on these latter experiments, which were conducted on Eremito Ridge at the northern boundary of Amboseli National Park, southern Kenya.

Geographical setting

The geographical setting of the experimental sites has been described in detail by Dunne and Dietrich (1980a) and Western and Dunne (1979). The hillslopes, which are remnants of the late-Tertiary erosion surface mapped by Williams

(1972), are long (600–3000 m) and gently convexo-concave. Their average gradients lie between 0.013 and 0.073, and the maximum local gradient at the inflection point ranges between 0.022 and 0.270. The hillslopes are developed on Precambrian gneisses, schists, and amphibolites, which have weathered to poorly aggregated, kaolinitic, sandy clay loam ferric-luvisols. Soil thickness ranges up to approximately 1 m, depending on local gradient (Fig. 2.1). Where soils are thin, gravel and cobbles, often originating in quartz veins, emerge at the surface and spread downslope as a thin, surface layer. Elsewhere, these coarse particles are dispersed widely over the surface and particularly along rills. The soil surface is covered by a discontinuous and poorly sorted lag of sand, with a mode in the range 0.25–0.5 mm.

The ridge receives an average of approximately 300 mm of rainfall per year during two short rainy seasons, April–June and November–December. Most rainstorms are short and intense. Sparse, uncertain rainfall and heavy grazing by cattle, sheep, goats, and wild herbivores leaves a thin groundcover of grass and low bushes. The density of this cover varies radically between and during seasons because of large inter-annual and seasonal variations of rainfall, and because of the strongly seasonal grazing pressure. At the end of an average rainy season, local ground cover, measured over the area of our 15 m^2 plots, varies from 0 to 80 percent, and basal cover (measured after clipping the leafy, aerial parts of plants) varies from 0 to 35 percent. After a long drought, cover is at its basal value, which is nowhere greater than 10 percent.

Despite erosion rates of several millimeters per year indicated by root exposures on bushes (Dunne et al., 1978), the hillslopes are unrilled within 500–600 m of the divide. Beyond this distance the rills rapidly develop widths of 0.5 m and depths of 0.1 m and increase to widths of 1.0 m and depths of 0.25 m on hillslopes one kilometer in length.

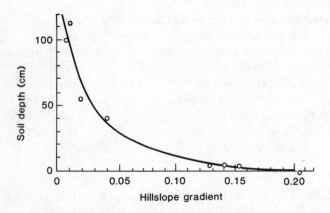

Figure 2.1 Variation of soil depth with gradient on Eremito Ridge.

Experimental methodology

Plots 6 m long and 2.5 m wide were bounded by sheet-metal walls and a trough cemented into the soil. Some plots were located around the heads of rills, and consequently had converging side boundaries and a concrete throat at a step excavated in the channel to allow the collection of runoff.

Two types of experiments were conducted on the plots. In the first type, artificial rainfall was applied to the plots at rates of 60–140 mm/h for 60 minutes. The rainfall was generated from a simulator similar to that described by Dunne et al. (1980), except that it included two movable nozzles spraying from a height of 5 m above the soil, and the nozzles used were the Veejet-100 and Veejet-150 models, manufactured by Spraying Systems Inc. of Des Moines, Iowa. The former was used by Meyer (1958) in the USDA "rainulator," where it operates at a height of approximately 2.5 m above the ground surface. In our experiments the nozzles operated at a pressure that should produce median drop sizes of 2.2 mm and 2.8 mm, according to the manufacturer's specifications. The experiments with rainfall included measurements of rainfall intensity and runoff, from which local rates of discharge (m^3/m s) at any distance along the plot could be calculated under the assumption of spatially uniform infiltration. Average flow depths were measured with a thin ruler at 0.1- or 0.25-m intervals across the plot and 0.5-m intervals along the plot. From the average discharge and flow depth at a cross-section, the average flow velocity could be computed. Average flow depths and velocities were also measured in flow concentrations. The depths and velocities were used together with gradient to compute Darcy–Weisbach friction factors, as described by Dunne and Dietrich (1980b). These hydraulic measurements were used as the basis for modeling the flow and checking results, as described later.

The second type of experiment involved spreading water across the upper ends of the plots without rainfall. Water was applied to the plot by pumping it from two pipes onto a sloping board at the upper boundary. It then flowed onto a rumpled canvas, which ponded it briefly before allowing drainage onto the plot at low velocity. As the water spread downslope, infiltration reduced the discharge, which therefore declined more or less linearly downslope in contrast to the normal case when rainfall excess increases the discharge along the plot. As all experiments were preceded by an experiment involving artificial rain one day earlier, the initial soil-moisture content was between field capacity and saturation. Therefore, infiltration quickly reached a constant value. Runoff typically reached the trough within 1–4 minutes of application, and the runoff rate was monitored by continually timing the filling of wide-mouthed sample bottles. The measurements of runoff varied by no more than 5 percent once a steady rate had been established. The same procedure was used to measure rates of sediment transport every two minutes. The same hydraulic measurements and computations were made as for the rainfall experiments.

Before and after each experiment, plot microtopography was surveyed with a level to the nearest millimeter of elevation at intervals of 0.5 m along the plot

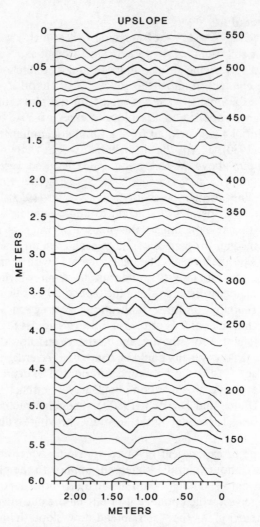

Figure 2.2 Contour map of plot KR-11 before the experiments. Numbers on the right-hand side of the figure are elevations in millimeters above an arbitrary datum. Contour interval is 10 mm.

and 0.1 m or 0.25 m across the plot. Contour maps of the plot surfaces were later constructed with a computer program that fits a bivariate spline surface to the measured elevations (Fig. 2.2). The same program was also used to interpolate a 10-cm-square grid of elevations for numerical computations of runoff, sediment transport, and erosion. These computations are described later.

The characteristics of the plots used for studying sheetwash without rainfall

Table 2.1 Characteristics of plots used for studying sheetwash without rainfall.

Plot	Average gradient	Vegetation cover (percent)	Basal cover (percent)	Rill head?
KR-1	0.020	35	19	
KR-2	0.017	75	35	
KR-5	0.011	~1	~1	yes
KR-7	0.028	~1	~1	yes
KR-8	0.011	~1	~1	
KR-9	0.040	5	5	
KR-10	0.036	~1	~1	
KR-11	0.074	5	5	
KR-12	0.045	5	5	yes
KR-13	0.046	9	9	

are listed in Table 2.1. The range of plot gradients extends only to 0.074, but Figure 2.1 indicates that this includes an important fraction of the land with a soil cover. Steeper slopes expose bedrock or a mantle of gravel.

Results

Experiments with rainfall

None of the preceding experiments with raindrops (median diameter = 2.2 mm) produced rills on previously unrilled plots, despite the fact that the one-hour artificial rainstorms had intensities between 60 and 110 mm/h and recurrence intervals between 5 and 25 years (Fiddes et al., 1974). On the steepest plot, KR-11, one could observe during rainfall the recession of 5-mm-high scarps or broad knickpoints at speeds of about 5 mm/min, anastomosing streams of concentrated sand transport 1–2 cm wide, and streaks of heavy minerals on the surface. However, there was no incision of rills. On the plots with gentler slopes we did not observe such concentrations of sediment transport. Measurements of the runoff and sediment transport from these rainfall experiments will be reported elsewhere.

On the plots around the heads of the three previously established rills (KR-5, KR-7, and KR-12), the intense rainsplash combined with sheetwash caused rapid rates of sediment transport from the rill divides, which were lowered. For example, the slopes bounding the rill at the lower end of the plot KR-5 were lowered 5–30 mm by three rainstorms with a combined duration of 155 minutes and intensities of 80–110 mm/h. Depressions in the profile were also filled by sediment transported from upslope, smoothing the profile (see Fig. 2.3).

Sediment entered the rills, where the coarse and medium sand fractions traveled as bedload. In a natural rainstorm of such duration and intensity, the

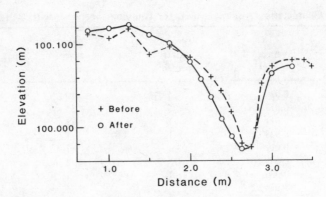

Figure 2.3 Cross-sections at a distance of 8.5 m from the top of plot KR-5 before the rainfall experiments (crosses) and after three rainstorms (circles) with durations of 35, 60, and 60 minutes, intensities of 96, 80, and 110 mm/hr, and median drop diameters of 2.2, 2.2, and 2.8 mm, respectively. Note the lowering of sideslopes and the aggradation and smoothing of the left-hand divide by sediment carried from upslope.

large drainage area would supply high rates of discharge. For example, a strip of hillside 2.5 m wide and 600 m long with an average infiltration capacity of 30 mm/h (Dunne and Dietrich, 1980*a*, p. 49) draining to plot KR-5 in a 110 mm/h, 60-minute rainstorm would generate a rill discharge of 0.034 m^3/s. Instead the peak discharge measured during a rainstorm of this intensity was 0.0005 m^3/s. Even if the runoff from the natural storm were originally loaded to capacity, its convergence from a width of 2.5 m into the head of the approximately 0.5-m-wide rill (see Fig. 2.3) would have increased its transport capacity by approximately $(2.5/0.5)^n$, as indicated by (3). Since n is commonly accepted to be $\geqslant 2.0$ (Kirkby, 1971; Smith and Bretherton, 1972), the increase in transport capacity due to the convergence would be large. Figure 2.3 shows that even the small experimental discharge could carry away all of the sediment splashed from the divides, so the natural discharge could have transported more than this amount, and would have scoured the sand bed of the rill, tending to maintain or increase the amplitude of the cross-section. Instead the rill became shallower during the rainfall experiments.

Experiments without rainfall

After approximately one day of drainage, the plots were subjected to sheetwash without raindrops. After a constant rate of outflow was established at the trough, specific discharges ranged up to 400×10^{-6} m^3/m s, and sediment flux rates ranged up to 7×10^{-3} kg/m s. Average flow depths varied slightly as discharge declined along each plot, and across the plot in response to microtopography. Typical flow depths averaged 7–11 mm (maximum local depth 13 mm) along KR-2 with a gradient of 0.017 and a cover density of 75 percent, and 1–2 mm (maximum local depth 8 mm) along KR-11 with a gradient of 0.074 and a cover of 5 percent.

Substitution of these values into the Shields criterion (Raudkivi, 1976, p. 30) for transport of 0.25–0.50 mm sand (a range which included the mean grain sizes of the armor layers on all plots) predicted transport everywhere on the almost-bare plots once discharge attained a steady state. This criterion is less easy to apply on densely vegetated plots because most of the total shear stress is there taken up by plant stems. However, the criterion predicted transport on the bare patches between clumps of ground cover.

On plots with gradients less than 0.05, transport rates were low and they declined rapidly within the first 10 minutes of each experiment as finer particles, stranded at the end of the preceding rainstorm, were washed from the plot (Fig. 2.4A). Throughout most of the experiment sediment transport was confined to bedload in the sandy armor layer, and to deeper filaments of flow between vegetation clumps or microtopographic highs.

In these zones of flow convergence, narrow zones of coarsened bed developed and were lowered, as Horton suggested. Scouring of these zones occurred very slowly because of the low rates of sediment transport, and most of the scoured sediment was redeposited in slower, deeper or diverging flow downslope. An integrated network of scoured zones did not develop on these gently sloping plots during the 35- to 60-minute experiments. One or two scour zones, 10–15 cm long, developed on gradients of 0.02 or less, and two or more zones were scoured on gradients of 0.02–0.05. On these steeper gradients, small arcuate steps, 3–4 mm high, developed and eroded slowly headward, widening and degrading. The greater density and integration of scoured zones on the

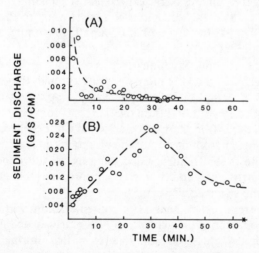

Figure 2.4 (A) Sediment discharge versus time for a sheetwash experiment without raindrops which produced no rills. (B) Sediment discharge versus time for a sheetwash experiment without raindrops which incised rills on plot KR-11.

gradients of 0.02–0.05 presaged the formation of a channel network, but the magnitude and duration of flow were insufficient to extend and deepen the scoured zones into a dendritic network. At the end of the experiments, a few scour marks led into small sand splays downslope. It seems reasonable to conclude that if the plots had been receiving runoff (without raindrops) commensurate with their position several hundred meters from a divide during natural storms, the sediment transport, scour, and network integration would have occurred much faster, but we were not able to reproduce these conditions on gradients less than 0.05. Each of the flow concentrations was quickly filled and eradicated by a succeeding artificial rainstorm.

On the steepest plot, KR-11, we were able to generate sufficient shear stress and sediment transport to integrate the scoured sections and form a rill network. In this case the arcuate steps grew to heights of as much as 5 cm as they quickly eroded headward. Their retreat produced well-defined rills up to 3.5 cm deep and 25 cm wide with vertical banks and headscarps. They intersected one another by deepening upslope and downslope, and branched as they eroded headward around clumps of vegetation or mounds.

The temporal variation of sediment transport from KR-11 (Fig. 2.4B) was strikingly different from that on unrilled plots (Fig. 2.4A). The sediment transport in flow concentrations was sufficiently intense to breach the sandy armor layer and exploit the sandy clay loam beneath. Thus, the sediment transport rate increased during the first 30 minutes. After that time the slowing of rill growth and the armoring of the channel beds with coarse sand and small gravel caused a decline in sediment flux. This decline was most pronounced for the finest proportion of the grain-size frequency distribution. The 16th percentile of the distribution of grain sizes washed from the plot declined sharply in the first few minutes of the experiment from 0.040 mm to 0.016 mm as the armor layer was breached, and then rose steadily to that of the sandy armor layer (0.088 mm) after 30 minutes of runoff, remaining at this value for the rest of the experiment (Aubry, 1984).

At the end of the one-hour experiment an anastomosing network of rills had formed. It had become gradually more dendritic through the experiment, but integration was not complete before the water supply was exhausted. We recorded the distribution of the rills on overlapping "aerial photographs" from the top of the simulator (Fig. 2.5). The density of rills declined from 3.9/m of cross-section one meter from the top of the plot to 2.2/m four meters from the top. Figure 2.6 shows the frequency distributions of width and depth of the rectangular channels. We could find no consistent changes of channel geometry along the short network.

After the rill-forming experiment a 60-minute rainstorm with an intensity of 139 mm/h and a median drop diameter of 2.8 mm was applied to the plot. Despite the radical increase in soil loss over that during the sheetflow experiment, rills were filled and eradicated. At the beginning of the storm, water concentrated in the channels, leaving their angular margins exposed to raindrop impact. Equation (5) predicts that the lowering of these corners

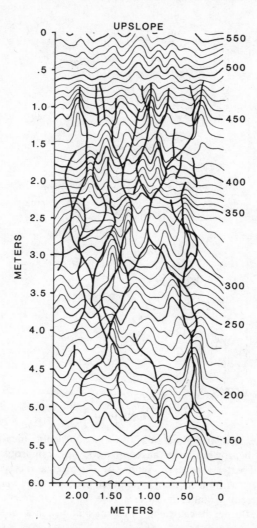

Figure 2.5 Predicted contour map of plot KR-11 after simulation of bed erosion. The dark lines are the observed locations of rills developed during a one-hour sheetflow experiment. Both the predicted and observed locations of rills originated on the same surface, shown in Figure 2.2, which was used as input to the erosion model. Contour interval is 10 mm.

should be particularly rapid. They were replaced gradually by rounded shoulders and then by more or less horizontal surfaces as the rills were filled and their margins lowered. Some rills were not completely eradicated by the end of the rainstorm, but their depths had decreased and widths had increased, as shown in Figure 2.6. The shallow, trough-shaped cross-sections continued to flatten until the end of the experiment.

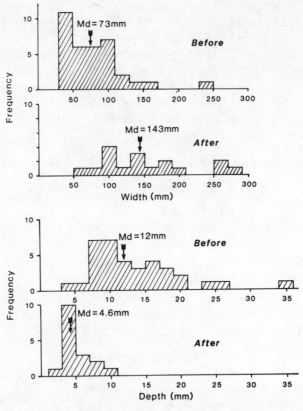

Figure 2.6 Frequency distributions of rill widths and depths measured every meter along plot KR-11 before and after an experiment with raindrops onto the micro-topography and rill pattern shown in Figure 2.5. The original rills were incised during an earlier experiment without raindrops, and their cross-sections were rectangular. At the end of the succeeding rainstorm the rills were fewer in number and had degraded to shallow troughs; the depths shown here were maximum values for the cross-sections.

Review of hypotheses

The three hypotheses proposed earlier were vindicted by the experiments. Altering the balance of rainsplash and flow transport degraded the form of even large rills originally formed by copious runoff from long hillsides in natural storms. Sheetwash without raindrops became unstable due to small heterogeneities in the sediment transport field that caused incision and set up a positive feedback whereby the convergence of water toward the incision extended the downcutting process, forming a channel network. The addition of raindrops countered this incision by diffusing sediment from the margins of incisions and filling them. Where the intensity of rainsplash was high relative to the capacity of sheetwash, rills were eradicated. However, even under rainfall impact, plot KR-11 had a gradient close to that required for rill formation 5 m from a divide. It is easy to understand, therefore, why hillslope segments steeper than 0.074 have surfaces of bedrock or a thin lag of gravel. Maasai pastoralists who live on these hillslopes indicated to Western and Dunne (1979, p. 868) that they avoid placing their settlements on gradients steeper than 0.07–0.08 partly because the shallow soils make it difficult to implant house poles, and the frequency of stones larger than 100 mm on these gradients cause discomforts and injury to cattle.

Modeling rill initiation by sheetflow

In order to generalize from the results of the sheetwash experiments it is necessary to construct a mathematical model of the flow and sediment transport processes that were measured. The remainder of this chapter describes the progress made so far and the problems encountered in the modeling of sheetwash without raindrop impact. The purpose of the model is to compute the spatial pattern of erosion and thus the change in shape of a segment of hillside. The model needs to be based on the continuity equation (1) in its two-dimensional form because we are particularly interested in the formation of rills. Equation (2) indicates that the computation of sediment flux requires a calculation of water discharge everywhere on the plot. Thus, the erosion model is driven by a two-dimensional model of runoff. The initial condition is determined by the microtopography measured before the sheetflow experiments, although any initial non-planar geometry could be used.

Runoff model
A mesh of bed elevations at the center of all 10 cm × 10 cm cells is first interpolated from a contour map, such as Figure 2.2. The magnitude and orientation of the gradient at each mid-point is then computed from a least-squares planar fit to the mid-point and its eight neighbors. The algorithm for this gradient computation was developed by C. D. Tomlin of Yale University.

Water enters the upper row of cells at the chosen rate, depending on the intensity of excess precipitation and distance from the upper end of the hillslope. The discharge is reduced by infiltration, and outflow is then divided between the downslope and adjacent cells according to the orientation of the local gradient. A set of such calculations is made successively across and down the plot for a single steady-state discharge (Aubry, 1984). The result is a map of discharges out of the lower end of each cell.

Local flow depths are calculated for each cross-section at 10 cm increments of distances along the plot. The computer program divides the micro-topography into blocks between local maxima on the cross-profile and then calculates the total discharge between each pair of maxima. It assumes that the water surface is horizontal across each block and computes the flow depths required to pass the discharge. If the required depth causes sideways spill over a topographic high, the blocks are redefined and the calculation repeated. The depth–discharge relationship is obtained from the Darcy–Weisbach equation, incorporating local slope and the hydraulic roughness of flow, which is mainly a function of vegetation density and Reynolds number. The relationship between hydraulic roughness, vegetation density, and Reynolds number were defined as described earlier from simultaneous measurements of flow depth, velocity, and gradient during the plot experiments (see also Dunne and Dietrich, 1980b). Comparisons of predicted cross-sectional variation of flow depths and velocities with values measured during experiments showed that the locations and approximate values of the deeper and faster portions of the flow were reproduced well, but that the model predicted zero depths for microtopographic highs, even though the field measurements indicated that very thin flow from higher microtopography upslope could cover a protuberance. Depths and velocities were predicted closely for rills. Since most of the sediment transport occurred in deeper filaments of sheetflow and in rills, the predictions were judged to be usefully close to reality.

Several drawbacks to the model include its crude discretization of the runoff process, its use of the Darcy–Weisbach flow equation, which is strictly valid only for steady, uniform flow, and its inability to take into account convective accelerations (and the associated shear stresses), which would affect sediment transport.

Sediment transport model

Sediment transport by sheetflow has received little experimental study (e.g., Meyer and Monke, 1965; Parsons, 1971; Moss et al., 1979; Singer and Walker, 1983), and there are even fewer studies that can be extrapolated quantitatively to natural hillslopes. Foster and Meyer (1972) concluded that the Yalin (1963) bedload equation was the best one available for transport by sheetflow. They reviewed its performance in predicting soil loss from experimental plots and small channels and found that large (severalfold) over- and underpredictions of measured values are common. There has been little or no critical re-examination of the equation's suitability for sheetwash since that time, and the Yalin and other equations developed for much deeper flows are used routinely (e.g., Simons et al., 1982) in mathematical models of erosion. Our field measurements provided a chance to evaluate the usefulness of such equations.

Sheetflow without raindrop impact carried little or no suspended load during our experiments. Thus we confined the computations to predicting the intermittent saltation of sand, which we observed in the field. Results from two bedload transport equations were compared: the Yalin and Einstein equations. Both were initially developed for flows much deeper and more turbulent than sheetflow and were tested in flumes and applied to rivers. However, the Yalin (1963) equation is known to predict sediment transport accurately at low excess shear stresses. The Einstein equation (Einstein, 1950; Einstein and Chien, 1953) is a flexible, widely used, predictor of bedload transport in mixtures of sand. Both methods predict that bedload transport is extremely sensitive to local boundary shear stess, and they incorporate empirical factors to account for the effects of Reynolds number (and therefore thickness of the laminar sublayer), grain size, and, in the Einstein case, of the influence of other grain sizes.

The bedload equations were first used on the grain-size distributions of composite sediment samples scraped from the upper 2–5 mm of the plots. Sediment discharge from the last row of cells on each plot was computed using the flow characteristics predicted by the two-dimensional flow model. The results were compared with sediment flux rates measured in the first 30–60 seconds of each experiment, before coarsening of the surface armor could become important. The results showed that both equations overestimated measured rates of sediment transport in sheetflow by one-half to one order of magnitude, and that the predictions for rill transport were very close to measured values. The overpredictions may be due to occasional overpredictions of flow depths (and therefore shear stresses) in flow concentrations, and to the fact that form drag on large, immobile grains or mounds was ignored in the equations, despite its role in reducing the shear stress effective in sediment transport. It is clear that carefully controlled experiments and theoretical studies are needed to develop a better transport equation for shallow, laminar, and turbulent flows over very rough surfaces.

We also attempted to predict the effects of surface armoring on bedload by computing how selective transport would alter the grain-size distribution of a reservoir of sediment with the initial depth and texture of the armor layer, and would thereby alter sediment transport in the successive time step. The Einstein equation was used for these computations because it gave the closer approximation to the measured grain-size distribution of sediment leaving the plots. The results reproduced the observed general pattern of declining sediment transport, but the rate of decline was extremely sensitive to overpredictions of flux rate and to the assumed initial size of the reservoir of sediment subject to armoring. Improvements in this aspect of sediment transport prediction await a transport equation more appropriate to sheetflow.

Erosion model
The transport model reproduced general patterns but not the true rates of sediment flux from the plots, and so it could not be used to compute rates of erosion. However, we did use it as an index of the local intensity of sediment transport. With this caveat, we have used the Einstein equation to compute the spatial pattern of erosion and deposition from the divergence of q_s in (1).

During the experiments we noted that there was little transport in lateral directions because lateral gradients were small relative to those downslope. Therefore, to reduce computer costs we calculated sediment flux for one-minute periods in only the downslope direction for each cell across a row. These influxes to the second row were compared with effluxes from cells in that row, and differences in flux were converted to erosion or deposition within the cells. The procedure was extended through each row on the plot, after which a new topographic map and gradient matrix were computed and subjected again to the runoff and sediment transport models. The latter was applied in its simplest form, without grain-size interactions or armoring, as we were using the calculated flux rates only as an index. Successive alternation of these steps evolved a spatial pattern of erosion and deposition—that is, a change in the form of the plot surface.

The crudeness of the computation, and particularly the magnitudes of the increments used to discretize time and space caused an oscillating instability of the calculated surface as waves of sediment passed through each cell. Thus, the contour maps computed at various times during model runs showed linear strings of small depressions and protuberances that oscillated their positions between time steps. However, as erosion was allowed to proceed to a depth similar to that observed during the sheetflow experiment, these linear strings continued to define the position of the deepening incision. In order to reduce the costs of computation, these oscillations were smoothed graphically, producing maps such as Figure 2.5.

The microtopography of plot KR-11 was not changed to a measurable degree by the first rainfall experiment. Therefore, Figure 2.2 is an accurate representation of the surface before the sheetwash experiment, and the differences between the contours in Figures 2.2 and 2.5 indicate the changes predicted by the erosion model for sheetwash without raindrops. The initial cross-sections exhibited microtopography with an amplitude of approximately 1 cm, and with some low points that would concentrate flow locally, but each concentration gave way downslope to a protuberance or planar surface. There was no throughflowing channel. The erosion model predicted that erosion would be concentrated in some depressions but not in others, and that in other depressions the incision would shift laterally from the original local minimum because of the influence of water discharge and sediment load entering a flow concentration from relatively large or steep contributing areas. The model also predicted that some of the depressions would deepen and integrate into a channel system, with the density of channels per unit width of cross-section decreasing downslope.

Comparison with observations

The model predicted that sheetflow without raindrops would generate rills on the previously unrilled plot KR-11. The predicted locations of these rills are shown by the contours in Figure 2.5. The rills which formed during the field experiment had vertical margins, and so it was not possible to define them well by surveying. Instead, we measured their depths and widths with a ruler and recorded their positions by taking vertical photographs from the top of the rainfall simulator. To compare the predicted locations against field results, rills were traced from the photographs onto the predicted contour map of the eroded surface. Figure 2.5 indicates that many of the rills lie along the depressions predicted by the model, although some occupy small ridges and the sideslopes of depressions. Some of these "misplaced" rills developed early in the experiment and were later abandoned by most of their original flow. However, we could not record channel changes photographically during the experiments. Continued erosion may have lowered the surface enough to eradicate some of these short, small rills. It is also possible that subtle spatial variations of infiltration or erodibility, which were not included in the model, may have been responsible for localizing some of the misplaced rills. In general, the predicted and observed patterns of rills, both of which originated with the same microtopography shown in Figure 2.2, agree remarkably well, and both display the expected integration downslope. This network development continued until the end of the experiment.

Summary

Field experiments and a two-dimensional model of flow, sediment transport, and erosion based on them confirm the projections of Horton and Smith that an erosive sheetflow is inherently unstable and will separate into streams that will incise channels and integrate into a network. Field experiments also demonstrate that such a tendency can be countered or reversed by raindrop impact that diffuses sediment from protuberances and fills depressions. We have not yet generalized our field results on rainsplash into a model of sheetwash–raindrop interaction.

The widely used models of sediment transport by sheetflow are not accurate predictors, and they need to be improved on the basis of theoretical and experimental study. In this investigation they could be used only as indices of the local intensity of bedload transport. Nevertheless, their incorporation into the erosion model yields some interesting results about the nature of the relation between process and morphology on rilled and unrilled hillslopes.

Acknowledgments

The research reported here was supported by the National Science Foundation (Grants EAR-80/8286 and EAR-8313172) and by the Kenya Rangeland Ecological Monitoring Unit (Government of Kenya). E. K. Wahome of KREMU, L. M. Reid, N. F. Humphrey, and S. Fouty of the University of Washington assisted with the field experiments.

References

Aubry, B. F., Runoff, sediment transport, and rill formation by sheetflow, M.S. thesis, 108 pp, University of Washington, Seattle, July, 1984.

Collins, B. D., Erosion of tephra from the 1980 eruption of Mt. St. Helens, M.S. thesis, 181 pp, University of Washington, Seattle, 1984.

Collins, B. D., T. Dunne, and A. K. Lehre, Erosion of tephra-covered hillslopes north of Mt. St. Helens, Washington: May 1980–May 1981, *Zeitschrift für Geomorphologie, Supplement Band 46*, 103–121, 1983.

Cordova, J. R., I. Rodriguez-Iturbe, and P. Vaca, On the development of drainage networks, *International Association of Hydrological Sciences Publication 137*, 239–249, 1983.

Dunne, T., Formation and controls of channel networks, *Progress in Physical Geography, 4*, 211–239, 1980.

Dunne, T., and W. E. Dietrich, Experimental study of Horton overland flow on tropical hillslopes: 1. Soil conditions, infiltration and frequency of runoff, *Zeitschrift für Geomorphologie, Supplement Band 35*, 40–59, 1980a.

Dunne, T., and W. E. Dietrich, Experimental study of Horton overland flow on tropical hillslopes: 2. Hydraulic characteristics and hillslope hydrographs, *Zeitschrift für Geomorphologie, Supplement Band 35*, 60–80, 1980b.

Dunne, T., W. E. Dietrich, and M. J. Brunengo, Recent and past erosion rates in semi-arid Kenya, *Zeitschrift für Geomorphologie, Supplement Band 29*, 215–230, 1978.

Dunne, T., W. E. Dietrich, and M. J. Brunengo, Simple portable equipment for erosion experiments under artificial rainfall, *Journal of Agricultural Engineering Research, 25*, 161–168, 1980.

Einstein, H. A., The bed-load function for sediment transport in open channel flows, *U.S. Department of Agriculture Technical Bulletin 1026*, 1950.

Einstein, H. A., and N. Chien, Transport of sediment mixtures with large ranges of grain sizes, *Missouri River Division, Sediment Series 2*, 49 pp., University of California Institute of Engineering Research, Berkeley, 1953.

Ellison, W. D., Studies of raindrop erosion, *Agricultural Engineering, 25*, 131–136, 181–182, 1944.

Ellison, W. D., Soil erosion studies, *Journal of Agricultural Engineering, 28*, 145–146, 197–201, 245–248, 297–300, 349–351, 1947.

Fiddes, D., J. A. Forsgate, and A. O. Grigg, The prediction of storm rainfall in East Africa, *Transport and Road Research Laboratory Report 623*, 50 pp., United Kingdom Department of the Environment, London 1974.

Foster, G. R., and L. D. Meyer, Transport of soil particles by shallow flow, *Transactions of the American Society of Agricultural Engineers, 15*, 99–102, 1972.

Hirano, M., A study of a mathematical model of slope development, *Geographical Review of Japan, 39*, 324–336, 1966.

Hirano, M., Simulation of developmental process of interfluvial slopes with reference to graded form, *Journal of Geology, 83*, 113–123, 1975.

Horton, R. E., Erosional development of streams and their drainage basins: hydrophysical approach to quantitative morphology, *Bulletin of the Geological Society of America, 56*, 275–370, 1945.

Kirkby, M. J., Hillslope process–response models based on the continuity equation, *Institute of British Geographers Special Publication 3*, 15–30, 1971.

McCarthy, C. J., Sediment transport by rainsplash, Ph.D. thesis, 215 pp, University of Washington, Seattle, 1980.

Melton, M. A., Correlation structure of morphometric properties of drainage basins and their controlling agents, *Journal of Geology, 66*, 442–460, 1958.

Meyer, L. D., An investigation of methods for simulating rainfall on standard runoff plots and a study of the drop size, velocity, and kinetic energy of selected spray nozzles, *Special Report 81*, 43 pp., Purdue University Agricultural Experiment Station, West Lafayette, Ind., 1958.

Meyer, L. D., and E. J. Monke, Mechanics of soil erosion by rainfall and overland flow, *Transactions of the American Society of Agricultural Engineers, 8*, 572–580, 1965.

Moss, A. J., P. Green, and J. Hutka, Small channels: their experimental formation, nature, and significance, *Earth Surface Processes and Landforms, 7*, 401–415, 1982.

Moss, A. J., P. H. Walker, and J. Hutka, Raindrop-stimulated transportation in shallow water flows: an experimental study, *Sedimentary Geology, 22*, 165–184, 1979.

Parsons, D. A., The speeds of sand grains in laminar flow over a smooth bed, in: *Sedimentation*, edited by H. W. Shen, pp. 1–1 to 1–25, Fort Collins, Colo., 1971.

Raudkivi, A. J., *Loose Boundary Hydraulics*, 2nd ed., 397 pp., Pergamon, Oxford, 1976.

Schumm, S. A., Evolution of drainage systems and slopes in badlands at Perth Amboy, New Jersey, *Bulletin of the Geological Society of America, 67*, 597–646, 1956.

Simons, D. B., R.-M. Li, T. J. Ward, and L. Y. Shiao, Modeling of water and sediment yields from forested drainage basins, *U.S. Forest Service General Technical Report PNW-141*, 24–38, 1982.

Singer, M. J., and P. H. Walker, Rainfall–runoff in soil erosion with simulated rainfall, overland flow, and cover, *Australian Journal of Soil Research, 21*, 109–122, 1983.

Smith, T. R., Conservation principles and stability in drainage basin evolution, Ph.D. thesis, 80 pp., The Johns Hopkins University, Baltimore, 1971.

Smith, T. R., Set-determined process and the growth of spatial structure, *Geographical Analysis, 8*, 354–375, 1976.

Smith, T. R., and F. P. Bretherton, Stability and the conservation of mass in drainage basin evolution, *Water Resources Research, 8*, 1506–1529, 1972.

Western, D., and T. Dunne, Environmental aspects of settlement site decisions among pastoral Maasai, *Human Ecology, 7*, 75–98, 1979.

Williams, L. A. J., Geology of the Amboseli area, *Geological Survey of Kenya Report No. 90*, 86 pp. Nairobi, 1972.

Yalin, M. S., Expression of bedload transportation, *Proceedings of the American Society of Civil Engineers, Journal of the Hydraulics Division, 89*, 221–250, 1963.

3
Erosion processes and sediment properties for agricultural cropland

L. D. Meyer

Abstract

Soil erosion and sedimentation by water involve the processes of detachment, transport, and deposition of soil materials by rainfall and runoff. Major sources of sediment eroded from sloping land are interrill areas, rills, and gullies, with the relative contribution from each varying widely depending on the specific situation. Eroded sediment generally consists of both primary soil particles and aggregates, which greatly affect the size distribution, density, and stability of the sediment and, consequently, sediment transportation and deposition. The rate of soil erosion depends on various climatic, soil, topographic, and cultural factors. By identifying and quantifying such factors, erosion rates can be predicted more accurately and better soil conservation practices can be developed.

Introduction

Productive agricultural soil constitutes only a small part of our Earth's mantle, yet it is one of humanity's essential natural resources. It provides most of the food we eat, the clothes we wear, the environment in which we live, and various other aspects of our existence that make possible "the good life." Our present-day soil has resulted from eons of physical, chemical, and biological processes, including massive natural erosion and sedimentation, without which our Earth might be a virtually sterile planet. However, recent human activity, particularly some agricultural practices, has accelerated soil erosion rates on much of our more productive land. Erosion exceeds agriculturally tolerable rates on about one-fourth of our nation's cropland, and annual rates above 100 t/ha occur on some areas (CAST, 1982). The effects of a few years of such rapid erosion may exceed those of centuries of natural erosion, so the time-frame of landscape changes may be relatively short when land is used intensively.

Excessive soil erosion not only reduces the productive potential of cropland, but it also may harm the lands and waters that receive the eroded material. Thus, soil erosion is a menace to the continued well-being of present and future civilizations that needs to be limited to tolerable rates. This paper discusses current knowledge concerning processes involved in soil erosion and sedimentation by water, physical characteristics of eroded sediment, and some of the major factors that affect soil erosion rates on sloping land in general, agricultural cropland in particular.

Processes involved in soil erosion and sedimentation by water

The companion phenomena of soil erosion and sedimentation by rainfall and runoff involve detachment of soil material from the soil mass, transportation downslope, and eventual deposition (Bennett, 1939; Stallings, 1953; Guy, 1970; Meyer, 1979). Soil erosion is the detachment of soil particles (both primary and aggregated) from the soil mass by raindrop impact or runoff shear and their transport by splash or by flowing water. Eroded soil in transport is sediment, and sedimentation occurs when eroded materials are deposited. Individual particles may undergo this series of events many times within a rainstorm and during subsequent storms, so final deposition may be far beyond the original source.

Rainfall and runoff
For soil erosion by water, impacting raindrops and flowing runoff are the erosive agents that work against the resisting forces of soil cohesion and gravity to detach and move soil material. Soil erosion on sloping land usually begins when raindrops strike exposed soil. Raindrops vary from mist sized to about 7 mm in diameter, with the average size for most rainstorms between 1 and 3 mm (Laws and Parsons, 1943). Several trillion raindrops annually bombard each hectare of land in the humid regions of the world at impact velocities up to 9 m/s. In areas where the annual rainfall is about one meter, the water falling on just one hectare has a volume of $10\,000\ m^3$, a mass of about $10\,000\,000$ kg, and falls with an impact energy of 200 to 300 MJ (Meyer, 1979). Unless the soil surface is protected by vegetation or other cover, these impacting raindrops can detach tremendous quantities of soil from the soil mass. The detached particles are splashed in all directions from the impact points with some net movement downslope.

Most soil eroded by water is transported downslope by surface runoff. Runoff does not begin, however, until the rain intensity exceeds the infiltration rate of the soil and the ponding capacity of the land surface has been satisfied. Thus, soil conditions having high intake rates or large surface-ponding capacities may appreciably delay runoff and reduce subsequent runoff rates. Runoff not only transports soil that is detached by rain, but it can also detach and transport additional soil. Only 10 mm of runoff from one hectare amounts to 100 000 liters with a mass of 100 000 kg. In the United States, annual runoff varies from very little in the Great Plains and desert areas to a depth of more than 500 mm totaling more than 500 000 000 l/km^2 near the Gulf of Mexico, across the New England states, and in several mountainous areas (Geraghty et al., 1973).

Detachment, transport, and deposition

Detachment of soil particles from the soil mass is the initial process during soil erosion. Since rainfall occurs relatively uniformly over a farm field or similar area, detachment by raindrop impact may occur anywhere that soil is exposed to the shear forces resulting when these generally spherical drops flatten at impact. In contrast, detachment by runoff is generally limited to a small portion of the land area where the runoff concentrates at erosive velocities. Detachment by concentrated runoff causes rills and gullies that are quite visible, whereas detachment by rainfall is seldom noticeable unless rocks or other surface covers are present to protect some of the land surface from raindrop impact and permit soil pedestals to form beneath them.

Transportation of detached soil results from both raindrop splash and overland flow. Net downslope movement by splash increases as the slope steepens (Ekern, 1950). The net rate of downslope transport by splash is usually quite small relative to that by runoff (Young and Wiersma, 1973; Lattanzi et al., 1974), but a considerable amount of soil may be splashed short distances to rills and gullies where concentrated runoff can transport it. The quantity and size of the material that can be transported by runoff is a function of runoff velocity and turbulence (ASCE, 1975), and these increase as the slope steepens and as the depth of flow increases. Velocity is usually quite slow until the flow accumulates and concentrates on a limited portion of the land surface.

Deposition of sediment being transported by runoff may occur when the transport capability of runoff is reduced, such as when the flow velocity is decreased by vegetation or by decreased slope steepness. Deposition is a selective process with the largest and most dense materials settling first, whereas very fine colloidal material may remain in suspension for long periods. Therefore, the size distribution of sediment leaving an eroding area may not be representative of the material detached and transported on that area if selective deposition has occurred upstream of the point of measurement.

The interrelationships among soil detachment, transportation, and deposition were described many years ago by Ellison (1947). He defined soil erosion as "a process of detachment and transportation of soil materials by erosive agents" (Ellison, 1947, p. 145). For erosion by water, these agents are rainfall and runoff. Ellison pointed out that each has both a detaching and a transporting capacity, and that these should be studied separately. Similarly, the erodibility of a soil may be separated into its detachability and transportability components. Meyer and Wischmeier (1969) incorporated these components into the framework of a mathematical model that considers (1) soil detachment by rainfall, (2) transport by rainfall, (3) detachment by runoff, and (4)

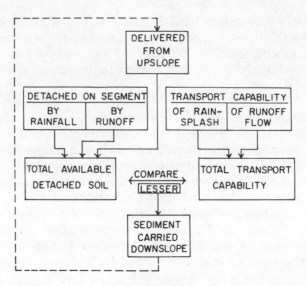

Figure 3.1 Approach used to simulate the process of soil erosion by water on each successive slope-length segment. The four subprocesses are evaluated for each segment, and sediment movement is routed downslope as illustrated. (From Meyer and Wischmeier, 1969).

transport by runoff as separate but interrelated phases of soil erosion by water (Fig. 3.1). Each phase then may be described by appropriate mathematical relationships plus parameters to characterize the soil properties that influence erosion. In this simulation approach the soil erosion rate for each successive slope-length segment is determined by the limiting component, either available detached soil or transport capability. If a situation occurs where available soil in transport exceeds the transport capability, part of the sediment load is deposited.

Sources of eroded sediment

Sediment that is eroded from sloping land comes from three sources: (1) *interrill areas*, the areas between channels where flow has not yet concentrated; (2) *rills,* small transitory, usually annual, channels eroded by concentrated run-off; and (3) *gullies*, larger, more persistent channels eroded by concentrated runoff.

Interrill erosion[1] is a relatively uniform removal of soil, so it is usually inconspicuous compared with the more obvious rill and gully erosion. It results primarily from raindrop impact on soil surfaces that are not protected by cover. Interrill erosion includes both movement by splash and transport of raindrop-detached soil by thin-film runoff, enhanced by raindrop-impact turbulence. The rate of interrill erosion is only slightly affected by the steepness of the interrill surface (Lattanzi et al., 1974; Harmon and Meyer,

1978) and it varies little due to location on a hillside, as raindrop impact is relatively uniform all over a slope. Interrill areas are not rilled because runoff has not attained a tractive force sufficient to detach soil particles by scour or because sediment is being detached so rapidly by rainfall that it satisfies the flow's transport capability.

Rill erosion results primarily from soil detachment by concentrated runoff and thus occurs on only a small proportion of the land surface. Rills may develop where runoff concentrates due to minor topographic variations, tillage marks, or random irregularities on the land surface. However, rill erosion (or rilling) does not begin until the flow's erosiveness exceeds the soil's ability to resist detachment, so runoff may flow for a considerable distance downrow or downslope before rilling starts. Once rilling begins, it may increase rapidly as flow accumulates (Meyer, Foster and Römkens, 1975b), so rill erosion increases with the length of slope. It also increases with slope steepness. Rills typically progress upslope by a series of intensely eroding headcuts or knick-points (Meyer, Foster and Römkens, 1975a).

Gully erosion results from soil removal by concentrated runoff also, but it is larger in scale and affects the landscape more than rill erosion. Gullies develop and persist due to major features of land topography, whereas rills may change locations from time to time due to farming and other land-use operations. Slumping of gully sidewalls from gravitational forces at times of high soil-water content often provides much of the soil debris removed by subsequent runoff (Bradford and Piest, 1980). Gully erosion characteristically proceeds upslope as a series of large headcuts (Piest and Bowie, 1974; Leopold, et al., 1964). Gullies that develop on intensively cropped agricultural land but are periodically smoothed or obliterated by cultivation have been given the special designation of ephemeral gullies. They form in major water-courses where other gullies form, but they are periodically filled with soil from beside them to make them crossable for farming or other uses.

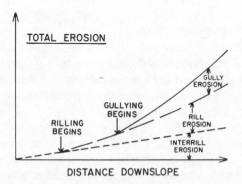

Figure 3.2 Typical contributions of interrill erosion, rill erosion, and gully erosion to total erosion from sloping land. Note that rilling begins some distance downslope and that gullying begins farther downslope.

The relative contributions to the total erosion from interrill, rill, and gully erosion vary depending on slope length and steepness, climatic patterns, land cover, and soil characteristics. Interrill erosion is relatively constant all over an area of sloping land, unless soil characteristics or land cover change appreciably. Rill erosion varies from none near the top of the slope (where all sediment originates from interrill areas) to major amounts downslope (where most runoff has concentrated). Severe gully erosion generally does not begin until major quantities of runoff have accumulated. The sediment originating from each of these three sources for increasing length of slope is illustrated in Figure 3.2. Interrill erosion dominates for the shorter lengths, rill erosion is greatest of the three sources at moderate lengths, and gully erosion may equal or exceed the other sources on very long slopes. The proportion of sediment from rill and gully erosion sources relative to that from interrill erosion may be expected to increase with steeper slopes, more intense rain, greater soil cover, and less permeable soil.

Physical characteristics of eroded sediment

Sediment eroded from cohesive soils consists of both individual (primary) soil particles and aggregates of soil material (Kemper and Koch, 1966; Gabriels and Moldenhauer, 1978; Alberts et al., 1980; Meyer et al., 1980, Young, 1980). Few if any major soils erode entirely as primary particles. Thus, the size distribution of sediment as it erodes from an upland field may be quite different from that of the dispersed soil or the dispersed sediment. Furthermore, the density of sediment aggregates is less than that of the primary particles of which they are composed. Both sediment size and density greatly affect the transportability of sediment by runoff and the probability of its subsequent deposition downslope. Other sediment characteristics such as the stability of sediment aggregates also may affect sediment transport.

Size distribution of eroded sediment

The size characteristics of sediment eroded from a specific soil depend on the size distribution of primary soil particles, the extent of aggregation of the soil materials, and the stability of the aggregated material during the erosion process. Generally, aggregation increases with the percentage of colloidal material in the soil, and this aggregation is further enhanced by soil organic matter and iron content (Hartman, 1979).

Size distributions of sediment eroded from soils of different textures are shown in Figure 3.3. These and additional data (Meyer et al., 1980; Meyer, 1985) were obtained by analyzing samples of runoff collected from crop row sideslopes during intense simulated rainstorms. They illustrate the wide range in sediment size distributions that may occur among different soils and the lack of a consistent relationship between soil texture and sediment size. The coarse

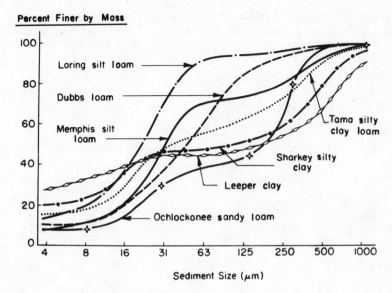

Figure 3.3 Size distributions of undispersed interrill sediment as eroded from the row sideslopes of soils with various textures.

sediment that resulted from interrill erosion of the sandy loam and loam soils was logical, but even coarser sediment resulted from erosion of some of the clay and silty clay soils. This coarse sediment from these fine-textured soils meant that most of it was aggregated because these soils contained a very small percentage of sand-sized primary particles. The medium-textured silt loam soils generally had finer sediment than either the coarse-textured or fine-textured soils, although some silt loams had considerably coarser sediment than others because of the wide range in aggregation and aggregate stability among them.

The difference in size distributions between the primary soil particles and the undispersed sediment that eroded from 17 soils is summarized by soil textures in Table 3.1. This difference was great for the predominantly fine-textured and high organic matter soils, so their sediment consisted mostly of aggregates. However, the predominantly medium- and coarse-textured soils had only small differences, indicating that only a small part of their sediment was aggregated.

Results from other research indicate that the sediment eroded from rills by concentrated flow is coarser and therefore more is in the form of aggregates than that eroded by rainfall from interrill areas of the same soils. Alberts et al. (1983) found that the median size of interrill plus rill sediment was almost twice that for sediment from interrill erosion only. In a study of row furrow gradients (Meyer and Harmon, 1984), erosion during a 70-mm/h rainstorm from the row sideslopes (interrill areas) to the furrow for a ridged (or bedded)

Table 3.1 Differences between the size distributions of primary soil particles and the undispersed sediment as eroded from row sideslopes for soils of various textures.

Soil texture		Percent finer by mass than			
		4 μm	16 μm	63 μm	250 μm
clays (2)[a]	dispersed soil	51	67	83	97
	undispersed sediment	23	33	42	61
	difference	28	34	41	36
silty clays (2)	dispersed soil	48	75	98	99
	undispersed sediment	20	34	43	52
	difference	28	41	55	47
silty clay loams (2)	dispersed soil	32	53	97	99
	undispersed sediment	15	31	63	79
	difference	17	22	34	20
well aggregated silt loams (4)	dispersed soil	21	40	89	96
	undispersed sediment	7	23	71	85
	difference	14	17	18	11
poorly aggregated silt loams (4)	dispersed soil	13	30	91	97
	undispersed sediment	9	26	86	95
	difference	4	4	5	2
loams (2)	dispersed soil	21	31	65	89
	undispersed sediment	12	20	49	83
	difference	9	11	16	6
sandy loam (1)	dispersed soil	6	14	45	77
	undispersed sediment	6	14	38	67
	difference	0	0	7	10

[a] Number of soils given in parentheses.

row of a moderately aggregated silt loam soil totaled 15.8 t/ha h, with 4.7 t/ha h sand-sized (> 63 μm) sediment and 11.1 t/ha h finer sediment. For a 9.3 m furrow of 5.0 percent gradient with moderate rilling, total sediment loss from the end of the furrow (row sideslope erosion plus rill erosion) was 21.2 t/ha h, with 8.1 t/ha h sand-sized sediment and 13.1 t/ha h finer. Total erosion from a 6.5 percent furrow with serious rill erosion was 24.2 t/ha h, with 11.3 t/ha h sand-sized sediment and 12.9 t/ha h finer. Thus, the sediment loss in excess of that from the row sideslopes, which originated from rilling in the furrows, consisted mostly of sand-sized sediment, whereas only 30 percent of the interrill sediment was sand sized. The coarser sediment produced by rill erosion is not surprising because these particles have not been stressed repeatedly by raindrop impact, and because rill flow is generally capable of transporting larger sediment than the thin-film flow typical of interrill areas. The losses of sand-sized sediment from 2.0 and 3.5 percent furrows where no appreciable rilling occurred approximately equaled the rate of sand-sized sediment delivered from the row sideslopes. For a furrow of 0.5 percent gradient most of the coarser sediment deposited along the furrow and only 1.3 t/ha h of the total 7.4 t/ha h sediment loss was sand sized.

Size distribution of dispersed sediment

Another important characteristic of aggregated sediment is the size distribution of the primary particles of which the sediment is composed. The sediment samples that were analyzed for size distribution in their aggregated state (Meyer, 1985) were subsequently dispersed to determine the primary particle sizes of the coarse sediment ($> 250 \, \mu m$), medium sediment ($63-250 \, \mu m$), and fine sediment ($< 63 \, \mu m$). These results, plus the size distribution of the primary particles for all sizes of sediment combined, are summarized in Table 3.2. For well-aggregated soils such as Leeper, Sharkey, and Monona, much of the sand-sized medium and coarse sediment consisted of aggregates of clay and silt primary particles. For all soils except those with considerable primary sand, the percentage of silt and clay in the coarse sediment was similar to that of the total sediment or the original soil. For many of these soils, the percentage of clay in the coarse sediment was higher than its percentage in the total sediment. This enrichment of clay in large aggregates was also found by Alberts et al. (1983). Because coarse sediment is least easily transported and most readily deposited, the coarser sediment from well-aggregated soils that deposits in sediment basins, where slopes flatten, or in vegetated filter strips may contain large percentages of very fine-textured material. The oft-made statement that eroded clay and fine silt are transported the farthest may be misleading for those soils where much of the sediment erodes in the form of coarse aggregates.

Density of aggregated sediment

Most sediment that erodes as primary soil particles has a density of about 2.65 g/cm^3, but sediment that erodes as aggregates has a lower density. Some information is available on the density of dry aggregates, but wet aggregates are denser because eroding sediment usually has been subjected to prolonged wetting by rainfall and thus most of the aggregates' pore space is filled with water rather than soil air.

The average density of coarse 250 to 1000 μm sediment eroded from different soils may vary widely depending on the relative proportion of aggregates and primary sand. For soils ranging from silty clay to sandy loam, average density varied from slightly less than 2.0 to about 2.6 g/cm^3 as the percentage of primary sand in the sediment increased (Rhoton et al., 1983a). Without the sand, the density of the remaining wet aggregates was between 1.9 and 2.1 g/cm^3 for all soils. These results suggest that a density of about 2.0 g/cm^3 is appropriate for coarse, wet aggregates that are eroded from typical agricultural soils. Slightly different densities may be appropriate for finer wet aggregates, but in all cases the densities of wet aggregates may be

Table 3.2 Size distributions of the primary particles in coarse ($> 250\ \mu m$), medium ($63-250\ \mu m$), and fine ($< 63\ \mu m$) sediment that eroded from various soils.[a]

Soil type	Eroded sediment		Primary particles (percent by mass)		
	Size class	Percent in size class	$> 63\ \mu m$ (sand)	$4-63\ \mu m$ (~ silt)	$< 4\ \mu m$ (~ clay)
Alligator	coarse	27	14	26	60
clay	medium	33	61	12	27
(23% sand)	fine	40	—	46	54
	all sediment	100	24	29	47
Leeper	coarse	48	12	34	54
clay	medium	12	42	25	33
(13% sand)	fine	40	—	35	65
	all sediment	100	10	34	56
Sharkey	coarse	39	3	41	56
silty clay	medium	13	8	37	55
(2% sand)	fine	48	—	49	51
	all sediment	100	2	44	54
Monona	coarse	14	9	51	40
silty clay	medium	15	11	56	33
loam	fine	71	—	75	25
(6% sand)	all sediment	100	3	69	28
Tama	coarse	28	6	53	41
silty clay	medium	19	7	55	38
loam	fine	53	—	69	31
(5% sand)	all sediment	100	3	62	35
Arkabutla	coarse	15	43	36	21
silt loam	medium	22	74	16	10
(27% sand)	fine	63	—	81	19
	all sediment	100	23	60	17
Memphis	coarse	19	2	66	32
silt loam	medium	9	4	66	30
(2% sand)	fine	72	—	78	22
	all sediment	100	1	74	25
Ora	coarse	10	64	26	10
silt loam	medium	17	75	18	7
(27% sand)	fine	73	—	85	15
	all sediment	100	19	68	13
Clarion	coarse	27	74	12	14
loam	medium	36	64	17	19
(49% sand)	fine	37	—	62	38
	all sediment	100	43	32	25
Ochlockonee	coarse	33	98	1	1
sandy loam	medium	29	96	3	1
(58% sand)	fine	38	—	88	12
	all sediment	100	60	35	5

[a] Only those soils having at least 9 percent of the total sediment in each size class are listed.

expected to be greater than those of dry aggregates and less than those of primary soil particles.

Stability of aggregated sediment

Aggregate stability generally increases as aggregates break down into smaller sizes. Once aggregated sediment has been detached from the soil mass, it is subject to further stresses from raindrop impact and flowing runoff. Sediment detached and eroded from interrill areas usually has been intensively stressed by impacting raindrops which tend to break weak aggregates. Such sediment will be further stressed as it is transported through the runoff system, and any additional breakdown will effect its subsequent transportability. To determine whether aggregated sediment from interrill areas will break down further during transport by concentrated runoff, soils varying from clay to loam were studied to determine the stability of their interrill sediment (Rhoton et al., 1983*b*). Sediment samples were placed with water in a tilted square channel that was rotated at 5, 10 and 20 rpm to simulate stresses caused by rill flow. Five minutes of rotation caused very little further breakdown of sediment coarser than 63 μm. Only when the highest stress condition (equivalent to flow greater than 0.5 m/s) was continued for 60 minutes did significant additional breakdown of the aggregated sediment occur, but seldom will sediment travel that long at such high flow velocities through upland flow systems. Therefore, this research suggests that sediment originating from most soils is quite resistant to further breakdown if it has been detached by erosive raindrops. As this study showed little evidence that any serious additional breakdown occurs due to upland runoff stresses, the size distribution of sediment eroded from interrill areas may be used in analyzing its transport through the upland flow systems. Breakdown of sediment detached by rilling has not been studied, but the larger aggregates from rills and gullies are probably less stable and more likely to suffer some further breakdown during transport.

Prediction of sediment sizes

Field experiments to determine the size distributions of sediment eroded from different soils are time consuming and costly. Yet, sediment-size data are needed for accurate evaluations of sediment movement, particularly for the sediment size distribution components of recently developed erosion/sedimentation models. For the CREAMS model a sediment size distribution consisting of three primary size groups and two aggregate size groups is estimated from the primary particle size distribution (Foster et al., 1981). An experimental method for simulating sediment size distribution in the laboratory was developed that predicts ten size groups (Rhoton et al., 1982). Bulk field samples of surface soil are wetted in a specific way and agitated in an orbital shaker for a specific time to stress them like rainstorms do. The size distribution data obtained using this procedure were mostly within ± 1 standard deviation of the distribution curves for sediment eroded during field experiments. The greatest similarity was for soils high in sand and silt, whereas soils high in clay and organic matter were more variable. Results from the procedure of Rhoton et al. (1982) gave much better estimates of the size distributions of sediment from interrill areas than either their primary particle size distributions or distributions obtained by gentle agitation procedures (Kemper and Koch, 1966) that are often used during conventional aggregate-index analyses.

Factors that affect soil erosion rates

The rate of soil erosion that occurs for any specific situation depends on many factors, some that are part of the natural environment and others that are affected by land use and management. Starting about 1940 researchers began to develop equations and tables to help quantify erosion rates (Smith and Wischmeier, 1962; Meyer, 1984). The best-known equation resulting from these efforts is the Universal Soil Loss Equation (USLE) (Wischmeier and Smith, 1965, 1978), which estimates average annual soil loss caused by rill and interrill erosion (only). The USLE uses six major factors to estimate upland soil erosion by water: erosiveness of rainfall and runoff (R), soil erodibility (K), slope length (L), slope steepness (S), cropping and management techniques (C), and supporting conservation practices (P). Since its introduction in the early 1960s, the USLE has contributed greatly toward improved land management planning and use of effective soil conservation practices. The statistical/empirical approach used for its development was necessary because at that time considerable erosion data were available, but the physical processes involved were not well understood. Recent advances in understanding fundamental principles and processes of soil erosion and sediment movement have provided knowledge for the development of physically based models that incorporate many basic erosion and sedimentation processes plus interrelationships among them (Bennett, 1974; Foster et al., 1977, 1981). The better understanding of soil erosion processes also provides a basis for improving erosion control practices.

Climate

Among the various aspects of climate that affect the rate of upland erosion, rainfall characteristics such as amount, intensity, and drop size usually have the greatest effect. Total runoff and runoff rate, and consequently soil erosion, increase directly with rain amount and intensity. Also, very low intensity rains often consist of small drops that fall slowly with little energy to detach soil or seal soil surfaces. Conversely, most intense rains fall as mostly large drops having great amounts of energy. The erosive potential of rainfall for much of the United States is a function of rainfall energy and intensity. The average annual potential varies over the United States from its highest total along the Gulf of Mexico to much less along the northern tier of states and west from the Great Plains (Wischmeier and Smith, 1978). The distribution of erosion potential through the year also varies from relatively uniform in the southeast states to occurring very dominantly in late spring and summer in certain other areas. The interrill erosion rate for high silt and sand soils is proportional to the square of rain intensity (Meyer, 1981). For high-clay soils, the exponent of intensity decreases from 2.0 as the clay content increases.

Infiltration characteristics of a soil

Soils vary greatly in their potential for water intake. Coarse-textured soils may have hydraulic conductivities of several centimeters per hour whereas fine-textured soils may be limited to a few millimeters or less per hour. However, soils that have relatively rapid hydraulic conductivities may be restricted in their water intake for other reasons. Some soils seal rapidly and severely from raindrop impact, and these surface seals may greatly impede water entry through the soil surface (Römkens and Prasad, 1985). Other soils may have restricting horizons such as genetic fragipans or induced traffic pans that severely reduce water intake once the storage above them is filled.

Soil erodibility

The susceptibility of a soil to erosion is termed its erodibility. Fine-textured soils are usually cohesive and difficult to detach, but once detached, their fine sediment is easily transported unless, as is often the case, it consists predominantly of sizeable aggregates. Coarse-textured soils are usually easier to detach, but their large sediment is difficult to transport and they often produce less runoff than fine-textured soils. Medium-textured soils such as loams and silt loams usually are both readily detached and transported, so they are generally quite erodible (Wischmeier and Mannering, 1969). The USLE K-factor is an average annual value for soil erodibility that can be estimated based on soil texture, organic matter content, permeability, and structure (Wischmeier et al., 1971). However, the K-factor combines two quite different types of erosion, rill erosion and interrill erosion, and it also includes runoff and transportability effects.

The interrill erodibilities of soils may vary greatly. For 18 soils ranging in texture from clay to sandy loam, erosion from about 100 mm of 71-mm/h rain varied from 12 to nearly 100 t/ha (Meyer and Harmon, 1984). The poorly aggregated silt loams were most erodible, and the high-clay soils were least erodible. Erosion from the 100 mm of rain averaged 66 t/ha for the nine silt and silt loam soils, 45 t/ha for the three loam and sandy loam soils, and 24 t/ha for the six clay, silty clay, and silty clay loam soils. Iowa soils were about half as erodible as Mississippi soils of the same texture, apparently due to the relatively high organic matter content and excellent structure of the former. Several soil properties individually explained over half of the variation in interrill erodibility, specifically clay ($< 2\,\mu$m) percentage and four closely related soil properties: 1500 kPa water, exchangeable calcium, sum of exchangeable bases, and cation exchange capacity. The next highest correlated property was organic matter content. All the above properties were negatively correlated with erosion. Complementary studies of the susceptibility of various soils to erosion by concentrated runoff (rill erodibility) are an important research need.

Soil surface cover

Surface cover dissipates raindrop impact energy, reducing soil detachment, transport by splash, rate of soil sealing, and consequent runoff. Cover also slows runoff velocity so that the flow is less erosive and has reduced capacity to transport sediment. Vegetation is usually quite effective in reducing erosion. The percentage of the surface covered by plant canopy increases with plant population and growth for most crops. Some crops such as cotton intercept raindrops, but large erosive drops of water may subsequently drip from their leaves. Studies of various surface mulches, including crop residues, stones, and woodchips have shown that even small amounts of cover can greatly reduce erosion and that, unless the slope is long or steep, complete cover can almost eliminate erosion (Meyer et al., 1970, 1972).

Tillage

Most agricultural crops require cultivation of the soil to prepare the seedbed and combat weeds. Such cultivation usually exposes additional soil to rainfall and runoff forces and makes cropland more susceptible to soil erosion. For most crops, the time when land is most susceptible to erosion is just before and after planting, but different crops are planted at different times of the year. Most wheat is planted in the autumn, whereas corn, soybeans, and cotton are planted in the spring. Cropping systems that minimize soil tillage during periods when the rainfall erosion potential is great reduce the erosion hazard. Various types of conservation tillage have been developed that leave more surface cover and thus minimize the exposure of erodible soil throughout the year. Practices such as no-till and wheat–soybean doublecropping usually decrease the potential for soil erosion while allowing intensive cropping (SCSA, 1983).

Slope steepness

Land gradient or slope steepness probably has the greatest effect on soil erosion of the several topographic characteristics. Steepness especially affects runoff flow velocity, a major factor in detachment and transport capability. The USLE slope-steepness factor

$$S = 65.41 \sin^2 \theta + 4.56 \sin \theta + 0.065$$

where θ is the angle from horizontal (Wischmeier and Smith, 1978), indicates that rill plus interrill erosion doubles as steepness increases from 2 to 4 percent, more than doubles again from 4 to 8 percent, and more than doubles again by 14 percent steepness. However, although considerable interrill erosion can occur on nearly flat areas, it only doubles as slope steepness increases from 2 to 20 percent (Lattanzi et al., 1974). This slight effect of steepness on interrill erosion as compared with its major effect on rill plus interrill erosion indicates the great effect of slope steepness on rill erosion. Furthermore, fundamental analyses show that sediment losses from interrill areas are essentially independent of slope steepness but that rill losses are proportional to slope steepness (Foster and Meyer, 1975).

Slope length

Slope length affects erosion because rill erosion increases as runoff accumulates downslope. However, erosion from interrill areas should not be influenced much by field length because the small interrill areas are similar everywhere on the slope. Hence, the rate at which total soil loss increases with distance downslope depends on the relative contributions of interrill erosion and rill erosion (Meyer et al., 1976). Soil losses from soils that are susceptible to rilling increase more rapidly with distance downslope than do losses from less-susceptible soils. Also, rills on susceptible soils usually begin nearer the top of the slope.

In the USLE the slope length factor

$$L = (\lambda/22.1)^m$$

where λ is slope length in meters (Wischmeier and Smith, 1978). For slopes of 5 percent and steeper, a value of 0.5 is recommended for the exponent m when computing average rill plus interrill erosion per unit of area. For lesser steepnesses, lower values of m are suggested. Analyses of fundamental erosion principles support a variable slope length exponent that increases as the contribution from rill erosion relative to that from interrill erosion increases (Foster and Meyer, 1975). The exponent $m + 1$ is appropriate when expressing accumulated soil loss or sediment load, and it ranges from 1 when essentially all sediment originates from interrill areas to 2 when rill erosion dominates and interrill erosion is negligible. When computing erosion per unit of area, m is the appropriate exponent. It ranges from 0 to 1, values that encompass the recommended USLE values (Wischmeier and Smith, 1978). These fundamental analyses suggest that higher values of m than used in the USLE are merited for steep, long slopes where the large majority of sediment originates from rill erosion.

Slope shape

The hillside shape of sloping land may affect soil erosion through the influences of slope length and slope steepness and their interrelationships (Meyer and Kramer, 1969; Foster and Wischmeier, 1974). On either uniform or irregular slopes for which both length and steepness are great, erosion will be great; where both are small, erosion will be relatively small. On concave slopes where slope steepness decreases as slope length increases, these effects tend to offset each other, so erosion along the slope and sediment losses from the toe of the slope will both be relatively small. However, on convex slopes, erosion is less than for other shapes near the top, but it increases very rapidly toward the toe of the slope. Consequently, sediment loss will be great because the greatest runoff occurs where the greatest slope steepness occurs. On complex slopes with the upper half convex and lower half concave, erosion will be great along the steep mid-portion, but the losses from the toe will be less than for a uniform slope because of deposition along the concave section. As erosion progresses through successive erosion cycles, all shapes will become more concave, erosion along the slopes will diminish, shape will change more slowly, and the sediment loss from the toe will decrease. Analysis of erosion and sedimentation rates along different slope shapes not only indicates how the shapes will evolve in the future, but it might also be used to suggest slope shapes that existed at earlier times.

Transport capacity of runoff

Essentially all sediment lost from sloping land during rainstorms is transported by concentrated runoff along furrows and rills. For concentrated runoff, either the sediment-transport capacity of the flow or the transportability (size, shape, and density) of the sediment may limit sediment transport and determine whether or not individual particles will be moved. Sediment transport capacity of runoff increases rapidly as the flow channel gradient increases (Meyer et al., 1983). Transport capacity of sand-sized sediment by shallow concentrated flow is very low at 0.2 percent, but 10 to 100 times as much sediment can be transported along a 1 percent furrow. Transport capacity increases another 5 to 50 times between 1.0 and 2.5 percent, and it doubles again between 2.5 and 5.0 percent. Sediment transport capacity also increases with increased flow, but the effect is less than that of gradient. For shallow-flow conditions the transport capacity is much greater for fine sand than for coarse sand. If the delivery of interrill sediment is at rates greater than the runoff in furrows or rills can carry, some of the sediment will deposit. If the runoff can transport more sediment than is delivered to furrows and rills, they may degrade by scour. A considerable range of flows and gradients may exist for cohesive soils where neither deposition nor scour will occur (Meyer and Harmon, 1985).

Transportability of sediment

Very fine, near-colloidal-sized sediment will generally be transported as far as the runoff flows. However, the transportability of larger sediment depends on its size and density characteristics as discussed earlier. Fine sediment is the easiest to transport and the least likely to deposit, whereas coarse sediment is more difficult to move and deposits more readily. However, fine silt and clay primary particles that are eroded as sizeable aggregates are much less transportable than in their primary state. Sediment that is in transport often deposits selectively, with the larger and denser material settling out first and the finer material being carried farther (Meyer and Harmon, 1985).

Threshold or critical erosion conditions

For a given situation, the erosion rate normally increases modestly per unit of increased slope length, slope steepness, or other erosion hazard until the resistance of the soil breaks down. Thereafter, erosion increases much more rapidly. For example, interrill erosion usually occurs anywhere soil is exposed, but major rill erosion does not begin until quite some distance from the top of the slope where runoff exceeds a critical condition. Horton's (1945) "belt of no erosion" was probably the area above the start of rilling where no erosion was apparent. However, most such areas, if they include exposed soil, would likely suffer serious interrill erosion even near the watershed divide. Data from a study of different types of mulches applied at various rates (Meyer et al., 1972, 1976) showed that the start of rilling caused a major increase in erosion rate per unit of slope length (Fig. 3.4). The average straight line trend, A, on this log–log plot does not adequately describe the effect of slope length on erosion for such data, so extrapolation well beyond the range of data can lead to seriously wrong conclusions. These data indicate that the lines of best fit will have slope B before rilling begins and slope C after rilling begins. For each set of data, the critical condition is the point where B and C intersect. This point of intersection is at a different slope length for each of these sets of data.

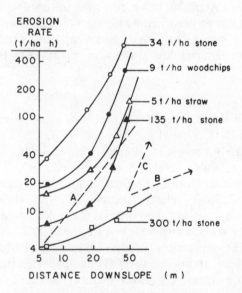

Figure 3.4 Effect of slope length on erosion rate for several mulch types and application rates. No rilling occurred at any of the lengths studied for the bottom curve, but rilling began before the shortest length of the top curve. The lengths where the other three curves changed from trend B to trend C were those at which serious rilling began. A straight line fitted to such data (illustrated by line A for the 135 t/ha stone data) indicates the type of error introduced by forcing a linear relationship. (From Meyer et al., 1976.)

Other illustrations of critical conditions are the threshold velocity in wind erosion and the critical tractive force in sediment transport hydraulics. Such thresholds were the theme of the Ninth Annual Geomorphology Symposium and the resulting book, *Thresholds in Geomorphology* (Coates and Vitek, 1980).

Sediment load versus detachment rate

Considerable evidence suggests that the rate of detachment by flowing water decreases as the sediment load increases (Willis, 1971; Meyer and Monke, 1965) or that the rate of deposition is directly proportional to the difference between the sediment concentration in the flow and the equilibrium concentration for that flow condition (Einstein, 1968). This concept may be expressed by the relationship (Foster and Meyer, 1975):

$$\frac{\text{detachment rate by flow}}{\text{detachment capacity of flow}} + \frac{\text{sediment load of flow}}{\text{transport capacity of flow}} = 1$$

This equation indicates that rilling will be greatest when the sediment load is low, but that little or no rilling will occur when the transport capacity of flow is satisfied with sediment produced by raindrop-impact erosion. It also shows that deposition may occur when the transport capacity of sediment-laden flow is decreased due to reduced flow gradient or increased hydraulic roughness.

Erosion control practices

The control of soil erosion and related problems is the ultimate goal of most efforts to better understand and quantify soil erosion. Since soil erosion can seldom be completely prevented, erosion control implies efforts to reduce erosion rates to tolerable levels. For continued agricultural production, tolerable rates of annual erosion (*T*-values) range from 5 t/ha for shallow soils to 11 t/ha for deep soils (Wischmeier and Smith, 1978). However, even lower limits may be necessary where the potential for serious offsite damages from eroded sediment exists downstream (Stewart et al., 1976).

Most erosion control practices (Barnes, 1971; SCS, 1984) reduce soil losses in one or more of the following ways: (1) dissipate raindrop impact energy and/or runoff scour forces; (2) reduce quantity or rate of runoff; (3) slow runoff velocities; or (4) improve soil characteristics that resist erosion. Dense vegetation is effective in all four of these ways, whereas terraces primarily slow runoff velocities. The difficulty of achieving erosion control generally increases as the erosion hazard of the site and the intensity of land use increase. Combinations of practices are often required to control erosion. For some situations, adequate erosion control and intensive land use are incompatible, so the land needs to be kept in permanent vegetation instead.

Conclusion

Erosion prediction and control technology has progressed tremendously since a concerted soil conservation thrust began about 50 years ago in the United States. A balanced research program of basic and applied research during this period has greatly improved our understanding of fundamental erosion and sedimentation processes, properties of eroded sediment, and control methods for most land-use situations. Were it not for these past efforts of many capable and dedicated conservation leaders and scientists, our nation's soil and water resources would be in a much poorer condition than exists today.

Most erosion research has been conducted with the goal of maintaining the productive potential of our agricultural land, but the findings are largely applicable for other uses. Many of the principles and processes involved in erosion of agricultural land also apply to other land, and major factors that affect soil detachment, transportation, and deposition operate wherever rain falls and runoff results. The greatest effect of modern agriculture is probably the relatively rapid rate at which erosion progresses on intensively cropped land. Other effects include tillage of the soil, introduction of vegetation that is exotic to the natural environment, and imposition of different drainage systems. For scientists, however, this accelerated erosion provides an opportunity to study significant landscape changes during their professional lifetimes.

Productive soil is one of our most basic natural resources, and it is essentially nonrenewable in an agricultural time-frame. Wise soil and water management to maintain soil productivity and environmental quality while enabling profitable agricultural use are noble goals for all who work with the land and its resources. Our well-being and that of many future generations depends on our ability to properly manage the precious soil and water resources entrusted to our care.

Note

1. Another term, 'sheet erosion,' has often been used to include all erosion that can be obliterated by normal tillage, thereby encompassing both interrill and rill erosion. However, true "sheet erosion" occurs only before runoff has sufficient detaching capacity to cause rill erosion. Use of the terms "interrill" and "rill" erosion eliminates the need for the confusing term sheet erosion because interrill erosion is literally sheet erosion, and rill erosion implies the nonsheet, locationally selective removal caused by runoff detachment (Meyer et al., 1975*b*).

References

Alberts, E. E., W. C. Moldenhauer, and G. R. Foster, Soil aggregates and primary particles transported in rill and interrill flow, *Soil Science Society of America Journal, 44,* 590–595, 1980.

Alberts, E. E., R. C. Wendt, and R. F. Piest, Physical and chemical properties of eroded soil aggregates, *Transactions of the American Society of Agricultural Engineers, 26,* 465–471, 1983.

American Society of Civil Engineers, Sedimentation engineering, edited by V. A. Vanoni, *American Society of Civil Engineers Manual 54,* 1975.

Barnes, R. C. Jr., Erosion control structures, in *River Mechanics Vol. II,* edited by H. W. Shen, p. 28–1 to 28–26, Fort Collins, Colo., 1971.

Bennett, H. H., *Soil Conservation,* 993 pp., McGraw-Hill, New York, 1939.

Bennett, J. P., Concepts of mathematical modeling of sediment yield, *Water Resources Research, 10,* 485–492, 1974.

Bradford, J. M., and R. F. Piest, Erosional development of valley-bottom gullies in the upper midwestern United States, in *Thresholds in Geomorphology,* edited by D. R. Coates and J. D. Vitek, pp. 75–101, Dowden and Culver, Stroudsburg, Pa., 1980.

Coates, D. R., and J. D. Vitek (Eds.), *Thresholds in Geomorphology,* 498 pp., Dowden and Culver, Stroudsburg, Pa., 1980.

Council for Agricultural Science and Technology (CAST), Soil erosion: Its agricultural, environmental, and socioeconomic implications, *Council for Agricultural Science and Technology Report 92,* 1982.

Einstein, H. A., Deposition of suspended particles in a gravel bed, *Proceedings of the American Society of Civil Engineers, Journal of the Hydraulics Division, 94;* 1197–1205, 1968.

Ekern. P. C., Raindrop impact as the force initiating soil erosion, *Soil Science Society of America Journal, 15,* 7–10, 1950.

Ellison, W. D., Soil erosion studies, *Agricultural Engineering, 28,* 145–146, 197–201, 245–248, 297–300, 349–351, 402–405, 442–444, 1947.

Foster, G. R., and L. D. Meyer, Mathematical simulation of upland erosion by fundamental erosion mechanics, *Proceedings of the 1972 Sediment Yield Workshop,* ARS-S-40, 190–207, 1975.

Foster, G. R., and W. H. Wischmeier, Evaluating irregular slopes for soil loss prediction, *Transactions of the American Society of Agricultural Engineers, 17,* 305–309, 1974.

Foster, G. R., L. D. Meyer, and C. A. Onstad, An erosion equation derived from basic erosion principles, *Transactions of the American Society of Agricultural Engineers, 20,* 678–682, 1977.

Foster, G. R., L. J. Lane, J. D. Nowlin, J. M. Laflen, and R. A. Young, Estimating erosion and sediment yield on field-sized areas, *Transactions of the American Society of Agricultural Engineers, 24,* 1253–1261, 1981.

Gabriels, D., and W. C. Moldenhauer, Size distribution of eroded material from simulated rainfall: effect over a range of texture, *Soil Science Society of America Journal 42,* 954–958, 1978.

Geraghty, J. J., D. W. Miller, F. VanDerLeeden, and F. L. Troise, *Water Atlas of the United States,* Plate 21, Water Information Center, Inc., Port Washington, N.Y., 1973.

Guy, H. P., Fluvial sediment concepts, in *U.S. Geological Survey Techniques of Water Resources Investigations, Book 3,* 1970.

Harmon, W. C. and L. D. Meyer, Cover, slope, and rain intensity affect interrill erosion. *Proceedings Mississippi Water Resources Conference,* Mississippi State University, Mississippi State, Miss., pp. 9–16, 1978.

Hartman, R., Aggregation, in *The Encyclopedia of Soil Science, Part I,* edited by R. W. Fairbridge and C. W. Finkl, Jr., pp. 24–27, Dowden, Hutchinson and Ross, Stroudsburg, Pa., 1979.

Horton, R. E. Erosional development of streams and their drainage basins, *Bulletin of the Geological Society of America, 56,* 275–370, 1945.

Kemper, W. D., and E. J. Koch. Aggregate stability of soils from western United States and Canada, *USDA–ARS Technical Bulletin 1355,* 31–57, 1966.

Lattanzi, A. R., L. D. Meyer, and M. F. Baumgardner, Influences of mulch rate and slope steepness on interrill erosion, *Soil Science Society of America Proceedings, 38,* 946–950, 1974.

Laws, J. O., and D. A. Parsons, Relation of raindrop-size to intensity, *Transactions of the Americal Geophysical Union, 24,* 452–460, 1943.

Leopold, L. B., M. G. Wolman, and J. P. Miller, *Fluvial Processes in Geomorphology,* 522 pp., Freeman, San Francisco, 1964.

Meyer, L. D., Water erosion, in *The Encyclopedia of Soil Science, Part I,* edited by R. W. Fairbridge and C. W. Finkl, Jr., pp. 587–595 Dowden, Hutchinson and Ross, Stroudsburg, Pa., 1979.

Meyer, L. D., How rain intensity affects interrill erosion, *Transactions of the American Society of Agricultural Engineers, 24,* 1472–1475, 1981.

Meyer, L. D., Evolution of the universal soil loss equation, *Journal of Soil and Water Conservation, 39,* 99–104, 1984.

Meyer, L. D., Interrill erosion rates and sediment characteristics, *Proceedings of the 1983 International Conference on Soil Erosion and Conservation,* in press, 1985.

Meyer, L. D., and W. C. Harmon, Susceptibility of agricultural soils to interrill erosion, *Journal of the Soil Science Society of America, 48,* 1152–1157, 1984.

Meyer, L. D., and W. C. Harmon, Sediment losses from cropland furrows of different gradients, *Transactions of the American Society of Agricultural Engineers, 28,* 448–453, 461, 1985.

Meyer, L. D., and L. A. Kramer, Erosion equations predict land slope development, *Agricultural Engineering, 50,* 522–523, 1969.

Meyer, L. D., and Monke, E. J., Mechanics of soil erosion by rainfall and overland flow, *Transactions of the American Society of Agricultural Engineers, 8,* 572–577, 580, 1965.

Meyer, L. D., and W. H. Wischmeier, Mathematical simulation of the process of soil erosion by water. *Transactions of the American Society of Agricultural Engineers, 12,* 754–758, 762, 1969.

Meyer, L. D., D. G. DeCoursey, M. J. M. Römkens. Soil erosion concepts and misconceptions. *Proceedings of the Third Federal Inter-Agency Sedimentation Conference,* Symposium 2, 1–12, 1976.

Meyer, L. D., G. R. Foster, and S. Nikolov, Effect of flow rate and canopy on rill erosion, *Transactions of the American Society of Agricultural Engineers, 18,* 905–911, 1975.

Meyer, L. D., G. R. Foster, and M. J. M. Römkens, Sources of soil eroded by water from upland slopes, *Proceedings of the 1972 Sediment Yield Workshop, ARS-S-40,* 177–189, 1975.

Meyer, L. D., W. C. Harmon, and L. L. McDowell, Sediment sizes eroded from crop row sideslopes, *Transactions of the American Society of Agricultural Engineers, 23,* 891–898, 1980.

Meyer, L. D., C. B. Johnson, and G. R. Foster, Stone and woodchip mulches for erosion control on construction sites. *Journal of Soil and Water Conservation, 27,* 264–269, 1972.

Meyer, L. D., W. H. Wischmeier, and G. R. Foster, Mulch rates required for erosion control on steep slopes, *Proceedings of the Soil Science Society of America, 34,* 928–931, 1970.

Meyer, L. D., B. A. Zuhdi, N. L. Coleman, and S. N. Prasad, Transport of sand-sized sediment along crop-row furrows, *Transactions of the American Society of Agricultural Engineers, 26,* 106–111, 1983.

Piest, R. F., and A. J. Bowie, Gully and streambank erosion, *Proceedings of the 29th Annual Meeting of the Soil Conservation Society of America,* 188–196, 1974.

Rhoton, F. E., L. D. Meyer, and F. D. Whisler, A laboratory method for predicting the size distribution of sediment eroded from surface soils, *Soil Science Society of America Journal, 46,* 1259–1263, 1982.

Rhoton, F. E., L. D. Meyer, and F. D. Whisler, Densities of wet aggregated sediment from different textured soils, *Soil Science Society of America Journal, 47,* 576–578, 1983*a*.

Rhoton, F. E., L. D. Meyer, and F. D. Whisler, Response of aggregated sediment to runoff stresses, *Transactions of the American Society of Agricultural Engineers, 26,* 1476–1478, 1983*b*.

Römkens, M. J. M., and S. N. Prasad, Surface sealing and infiltration—a symposia, *Proceedings of the 1983 Natural Resources Modeling Symposium,* Fort Collins, Colo., 1985.

Smith, D. D., and W. H. Wischmeier, Rainfall erosion, *Advances in Agronomy, 14,* 109–148, 1962.

Soil Conservation Service, *Engineering Field Manual for Conservation Practices,* 4th Printing, Soil Conservation Service, Washington, 1984.

Soil Conservation Society of America, Conservation tillage, a special issue, edited by Max Schnepf, *Journal of Soil and Water Conservation, 38,* 1983.

Stallings J. H., Mechanics of water erosion, *USDA Soil Conservation Service TP-118,* 1953.

Stewart, B. A., D. A. Woolhiser, W. H. Wischmeier, J. H. Caro, and M. H. Frere, Control of water pollution from cropland, Vol. II, *Cooperative ARS–EPA Manual: ARS-H-S-2 and EPA 600/2-75-026b,* 1976.

Willis, J. C., Erosion by concentrated flow, *ARS 41-179,* 1971.

Wischmeier, W. H. and J. V. Mannering, Relation of soil properties to its erodibility, *Soil Science Society of America Journal, 23,* 131–137, 1969.

Wischmeier, W. H., and D. D. Smith, Predicting rainfall-erosion losses from cropland east of the Rocky Mountains, *USDA Agricultural Handbook 282,* 1965.

Wischmeier, W. H., and D. D. Smith, Predicting rainfall erosion losses—A guide to conservation planning, *USDA–SEA Agriculture Handbook 537,* 1978.

Wischmeier, W. H., C. B. Johnson, and B. V. Cross, A soil erodibility nomograph for farmland and construction sites, *Journal of Soil and Water Conservation, 26,* 189–193, 1971.

Young, R. A., Characteristics of eroded sediment, *Transactions of the American Society of Agricultural Engineers, 23,* 1139–1142, 1146, 1980.

Young, R. A., and J. W. Wiersma, The role of rainfall impact on soil detachment and transport, *Water Resources Research, 9,* 1629–1636, 1973.

4

Plant cover effects on hillslope runoff and erosion: evidence from two laboratory experiments

R. P. C. Morgan, H. J. Finney, H. Lavee, Elaine Merritt, and Christine A. Noble

Abstract

Laboratory studies of detachment of soil particles from a sandy soil by rain-drop impact in storms of 50 mm/h and 61 mm/h for 5 min duration showed that the rate of detachment under Brussels sprouts decreased as the canopy cover increased from 0 to 15–25 percent but then increased again with further increases in percentage canopy cover. With 50 percent cover, the detachment rate equalled that on bare soil. Data for potatoes showed some support for a similar trend but those for sugar beet were less conclusive. Measurements of erosion on the same soil under a cover of 180 mm tall toy plastic oak trees in a 21-min three-stage storm of rainfall followed by rainfall with additional runoff and runoff without rainfall showed that a 15 percent canopy cover decreased erosion by 50 percent compared with that on bare soil. Erosion rates increased again, however, with further increases in canopy cover. The major difference in soil loss with different canopy percentages occurred in the first stage of the storm and was attributed to differences in soil detachment by rain-drop impact. The results lend further support to a small but increasing body of literature which shows that under certain circumstances plant covers are associated with higher rather than lower rates of erosion. The reasons for this and the circumstances involved remain imperfectly understood.

Introduction

Recent approaches to hillslope hydrology and erosion have concentrated on the separation of interrill from rill processes, the subdivision of interrill erosion into that by raindrop impact and that by overland flow, and the recognition of erosion as a two-phase activity comprising detachment of soil particles from the soil mass and their net transport downslope. An understanding of these aspects provides the basis for the mathematical modeling of hillslope runoff and sediment production. In current models the water erosion processes are represented by a series of partially physically based equations, relating to the mechanics of erosion, but the effects of soil, slope, and plant cover are often accounted for empirically (Meyer and Wischmeier, 1969; Kirkby, 1980; Foster et al., 1981). Plant cover effects are usually allowed for by coefficients that express the runoff or erosion expected under a given vegetation or crop as a proportion of that from bare soil. Since these coefficients are generally less than unity, they emphasize the protective role of the vegetative cover. This role is well-known theoretically and relates to the interception of rain by the canopy, increases in the infiltration rate and shear strength of the soil brought about by the root network, and reductions in flow velocity as a result of drag imparted by the leaves and stems. The values of the coefficients depend upon the height, percentage canopy area, ground cover, root density, and residue production of the plants.

Wischmeier (1975) and Foster (1982) argued that the coefficients decrease in value linearly with increasing percentage canopy cover and exponentially with increasing percentage mulch or residue cover. An exponential relationship has been observed for covers in direct contact with the soil surface, such as stones (van Asch, 1980), crop residues (Leflen and Colvin, 1981), and grass (Lang and McCaffrey, 1984). Hussein and Laflen (1982), however, found that increasing residue cover caused a linear decline in interrill erosion but an exponential decline in rill erosion. Elwell (1981) and Shaxson (1981) favor an exponential rate of decrease for both canopy and surface cover effects.

Some recent studies have unexpectedly failed to confirm either of the above relationships and have shown instead that a plant cover may result in greater erosion. De Ploey et al. (1976) recorded an increase in erosion by runoff with increasing grass cover. This was attributed to the flow concentrating between the plant stems and creating localized scour. A similar effect was recorded for stone covers by Yair and Lavee (1976). Morgan (1982) found that winter wheat, winter oats, and spring barley reduced rates of soil detachment by rainfall compared with bare ground but that the detachment at these lower rates was inversely related to rainfall energy. This finding implies that the crop covers afford better relative protection in rains of higher intensity. Assuming that rainfall energy at the ground surface decreases with increasing cover, it also implies that once detachment rates have fallen to their new level, they will increase again as the canopy grows further.

This chapter provides additional experimental evidence that plant covers can enhance erosion. Two studies are reported, one on soil detachment by rainfall and the other on interrill erosion.

Soil detachment by rainfall

Experimental design

A laboratory experiment was designed to investigate for Brussels sprouts, sugar beet, and potatoes the relationships between percentage canopy cover, the volumes and energies of the rain reaching the ground surface as throughfall and leaf drip, and the rate of soil detachment. Each crop was sampled throughout its growing season at approximately two-weekly intervals or whenever there appeared to be a significant change in canopy area. Data were collected for two seasons. As far as possible the experiment reproduced the field conditions of the Silsoe area, Bedfordshire, England.

A design storm of 5 min duration and an intensity of 50 mm/h in the first year of study and 61 mm/h in the second year was provided by a rotating disk rainfall simulator (Morin et al., 1967). A 5-min intensity of 52 mm/h has a return period of once a year and that of 72 mm/h a return period of five years in the Silsoe area (Natural Environment Research Council, 1975).

A 5-min duration was chosen for the storm as being sufficiently long to produce measurable quantities of detached soil but short enough to avoid the generation of surface runoff and minimize the effects of changes in soil surface properties through crusting and armoring. This allowed the effects of plant canopy on soil detachment by raindrop impact to be studied without interactions of soil and time.

A sample plant was taken from a local field and, after removal of the roots, positioned under the simulator in the middle of a 760×760-mm catch tray (Fig. 4.1). Rainfall received on the surface of the tray was channeled downslope through a triangular section into a pipe and collecting bucket. The tray was supported on an angle-iron base at an arbitrary slope of $10°$ in the first year and $5°$ in the second year, sufficient for the rainfall to drain off, and at a height of 1.3 m from the nozzle of the simulator. The plant was mounted in a plastic tube passing through a 75-mm diameter hole in the center of the tray. The tube served to hold the plant in a vertical position and to channel the stemflow into a cylinder for separate measurement.

Prior to rainfall simulation, a vertical photograph was taken of the plant canopy and catch tray. The effective canopy area for rainfall interception was measured from the photograph with a planimeter.

With the plant in position, three 5-min simulated rainstorms were applied. During the first storm the volumes of stemflow and total throughfall were determined, the latter being defined by all the rainfall collected on the catch tray. The difference in volume between the design storm and the sum of total throughfall and stemflow gives permanent interception. In the second storm a set of 15-mm diameter containers was placed beneath leaf drip points observed during the first storm to determine the volume of leaf drip. The difference in volume between total throughfall and leaf drip was assumed to be direct throughfall. It is recognized, however, that this component also includes raindrops striking the leaves and rebounding. In the third storm, soil

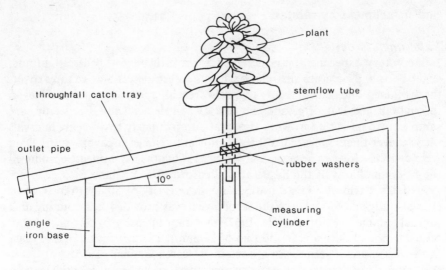

Figure 4.1 Experimental setup for the soil detachment study. (From Noble and Morgan (1983), reprinted by permission of John Wiley & Sons, Ltd.)

detachment was measured as the weight of soil lost from targets placed under the plant containing samples of an erodible sandy soil which had been previously oven-dried and passed through a 4-mm sieve to remove stones. The soil is from the Bearsted 1 Association, Typic Quartzipsamment. Prior to rainfall simulation, the soil was saturated and allowed to drain for 48 h.

In the first year of study, soil trays, 80 mm^2 in area, were used as targets and the number of trays was increased as the plant canopy became larger throughout the growing season. In this way an assessment of detachment by total throughfall under the plant canopy was obtained. Detachment by leaf drip was measured by placing smaller soil trays underneath the leaf drip points. In the second year the trays were replaced by circular splash cups, 30 mm in diameter, filled with soil. Six of these were placed beneath the canopy to assess total throughfall detachment and three were located beneath leaf drip points. The size of the soil targets was kept deliberately small so that detachment by raindrop impact could be measured without interference from runoff. The measured data relate to detachment in the strict sense of the breakdown of soil into its component particles. Since direction of particle movement is not considered, the data do not give a measure of soil loss or erosion.

Separate simulations were carried out without a plant cover to serve as a control using the soil trays and splash cups in the first and second year, respectively. The total detachment rate for each storm with a plant was determined by assuming that detachment occurred at the total throughfall rate for the area under the plant and at the bare soil rate for the area outside the plant. The sum of the two rates weighted by their respective percentage areas yields the storm detachment rate.

Rainfall energies were calculated from measurements of the drop-size distribution of the rain and estimates of the fall velocities of the raindrops. In the first year, drop-size distributions of the design storm, total throughfall, and leaf drip were obtained using the stain technique (Hall, 1970). The leaf drips were greater than 4 mm in size, whereas the design storm contained no drops larger than 5 mm. Rainfall energy under the plants was estimated by arbitrarily assuming that all drops ⩽ 4 mm in size derived from direct throughfall and had velocities of the design storm and that drops > 4 mm derived from leaf drip. The velocity v of the leaf drips was estimated as a function of height of fall h using the equation

$$v^2 = u^2 + 2gh$$

where u is the initial velocity of the drop, assumed here to be zero, and g is the acceleration due to gravity. This procedure ignores the effects of changes in drop shape and reductions in velocity resulting from air resistance. Since the maximum plant height was 870 mm and 80 percent of the plants tested were less than 500 mm high, these effects are likely to be small. For the second year, drop-size distributions were determined using the flour pellet method (Hudson, 1964). The drop-size distributions measured under the plant were divided into class groups and compared with the distribution for the design storm. Any increase in the percentage volume in a particular class was attributed to interception of the rainfall by the plant and the velocity of these additional drops was calculated as though they were leaf drips. The remaining volume in the size class was assumed to have the same drop velocities as that class in the design storm. Where the percentage volume of water in a given size class was equal to or less than that in the design storm, the velocities of all the drops were assumed to be the same as those of that class in the design storm. The velocities of the leaf drips were calculated using a computer program developed by Marchant (1977) that takes account of aerodynamic and gravitational forces.

The total energy of the rainfall at the ground surface was determined by summing the energy per unit area of the rainfall under the plant and the energy per unit area of the design storm weighted respectively by the proportions of the area of the catch tray under and outside the plant canopy.

The differences in procedure between the two years of study mean that the results are not always comparable in an absolute sense. Comparisons are therefore based on expressing the values measured in each year with a plant cover as a percentage of those obtained without cover for that year.

Rainfall interception

The effects of interception on the rainfall reaching the ground surface were reasonably consistent between the three crops and the two years of study

Table 4.1 Changes in rainfall properties and rates of soil detachment with percentage canopy cover[a].

Parameter	Control	Brussels sprouts	Sugar beet	Potatoes	Year
percentage canopy cover	0	3.8–49.9	5.4–13.6	5.5–21.1	1
		1.4–40.0	1.4–27.7	2.0–26.6	2
percentage volume in total throughfall	100	75.2–89.6	71.1–83.9	82.8–96.1	1
		71.6–97.4	43.2–97.0	72.2–95.3	2
percentage volume in leaf drip	0	0.3–17.0	11.3–24.2	0.0–7.8	1
		1.2–23.4	0.4–17.7	1.5–21.3	2
percentage volume in stemflow	0	0.0–2.9	3.9–11.8	0.0–1.4	1
		0.8–6.0	1.8–42.5	2.9–7.3	2
percentage design storm energy in total throughfall	100	19.8–89.5	24.2–80.2	64.2–85.2	1
		83.3–107.1	49.5–95.6	76.4–102.5	2
percentage design storm energy in leaf drip	100	0.0–39.2	24.3–53.4	0.0–30.5	1
		11.8–56.4	5.5–43.6	3.6–30.9	2
detachment by total throughfall as percentage of that with no cover	100	74.8–127.3	96.0–104.2	79.5–124.4	1
		68.2–105.6	43.6–95.0	68.3–101.6	2
detachment by leaf drip as percentage of that with no cover	100	0.0–59.5	15.4–44.1	0.0–17.7	1
		2.6–169.1	14.5–131.5	7.9–18.2	2

[a] Data denote the range of values recorded in the experiments.

(Table 4.1). The major difference is the higher percentage of stemflow and the lower percentage of total throughflow recorded under sugar beet, particularly in the second year. As the percentage canopy of the crops increased, the percentage volumes of leaf drip and stemflow increased (Table 4.2). Permanent interception increased in direct proportion to canopy percentage up to about 20 percent cover for both Brussels sprouts and potatoes. Further increases in canopy cover caused interception to increase at a decreasing rate so that, with 40 to 50 percent cover, only 22 percent of the rainfall was permanently intercepted by Brussels sprouts. Permanent interception by potatoes was only 21 percent with 27 percent canopy cover. The pattern for sugar beet was similar up to about 20 percent cover, but thereafter permanent interception decreased because of the increasing contribution of stemflow; with 28 percent cover only 14 percent of the rainfall was intercepted. Leaf drip accounted for about 20 percent of the rainfall reaching the ground under all three crops with canopy covers of 15 to 20 percent, except for potatoes during the first year of study when it was only about 8 percent.

Table 4.2 Linear correlation coefficients in soil detachment experiments.

Parameters	Brussels sprouts[a]	Sugar beet[b]	Potatoes[c]	Year[d]
% canopy area vs % volume	−0.760[e]	−0.044	−0.981[e]	1
in total throughfall	−0.949[e]	−0.926[e]	−0.542	2
% canopy area vs % volume	0.730[e]	0.584	0.910[e]	1
in leaf drip	0.970[e]	0.953[e]	0.738	2
% canopy area vs % volume	0.810[e]	0.562	0.962[e]	1
in stemflow	0.600	0.892[e]	0.989[e]	2
% canopy area vs % design storm	−0.817[e]	0.073	−0.733[f]	1
energy in total throughfall	0.846[e]	−0.926[e]	−0.812[f]	2
% canopy area vs % design	0.680[e]	0.671[f]	0.921[e]	1
storm energy in leaf drip	0.995[e]	0.955[e]	0.874[f]	2
% canopy area vs % detachment	0.707[e]	0.776[f]	0.692[f]	1
by total throughfall	−0.876[e]	−0.952[e]	−0.944[e]	2
	−0.071			1 (4–20%)
	−0.126		0.412	1 (4–15%)
			−0.857	1 (4–12%)
	0.518			2 (24–40%)
% canopy area vs % detachment	0.640[e]	0.724[f]	0.935[e]	1
by leaf drip	0.543	0.084	−0.563	2

[a] For Brussels sprouts $n = 49$ (year 1) and 8 (year 2).
[b] For sugar beet $n = 7$ (year 1) and 8 (year 2).
[c] For potatoes $n = 7$ (year 1) and 5 (year 2).
[d] Percentage canopy covers are over the range given in Table 4.1 unless otherwise stated.
[e] Significant at $p = 0.05$.
[f] Significant at $p = 0.10$.

Rainfall energy

Although there were some instances where the energy of the rainfall beneath a plant exceeded that of the design storm without a plant, generally the total throughfall energy decreased with increasing percentage canopy cover. Percentage energy reductions were usually greater than the percentage reductions in rainfall volume. The rate of reduction was highest under sugar beet where the total throughfall energy decreased to between 24 and 50 percent of that of the design storm with only 14 to 28 percent canopy cover. A decrease in rainfall energy to 20 percent of that of the design storm was not recorded under Brussels sprouts until the canopy cover was nearly 50 percent. Reductions in energy were even less marked for potatoes. The energy of the total throughfall was still 64 percent and 76 percent of that of the design storm when the canopy cover was 21 percent and 27 percent in the first and second year of the study, respectively.

The energy of leaf drip was much lower than that of the design storm, but it increased as the canopy cover grew, reaching 30 to 50 percent of that of the design storm with canopy covers of 14 percent or more. This increase in energy

Table 4.3 Median volume drop size (mm).

	Brussels sprouts	Sugar beet	Potatoes
design storms (no cover)	1.8–2.1	1.8–2.1	1.8–2.1
direct throughfall	1.8–2.9	1.6–3.3	1.5–2.4
leaf drip	4.5–6.3	4.6–6.2	4.7–5.9

is explained by a greater plant height and a larger median volume drop size (Table 4.3).

Soil detachment

Rates of soil detachment with a crop cover range from 44 to 127 percent of those with no cover (Table 4.1). Their relationship to increasing percentage canopy cover is inconsistent (Table 4.2). All the data are plotted in Figure 4.2 where regression lines are shown for those relationships over the full range of data which are significant at $p < 0.05$ and for some other selected relationships. The associated equations are given in Table 4.4. The relationships were selected partly on the strength of the correlation coefficients but, in addition, to draw attention to similarities in trend. Consistency of trend, even without high levels of statistical significance, is an important aspect of data interpretation when, as here, a series of essentially parallel experiments is carried out with small samples.

Detachment decreased with canopy cover for all three crops in the second year of the study but increased in the first year. More detailed analysis of the data for Brussels sprouts, however, shows that for canopy covers above 25 percent the detachment rate increased with canopy cover in the second year. Also, for covers up to 23 percent in the first year of study the detachment rate

Table 4.4 Linear regression equations for percentage soil detachment (Y) as a function of percentage canopy cover (X).

Number[a]	Equation	Year	Percentage canopy cover
1	$Y = 94.4 + 0.232X$	1	4–50
2	$Y = 99.2 - 0.197X$	1	4–20
3	$Y = 93.3 - 0.439X$	2	1–40
4	$Y = 52.4 + 0.749X$	2	24–40
5	$Y = 98.2 + 0.677X$	1	5–14
6	$Y = 96.7 - 1.85X$	2	1–28
7	$Y = 92.8 + 1.06X$	1	6–21
8	$Y = 120.5 - 3.18X$	1	4–12
9	$Y = 94.6 - 1.06X$	2	2–27

[a] see Figure 4.2.

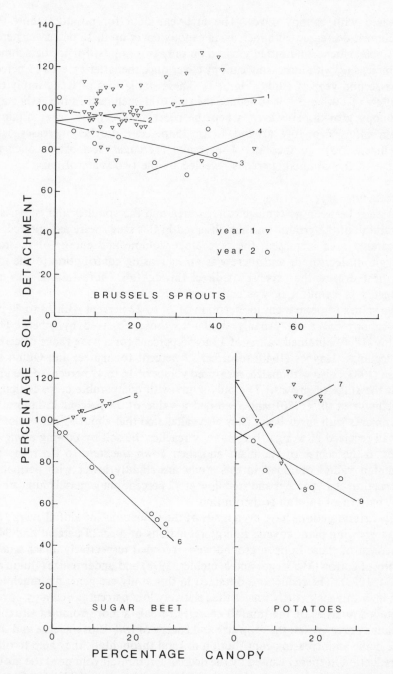

Figure 4.2 Relationship between percentage soil detachment (100 percent equals the bare soil value) and percentage canopy cover. Numbered lines denote best-fit regression lines (see Table 4.4).

decreased with canopy cover. The first-year data for potatoes show that detachment decreased with increasing canopy cover up to 12 percent. There is also some evidence, admittedly based on only two points, for the detachment rate increasing with increasing canopy cover once the latter exceeds 15 percent in the second year of study (Fig. 4.2). These similarities of trend imply that the effect of both crops is to reduce the rate of detachment in the early stages of canopy growth. However, when the percentage canopy cover attains a certain value, probably around 15 to 25 percent, further increases cause detachment rates to increase. The results for sugar beet are inconclusive because of the conflicting trends revealed by the two years of data.

Discussion

The trends between percentage canopy area and the volumes and energies of the rainfall at the ground surface observed in this study were as expected. As the canopy area increased, the percentage volume and energy of the total throughfall decreased, and there was an increasing contribution to the total from leaf drip at the expense of direct throughfall. More rainfall was also redirected as stemflow or was permanently intercepted.

The values of permanent interception of 14 to 22 percent with 14 to 50 percent canopy cover are broadly similar to those observed by others. Thus Wollny (1890) obtained values of 12 to 55 percent for a wide range of cereals and legumes, Haynes (1940) recorded 22 percent for maize, and Quinn and Laflen (1983), also with maize, measured values of 16 to 49 percent, depending upon the stage of growth. The only study with comparable crops is that by Appelmans et al. (1980) who recorded a value of 25 percent in laboratory experiments with sugar beet. They also calculated that about 55 percent of the rainfall received at a sugar beet canopy reaches the soil by flowing along the leaves to the center of the plant and then down the stem to the root. The maximum values obtained in this study are slightly lower with permanent interception at 17 percent and stemflow at 42 percent but generally indicate the same pattern of rainfall redistribution.

Few determinations have been made of the reductions in rainfall energy due to low-growing plant covers. Energy reductions of 5 to 25 percent and 34 to 62 percent of those in open ground were recorded respectively under a dense canopy of cotton (McGregor and Mutchler, 1978) and under maize (Quinn and Laflen, 1983). The reductions obtained in this study are generally much less. Although this may partly reflect the relatively low percentage canopy covers, a detailed analysis of the rainfall energies reveals a more complex situation. As canopy area grows, the drop size and energy of leaf drips increase and since these make an increasing contribution to total throughfall, they help to offset the reduction in energy caused by the decrease in rainfall volume at the ground surface. This effect was observed by Noble and Morgan (1983) for Brussels sprouts. Finney (1984), however, also found that although the percentage volume of direct throughfall decreased with increasing canopy cover, its energy per unit volume of rain reaching the ground surface increased. No physical explanation of this is yet available. Its net effect when combined with the greater role of leaf drip is to decrease the rainfall energy with increasing canopy cover more slowly than might otherwise be expected.

This pattern of changing rainfall energy with canopy growth may help to explain the somewhat unexpected results obtained for soil detachment. As canopy cover increases from 0 to 25 percent, detachment decreases because the percentage rainfall volume declines. With further increases in canopy, the reductions in volume are offset by the greater energy per unit rain penetrating the canopy. Thus detachment rates start to rise again. By the time 40 to 50 percent canopy cover is attained, detachment rates under Brussels sprouts, which are then 700 to 870 mm high, are similar to those recorded without a plant. Total energy levels of the throughfall continue to decrease, however, because of the decline in percentage rainfall volume so that the detachment rate becomes inversely related to total throughfall energy.

Interrill erosion

Experimental design

Erosion by rainfall and runoff combined was studied over a range of plant cover densities in a laboratory experiment using a rainfall–runoff rig. Rainfall was supplied from a nozzle simulator of the rotating disk type to a 1-m long and 0.6-m wide soil tray. The tray received runoff as overflow from a water tank at its top end. The rate of runoff was controlled by first passing the water through a constant-head tank and then through a gap meter. The tray had a wire mesh base covered with foam rubber and this was overlain by a 50-mm thick layer of soil from the Bearsted 1 Association. This arrangement allowed infiltration to take place but did not stimulate suction. A 0.2-m long wooden section with the test soil glued to it was placed between the water tank and the soil tray to provide an approach slope of similar roughness to that of the soil surface in the tray. Water and sediment passed from the tray through a plastic gutter and into collecting bottles. The tray was mounted on a car jack so that it could be set at a desired slope.

Prior to each experimental run the soil was saturated and allowed to drain for 1 h. The surface was leveled to a smooth condition and the tray set at a slope of $4°$. A design storm was applied consisting of nine minutes of rainfall at 50 mm/h, followed by seven minutes of rainfall at 50 mm/h with additional runoff at 2.5 l/min, and ending with five minutes of runoff only at 2.5 l/min to simulate a period of afterflow. Since plants could not be brought from the field, inserted in the soil tray and kept alive long enought to conduct the experiment, plant cover was simulated using 180 mm tall plastic oak trees manufactured by Britain Ltd (model 1822) and available from most toy shops (Fig. 4.3). This had the advantage of eliminating variability between individual plants. Experiments were conducted with two replications for 0, 3, 6, 9, and 12 trees, giving respectively 0, 15, 31, 46, and 61 percent canopy cover.

Figure 4.3 Interrill and microrill erosion on a sandy soil under a 31 percent canopy cover of toy plastic oak trees.

Flow velocity was measured by timing the movement of oil drops (Shell Vitrea Oil 22) over a 300-mm length of the soil tray. Flow depth was measured on the approach slope using a depth probe. It could not be measured on the soil because of the difficulty of judging when the probe touched the soil surface without sinking into it. Runoff and sediment were collected for 30-s periods at 0.5, 1, 1.5, 2, 2.5, 3, 6, 9, 10, 10.5, 11, 11.5, 12, 13, 16, 17, 17.5, 18, 18.5, and 21 min after the start of the storm. These sampling intervals were selected to be more frequent in the early stages of each phase of the storm in order to determine the immediate effects of changes in flow conditions.

Results

Flow depths ranged from 0.7 to 3.3 mm and velocities ranged from 72 to 372 mm/s. Velocities were generally rather high because microrills formed during all the experimental runs and the oil drops tended to follow their paths. As a result Reynolds and Froude numbers are also higher than those typical of overland flow in the field (Morgan, 1980). They are similar, however, to those produced in laboratory simulations by Emmett (1970) and Savat (1977). Reynolds numbers varied from 57 to 1018 but were generally between 200 and 600. Froude numbers ranged from 0.4 to 2.8 but were generally greater than 1.0. These values are typical for microrills without headcut development (Merritt, 1984). There were no discernible differences in flow conditions between the phases of the storms or with changes in canopy cover.

Neither the total runoff volumes nor those for the separate storm phases differ appreciably for different percentage canopy covers (Table 4.5). Applications of water to the plot were 4.5, 21.0 and 12.5 l, respectively for the three phases of the storm. With average runoff volumes of 2.8, 10.0, and 5.2 l, runoff represents respectively 62, 48, and 42 percent of the water applied. Runoff rose rapidly in the first two minutes of the storm before steadying at about 0.8 l/min (Fig. 4.4). With the onset of additional runoff from upslope, the hydrographs showed another rapid rise before leveling off at 2.8 to 3.2 l/min. A rapid fall to about 2.0 l/min occurred with the cessation of rain followed by a further rapid decline when all runoff ceased.

Total soil loss was highest with no canopy cover (Table 4.6). An increase in canopy cover to 15 percent approximately halved the soil loss, but further increases caused the soil loss to rise again. Despite the rise, however, the soil loss with 61 percent cover remained below that for bare ground. With the exception of the very low soil loss recorded in the second replication for phase 2 of the storm with 15 percent canopy cover, the greatest differences in soil loss with percentage canopy cover occurred in the first phase of the storm with

Table 4.5 Runoff (liters) recorded under different percentage canopy covers in the interrill erosion experiment.

| | Percentage canopy cover | | | | | |
	0	15	31	46	61	Replication
phase 1	2.76	2.75	2.31	2.90	2.61	1
	3.02	2.80	3.01	2.58	2.92	2
phase 2	9.60	10.39	10.23	10.38	10.20	1
	9.25	10.03	9.83	10.11	9.68	2
phase 3	3.81	5.54	5.51	5.69	5.18	1
	4.74	5.56	5.52	5.33	4.67	2
storm total	16.17	18.68	18.05	18.97	17.99	1
	17.01	18.39	18.36	18.02	17.27	2

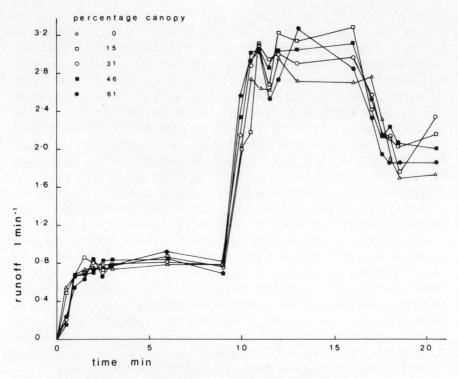

Figure 4.4 Hydrographs for the interrill erosion study.

rainfall alone and in the afterflow. The sedigraphs (Fig. 4.5) showed a similar pattern to the hydrographs with peaks occurring in the early stages of the first and second phases. The cessation of rainfall produced a sudden drop in soil loss followed by a slight rise in the afterflow before falling away to zero.

Table 4.6 Soil loss (g) recorded under different percentage canopy covers in the interrill erosion experiment.

| | Percentage canopy cover | | | | | |
	0	15	31	46	61	Replication
phase 1	16.87	7.17	12.14	9.21	14.24	1
	14.24	7.17	9.52	8.59	15.26	2
phase 2	31.36	20.66	22.31	32.28	28.76	1
	24.13	9.92	34.67	25.64	28.49	2
phase 3	6.73	5.44	4.36	4.48	4.61	1
	19.07	3.25	5.77	5.10	2.21	2
storm total	54.96	33.27	38.81	45.97	47.61	1
	57.44	20.34	49.96	39.33	45.96	2

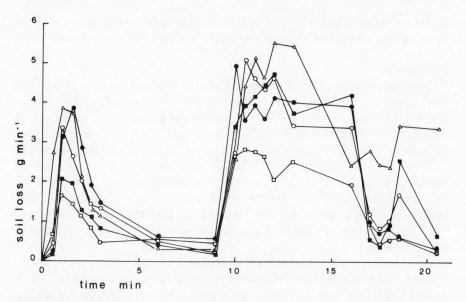

Figure 4.5 Sedigraphs for the interrill erosion study. Key to symbols appears on Figure 4.4.

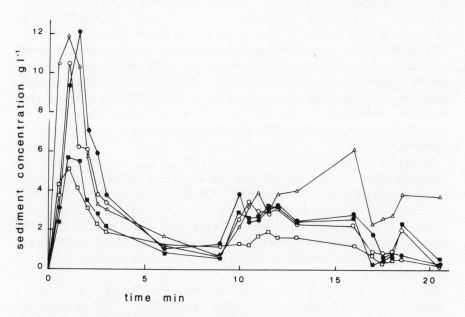

Figure 4.6 Patterns of sediment concentrations over time for the interrill erosion study. Key to symbols appears on Figure 4.4.

Sediment concentrations (Fig. 4.6) followed the same trends except that the peak values occurred in the first phase of the storm and not the second.

Discussion

As expected, increases in canopy cover had little effect on flow conditions, but their failure to reduce the volume of runoff was rather surprising and implies that the increases had little effect on permanent interception. Observations during the experiments suggest that permanent interception may be rather low. Some large single drops remained on the leaves, but most raindrops appeared to strike the plastic surfaces and rebound onto the soil. Despite the lack of runoff reduction, the plant cover, as expected, brought about a decrease in soil loss. The pattern of soil loss falling initially and then slowly rising again with increasing canopy cover was not anticipated, however, since it failed to conform with either the linear or exponential decreases in erosion recorded in the literature.

With the canopy cover having little effect on erosion by the combined rainfall and runoff in phase 2 of the storm and with the soil loss in the afterflow declining initially with increasing percentage canopy and then fluctuating about the reduced value, the pattern is best explained with reference to the first phase of the storm. Erosion in this phase is associated with a rapid increase in sediment concentration in the first few minutes. A similar increase has been recorded elsewhere by Yair and Lavee (1977, 1981). High sediment concentrations so early in a storm must be due to a high rate of soil detachment which provides a supply of loose particles for transport with a relatively low volume of runoff. Thus the increase in soil loss as canopy cover increases from 15 to 61 percent may be attributed to an increase in the rate of detachment in phase 1 of the storm. Since the flow velocities throughout the storm were generally sufficient to detach soil particles, it is not possible to attribute the detachment specifically to either raindrop impact or runoff generated by the rainfall. However, since runoff effects dominate phases 2 and 3 of the storm without producing differences in soil loss in relation to canopy cover, it seems reasonable to interpret these phase 1 differences in terms of detachment by raindrop impact.

Observations during the storm showed that the soil directly beneath the simulated plants was protected from erosion and that runoff became concentrated in microrills between the plants. Although this is similar to the effects of plant cover on flow described by De Ploey et al. (1976), there was no evidence in the study described here to show that it increased erosion. The scouring by microrills may, however, have been a contributory factor for the failure of erosion to decrease as canopy cover and therefore plant stem density increased.

The general pattern of soil loss and sediment concentration over time is similar for all percentage canopy covers and follows well-described and accepted trends. The high soil loss and sediment concentrations of the first phase relate to the detachment and removal of loose soil particles. Both soil loss and sediment concentration decline through phase 1 of the storm, presumably as the most erodible particles are removed. The rise in sediment concentration to a secondary peak in phase 2 most likely relates to detachment of soil particles by runoff. The increase in soil loss to a peak value for the storm during this phase is attributable to the greater runoff volume.

Implications

The results of these two studies fail to conform to many of the accepted ideas on the role of a plant cover in hillslope hydrology and erosion. The first study showed that permanent interception of rainfall by the canopy reduces the volume of rain reaching the ground. With the expected higher levels of infiltration caused by the root network opening voids in the soil, runoff in the field should decline compared with that on bare soil. Although the second study failed to show this effect, the conditions were rather special with a low level of permanent interception and no root system and so were not a realistic simulation of this component of the hydrological cycle.

Both studies yielded evidence for a pattern of erosion declining initially with increasing canopy cover and then rising again when the canopy exceeds 15 to 25 percent. The first study showed this pattern for detachment of soil particles by raindrop impact under Brussels sprouts and provided less convincing evidence for it under potatoes. No consistent trend of detachment with canopy cover was obtained for sugar beet. The results of the second study appear to be best explained through changes in the rate of soil detachment by raindrop impact with increasing canopy cover.

From the first study it is suggested that detachment increases, despite the reduction in the amount of rain penetrating the canopy, because of a greater contribution of leaf drip with its larger drop sizes and also, from Finney's (1984) observations, because of an unexplained increase in the energy per unit rain of the direct throughfall.

One mechanism commonly thought to explain the decrease in erosion with increasing plant canopy is the protection afforded to the soil surface against raindrop impact. This prevents the surface from becoming crusted and helps to maintain a high infiltration rate. As crusting is dependent upon the energy of the rainfall and the rate of soil detachment, the data presented here would not support this explanation. However, no observations were made on crusting in these experiments. Future studies should examine changes in the susceptibility of the soil to crusting in the presence of a plant cover.

Caution should prevail in a general application of the findings of these two studies because different plants are likely to behave in different ways. Plants with bladed leaves and closer proximity to the soil surface, such as wheat, barley and grass, may not affect soil detachment in the same way as Brussels sprouts. Also, the first study took no account of changes in soil properties during a storm which interact with the plant cover to influence detachability.

Neither study allowed for the effects of different soils, seasonal changes in soil properties, or the role of a root mat and litter layer.

Nevertheless, focusing attention on plant canopy effects has shown that an effective understanding of the role of the plant cover in hillslope hydrology and erosion requires a detailed analysis of the changes in rainfall energy, as well as in volume of runoff, brought about by the interception phase of the hydrological cycle. Present ideas on the role of canopy cover seem inadequate. These two studies have revealed aspects of erosion that cannot yet be properly explained. If an increasing canopy cover decreases the energy and volume of the rain reaching the ground surface, why should soil detachment rates increase? What is the mechanism by which raindrops become more effective detaching agents following interception by a plant cover? If it is through an increase in energy per unit volume of rain, what physical process is involved? How general are the findings of the studies reported here and, if they are not general, what are the circumstances that bring them about? Clearly more attention must be directed at the way a plant cover transforms the characteristics of the rainfall and how these characteristics, in turn, influence the detaching power of both leaf drips and direct throughfall.

Acknowledgments

The studies were carried out with the assistance of a Research Grant from the Natural Environment Research Council, United Kingdom.

References

Appelmans, F., J. van Hove, and L. De Leenheer, Rain interception by wheat and sugar beet crops, in *Assessment of Erosion*, edited by M. De Boodt and D. Gabriels, p. 227–235, Wiley, Chichester, 1980.

De Ploey, J., J. Savat, and J. Moeyersons, The differential impact of some soil factors on flow, runoff creep and rainwash, *Earth Surface Processes, 1*, 151–161, 1976.

Elwell, H. A., A soil loss estimation technique for southern Africa, in *Soil Conservation: Problems and Prospects*, edited by R. P. C. Morgan, pp. 281–292, Wiley, Chichester, 1981.

Emmett, W. W., The hydraulics of overland flow on hillslopes, *U.S. Geological Survey Professional Paper 662-A*, 1970.

Finney, H. J., The effect of crop covers on rainfall characteristics and splash detachment, *Journal of Agricultural Engineering Research, 29*, 337–343, 1984.

Foster, G. R., Modeling the erosion process, in *Hydrologic Modeling of Small Watersheds*, edited by C. T. Haan, H. P. Johnson and D. L. Brakensiek, pp. 297–380, American Society of Agricultural Engineers Monograph 5, 1982.

Foster, G. R., L. J. Lane, J. D. Nowlin, J. M. Laflen, and R. A. Young, Estimating erosion and sediment yield on field-sized areas, *Transactions of the American Society of Agricultural Engineers, 24*, 1253–1263, 1981.

Hall, M. J., A critique of methods of simulating rainfall, *Water Resources Research, 6*, 1104–1114, 1970.

Haynes, J. L., Ground rainfall under vegetative canopy of crops, *Journal of the American Society of Agronomy*, 32, 176–184, 1940.

Hudson, N. W., The flour pellet method for measuring the size of raindrops, *Research Bulletin 4*, Department of Conservation, Salisbury, Rhodesia, 1964.

Hussein, M. H., and J. M. Laflen, Effects of crop canopy and residue on rill and interrill soil erosion, *Transactions of the American Society of Agricultural Engineers, 25*, 1310–1315, 1982.

Kirkby, M. J., Modelling water erosion processes, in *Soil Erosion*, edited by M. J. Kirkby and R. P. C. Morgan, pp. 183–216, Wiley, Chichester, 1980.

Laflen, J. M., and T. S. Colvin, Effect of crop residue on soil loss from continuous row cropping, *Transactions of the American Society of Agricultural Engineers, 24*, 605–609, 1981.

Lang, R. D., and L. A. H. McCaffrey, Ground cover: its effects on soil loss from grazed runoff plots, Gunnedah, *Journal of the Soil Conservation Service, New South Wales, 40*, 56–61, 1984.

McGregor, J. C., and C. K. Mutchler, The effect of crop canopy on raindrop size distribution and energy, *USDA Sedimentation Laboratory Annual Report*, Oxford, Miss., 1978.

Marchant, J. A., Calculation of spray droplet trajectory in a moving airstream, *Journal of Agricultural Engineering Research, 22*, 93–96, 1977.

Merritt, E., The identification of four stages during micro-rill development, *Earth Surface Processes and Landforms, 9*, 493–496, 1984.

Meyer, L. D., and W. H. Wischmeier, Mathematical simulation of the processes of soil erosion by water, *Transactions of the American Society of Agricultural Engineers, 12*, 754–758, 762, 1969.

Morgan, R. P. C., Field studies of sediment transport by overland flow, *Earth Surface Processes, 5*, 307–316, 1980.

Morgan, R. P. C., Splash detachment under plant covers: results and implications of a field study, *Transactions of the American Society of Agricultural Engineers, 25*, 987–991, 1982.

Morin, J., D. Goldberg and I. Seginer, A rainfall simulator with a rotating disk, *Transactions of the American Society of Agricultural Engineers, 10*, 74–77, 79, 1967.

Natural Environment Research Council, *Flood Studies Report II, Meteorological Studies*, London, 1975.

Noble, C. A., and R. P. C. Morgan, Rainfall interception and splash detachment with a Brussels sprouts plant: a laboratory simulation, *Earth Surface Processes and Landforms, 8*, 569–577, 1983.

Quinn, N. W., and J. M. Laflen, Characteristics of raindrop throughfall under corn canopy, *Transactions of the American Society of Agricultural Engineers, 26*, 1445–1450, 1983.

Savat, J., The hydraulics of sheet flow on a smooth surface and the effect of simulated rainfall, *Earth Surface Processes, 2*, 125–140, 1977.

Shaxson, T. F., Reconciling social and technical needs in conservation work on village farmlands, in *Soil Conservation: Problems and Prospects*, edited by R. P. C. Morgan, pp. 385–397, Wiley, Chichester, 1981.

van Asch, Th. W. J., Water erosion on slopes and landsliding in a Mediterranean landscape, *Utrechtse Geografische Studies, 20*, 1980.

Wischmeier, W. H., Estimating the soil loss equation's cover and management factor for undisturbed areas, in *Present and Prospective Technology for Predictor Sediment Yields and Sources*, Report ARS-S-40, pp. 118–124, U.S. Agricultural Research Service, 1975.

Wollny, E., Untersuchungen über das Verhalten der atmosphärischen Neiderschläge zur Pflanze zum Boden, *Forschungen, Gebeit der Agrikulturphysik, 13*, 316–356, 1890.

Yair, A., and H. Lavee, Runoff generative process and runoff yield from arid talus mantled slopes, *Earth Surface Processes, 1*, 235–247, 1976.

Yair, A., and H. Lavee, Trends of sediment removal from arid scree slopes under simulated rainstorm experiments, *Hydrological Sciences Bulletin, 22*, 379–391, 1977.

Yair, A., and H. Lavee, An investigation of source areas of sediment and sediment transport by overland flow along arid hillslopes, *International Association of Hydrological Sciences Publication 133*, 433–446, 1981.

5
Sediment movement in ephemeral streams on mountain slopes, Canadian Rocky Mountains

James S. Gardner

Abstract

Sediment movement in, and some hydrological characteristics of, bedrock gullies on steep mountain slopes are described. Data are presented from a measurement and monitoring study at four sites in the Mt. Rae area of the Canadian Rocky Mountains between 1979 and 1984. Sediment movement was monitored using painted and numbered clasts arranged in transects. Observations and measurements of snowmelt- and rainstorm-generated discharge indicate intermittent surface and subnival flows capable of transporting sediment. Mean annual movement distances of marked clasts by transect ranged from less than 1 m to greater than 50 m. Mean values in these gullies exceed by an order of magnitude movements in such processes as debris shift on open slopes, but variability is generally the same as previously published variability measures. Large spatial and temporal variability in movement are suggestive of intermittent and spatially confined transport agents such as ephemeral streamflow. Mixing of channel bed sediments, imbrication in reworked channel bed deposits, and incision in debris pockets within the bedrock channels are also indicative of fluvial activity. Evidence presented in this chapter indicates that intermittent fluvial activity is an effective erosional agent in the high mountain, periglacial environment.

Introduction

Even on the steepest bedrock slopes in the Canadian Rocky Mountains, large amounts of debris are stored awaiting downslope transport. Mechanisms of transport include rockfalls, snow avalanches, icefalls, overland and stream flow, and combinations of the same. This chapter focuses on debris transport in ephemeral stream flow in a high mountain setting. During rainstorms and snowmelt, water flows off bedrock slopes into declivities, gullies, or couloirs. This phenomenon, which is a type of ephemeral streamflow, rarely has been described (e.g., Matthes, 1930; Behre, 1933; Bones, 1973). Usually it is discussed in the context of debris slope development and modification (Rapp, 1960; Church et al., 1979; Whitehouse and McSaveney, 1983). As a form of ephemeral and high mountain stream (e.g. Kellerhals, 1978; Day, 1972) the phenomenon has received no attention. In some circumstances these short-term ephemeral streams generate debris torrents or debris flows on lower debris-covered slopes, sometimes with striking geomorphic and disastrous economic consequences (Mears, 1979; Costa and Jarrett, 1981; Desloges and Gardner, 1984).

The objective of this chapter is to describe some of the hydrological characteristics and their geomorphic consequences in four ephemeral streams in the Mt. Rae area (50°35'N, 116°6'W) of the Canadian Rocky Mountain Front Ranges. The dynamic geomorphology of this area has been described by Gardner et al. (1983) and some aspects of slope erosion in the area have been described in various publications (e.g., Gardner, 1983a, b, c). In this context it was clear that runoff concentrated in gullies on the bedrock slopes is an important mobilizing, transport, and depositional agent. It is thought that the same generalization may apply to other temperate mountain regions.

The study described herein sought to measure ephemeral discharge, to monitor rates and patterns of debris movement in ephemeral channels, and to describe the erosional and depositional effects of the discharge.

The study area

The study area, like most of the Front Ranges, features prominent mountain ridges aligned on a northwest to southeast strike. Defined by extensive thrust faults along their northeastern to eastern margins, the ridges are composed of Paleozoic limestone and dolomite strata dipping steeply to the southwest. Intervening valleys, passes, and low ridges are composed of Mesozoic shales, sandstones, conglomerates, and coal deposits. The linear pattern of ridges and valleys is broken by cirques and through valleys oriented across the strike. Scarp slopes, dip slopes, and strike slopes are characterized by steep bedrock surfaces from 50 to 800 m high. Dip-slope bedrock surfaces have a flatiron morphology, whereas numerous strata of varying resistance exposed on scarp slopes produce a bench and cliff topography. Strike slopes are characterized by complexes of benches, cliffs, and flatirons (Fig. 5.1). Channels are super-imposed on these various bedrock topographies.

The study area displays evidence of classical alpine glaciation. Several cirques with northerly exposures contain remnants of small glaciers and moraines dating from the Neoglaciation. None of the ephemeral streams examined issue from glaciers or permanent snowpatches.

Figure 5.1 Site C is located on a northwestern exposure on a combination of strike and dip slopes. The photo was taken in late August and is indicative of the generally dry, snowless conditions in late summer and early autumn.

Climatological data for the immediate area are limited to short-time records from Highwood Pass (2100 m a.s.l.) and several experimental sites. At Highwood Pass, which is just below treeline, mean annual temperature is about $-1°C$. Snowfall accounts for over 60 percent of the mean annual precipitation of 800 mm, with April and May experiencing the heaviest snowfall. Although snowfall has occurred in all months, winter snowpacks persist from six to nine months, depending on elevation and exposure. Wind scour and direct sunlight limit snow accumulation on south- and west-facing slopes. Drifting and shading result in preferential deposition and long-lasting snowpacks on northern and eastern exposures. Snow is an important source of water for ephemeral streamflow and the channels or gullies accumulate snow preferentially by drifting and avalanching (Fig. 5.2).

Figure 5.2 Site A is developed on a western exposure. The photo was taken in mid-July and illustrates the presence of late-lying snow in the bedrock gully.

Table 5.1 Return periods for maximum one-day precipitation in the Kananaskis region.

Return period (years)	One-day precipitation (mm)
2	36
5	51
10	61
25	81
100	97

Source: Pollock and Gaye (1975).

The area is on the western margin of a Front Range/Foothills zone which experiences a high frequency of extreme, short-duration rainstorms during the summer. Starkel (1979) has attributed high-magnitude geomorphic change in high relief areas to storms of extreme intensity and to sudden thaws. Table 5.1 shows the return periods for one-day (24 h) rainfall extremes in the Kananaskis area. These values place the study area between 24-h precipitation extremes for tropical and arctic areas, where the geomorphic impacts of such storms have been studied (Rapp, 1974). The most intense storm observed and measured in the Mt. Rae area since 1975 yielded a 24-h value of 55 mm (return period $\simeq 8$ y). This storm was observed to produce overland flow on most bedrock surfaces, channelized flow in most gullies, and at least one debris flow.

Study sites

Four channels were chosen for detailed study. The choice of study sites was based on two criteria, the first of which related to suitable locational and morphological characteristics. The four sites include the important contrast between north and south exposures. The channels are formed on steep bedrock slopes, are composed of several tributaries, have small catchments (0.5 km^2), and feed onto prominent debris slopes that display evidence of fluvial and/or debris flow modification (Figs. 5.1 and 5.2). The second criterion for study site selection was purely logistical, relating to time available to carry on repetitive measurements over a period of years.

The four sites are located at or above treeline in an altitude range of 2300 to 2900 m. Figures 5.1 and 5.3 show sites C and B representing north and south exposures, respectively. The catchment at C is formed in part on dip slopes in Paleozoic limestones where the most pronounced channels coincide with bedding planes. Site B is formed in Paleozoic limestones exposed on the strike slope. Channels follow recessive units and bedding planes in the steeply dipping strata. The other two sites, A and D, have southwesterly and northwesterly exposures and are located on dip and strike slopes respectively.

Figure 5.4 shows profiles for the valley side slopes on which the study sites are located. Slope gradient data were taken from 1 : 50 000 topographic maps. Basal debris slope gradients range from 10° to 25° and the bedrock slopes range from 30° to over 50° in general. In detail, the bedrock channel gradients are stepped as a result of varying resistances to erosion of the bedrock units as well as their geomorphic development. A high gradient equivalent to pool and riffle sequences is present with steep (vertical to overhanging) plunge pools interspersed with flats or low-gradient reaches. Steep reaches are generally devoid of debris. Both high- and low-gradient sections of channel are abraded and smoothed presumably by water transported debris and perhaps by snow avalanches and rockfalls. Debris accumulates in low-gradient reaches through in-channel transport and mass wasting from surrounding slopes, as demonstrated by McPherson (1971) in high-gradient alpine streams.

Figure 5.3 Site B has a southern exposure. The photo was taken in early July and shows remnants of the winter snowpack supplemented by a late June snowfall. Discharge is present in the bedbrock channel.

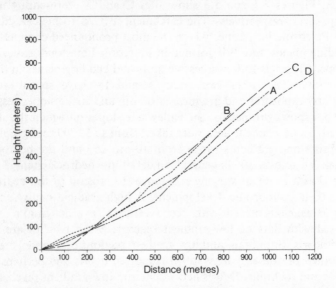

Figure 5.4 Slope profiles at the four study sites (A to D). These profiles are valley-side profiles extending from ridge crests to the toe of debris slopes.

Data collection

At each of the four sites the observation program consisted of periodic assessment and recording of snow conditions by photography, observation and measurement of channel discharge, and monitoring of debris movement in the channels. Assessment and recording of snow conditions was performed through late spring and summer (mid-June to late August) in 1979, 1980, and 1981. Channel discharge was assessed in two ways: the presence or absence of flowing water was noted on at least a weekly basis in the 1979, 1980, and 1981 field seasons, and velocity measurements were made on occasion using an Ott meter at known cross-sections. Obviously, velocity measurements in the turbulent flow of these streams are suspect. Measurements were made at bedrock channel cross-sections where turbulence appeared to be minimal. At best, the derived discharge values are approximations. Debris movement was monitored using painted and numbered clasts taken from in-channel debris accumulations. The marked clasts were representative of the size and shape variations at each site, excluding material in the coarse sand and smaller size range. Marked clasts varied in size from 1.5 to 20 cm along their intermediate axis. Fifteen transects of 25 clasts each were placed at various levels in the four channels in August 1979. Each transect was arranged across the channel gradient and was confined within the bedrock margins of the channel. The transects were located in areas of debris accumulation in the channels (i.e., low-gradient reaches). The positions of marked clasts were remeasured by tape in July 1980, August 1981, August 1982, and August 1984. Original transect positions were indicated by bedrock markers. Measurement error is thought to be about ± 5 cm, as in prior monitoring studies of debris shift (Gardner 1979).

Results

Observations obtained in this study are both qualitative and quantitative. Information relating to hydrology and modes of debris transport and deposition are largely qualitative, whereas data relating to rates and patterns of debris movement are largely quantitative.

Initially it was thought that snow would be an important factor in debris transport through the production of meltwater and possibly through snow avalanching. Observations of snow conditions at the four sites indicate that at B, winter snow is usually melted by early to mid-June as a result of a direct southerly exposure. Snowmelt at this site produces flow in the bedrock channel. Subsequent summer snowfalls result in short-term accumulations at B followed by rapid melt and runoff. Sites A, C, and D retain snow well into July and occasionally into August each summer (Fig. 5.2). Snow accumulates preferentially in the bedrock channels through drifting and avalanching. Channel snow patches usually remain long after snowpacks in the upper catchments have melted, producing a "melting-from-above" effect. In such cases, flow in the channels is subnival (Fig. 5.5) meaning that channel snow patches do not necessarily retard debris transport by meltwater. However, there is some evidence that channel snow does retard debris transport by snow avalanches. Channel snow is relatively stable and of a high density owing to wind accumulation and compaction, and accumulation by small, direct-action powder avalanches. Subsequent slab and wet snow or snowmelt avalanches rarely make contact with the channel bed and transportable debris, moving over the stable channel snow instead.

Discharge occurs as a result of spring/summer snowmelt, melt of occasional summer snowfalls, rainfall, and their combinations. Spring and early summer snowmelt was the most reliable discharge generator both in terms of volume and timing of discharge. Rainstorm-generated discharge was observed only

Figure 5.5 Late-lying snowpatch at the base of the bedrock channel at site D. Note the discharge in the channel and the fact that it becomes subnival. Subnival drainage is typical in the bedrock channels in early and mid-summer.

once at the four sites in three field seasons of observations. However, in the four field seasons prior to this study, casual observations in the same general area suggest that it occurs rather more frequently as a result of both long-duration heavy rain and short but very intense local convective storms. The apparent low frequency of rainfall-generated discharge is the result of a lack of a high-intensity storm (> 10 mm/h) in the area and the coarse nature of the temporal observations during the study. In addition, several long-duration frontal storms, which produced copious rain in the valleys, produced snowfall in the higher elevations of the basins studied. Thus runoff was delayed but discharge was usually observed as the snow melted.

Hydrographs are complex and show great variation in timing on an annual basis, and both timing and volume of flow on a seasonal and diurnal basis. Apart from occasional storm-generated flows, most discharge events are derived from snowmelt and thus are controlled primarily by incident radiation and available snow. The north- and south-facing sites (D and B, respectively) show a contrast in their respective hydrographs. Site B, the south-facing example, shows flow early in the annual melt period (late April to late May), followed by occasional flows through June and early July as late snowfalls are received and rapidly melt. By early July all winter snow has usually melted from the catchment of site B (Fig. 5.3). On one occasion in three field seasons, discharge was observed following an early August snowfall at site B. Generally, these isolated discharge events do not last longer than two to three days. Maximum estimated discharge at site B was $0.05 \text{ m}^3/\text{s}$.

At site D (north-facing), snowmelt flow began sporadically in early June. However, it was not until early July that runoff peaked and much of the bedrock channel became snow free. Maximum estimated discharge at site D occurred on July 4, 1981, when $0.27 \text{ m}^3/\text{s}$ was recorded. In 1981, snowmelt discharge from the winter snowpack continued until mid-August, a marked contrast from the south-facing site. Hydrological characteristics of sites A and C were intermediate between the north- and south-facing extremes. Site A generally retained snow in the bedrock channel until late July (Fig. 5.2), long after the snowpack had melted from the catchment. Thus discharge in the channel was in part subnival and tended to be of a very low magnitude. Maximum measured discharge at site A was $0.13 \text{ m}^3/\text{s}$. Site D (Fig. 5.1) experiences snowmelt discharge only until early July and periodic discharges thereafter in response to occasional summer rainstorms. This site lies in a low snow accumulation area as a result of extreme wind erosion of the winter snowpack. The catchment does not retain an extensive and deep snowpack and snowmelt discharge is relatively short-lived and of low magnitude. Maximum measured discharge at site D was $0.07 \text{ m}^3/\text{s}$.

Diurnal hydrographs at all sites differed somewhat in timing and volume. In the case of snowmelt-generated discharge, flow normally commenced in the morning and ceased by about 2200. Cessation of flow during the night resulted from the absence of incident solar radiation and night-time air temperatures at or near the freezing point at these high elevations. At site B (south-facing) flow commenced with incident solar radiation, usually between 0800 and 0930 and peaked between 1400 and 1600, assuming no major change in weather during the course of the day. At the other sites, flow was delayed due to shading and commenced between 1000 and 1200, reaching a peak between 1500 and 1800. Largely because of available snow and slightly larger catchments, peak discharges at sites A and D tended to be greater than at the other two sites. Rainstorm generated hydrographs were controlled entirely by the timing, duration, and intensity of the rain.

At no time during the study was there any evidence that discharge was generated from anything other than surface water sources.

Sediment transport was monitored using painted and numbered clasts taken from the channel bed. The rates of particle movement are summarized in Table 5.2. The values shown are arithmetic mean values for an entire transect. Within each transect variation in movement between the 25 marked clasts was large in most instances and is summarized by the coefficient of variation. Mean annual movement values for individual transects range from 0.01 to 114.3 m. The data must be used with caution as large mean values may result from different circumstances. They may reflect large movement by many clasts within the transect. Site B, transect 1, 1979–80 is a good example with a large mean value (13.13 m) and a relatively small coefficient of variation (0.36). Alternatively, they may reflect very large movement by just a few clasts. Site A, transect 4, 1980–81 is an example with a large mean value (26.72 m) and a relatively large coefficient of variation (2.04). Large mean values are also produced by situations where the number of available clasts is reduced and those few available moved long distances. Site D, transect 3, 1982–84 is an example where only one marked clast was available; thus the mean value is large but the coefficient of variation is 0.

The latter example illustrates a major shortcoming of the measurement procedure. This shortcoming is a decreasing sample size through time, as marked clasts are buried or broken (Table 5.2). Burial is inferred from the observation that many markers when found were partially buried and some that could not be found one year were found the following year. This problem has been noted previously with regard to measurement of talus shift (Gardner, 1979). Although it results in problems of representativeness, the fact that clasts disappear through burial is a valuable observation in its own right. Figure 5.6 illustrates the phenomenon of disappearing markers over the study period. The experiment commenced with a total of 375 markers. By August 1984, 114 markers remained. Sixty-nine percent of the original group could not be found. Disappearance progressed steadily from 1979–80 to 1984, while the number of clasts showing no movement declined steadily to the point where only 5 percent of the original sample showed no movement over five years. The number of clasts showing movement and disappearance increased through the study period until 1984. Approximately 10 percent of the original sample disappeared at some stage of the study but were found again during a later observation period, notably in 1984. These observations of disappearance and

Table 5.2 Mean movement values by slope, transect, and year.

Site	Transect	1979–80 N^a	$\bar{X}^b$ (m)	CV^c	1980–81 N	$\bar{X}$ (m)	CV	1981–82 N	$\bar{X}$ (m)	CV	1982–84 N	$\bar{X}/2^d$ (m)	CV
A	1	15	4.07	1.18	14	2.00	1.90	14	0.55	2.27	14	2.37	1.00
	2	20	13.90	1.55	17	1.12	2.63	18	2.03	3.80	18	8.52	1.33
	3	19	32.55	1.00	18	5.08	2.05	16	12.25	1.23	8	12.47	1.06
	4	11	37.79	1.35	10	26.72	2.04	9	11.63	1.89	4	54.39	0.21
	5	20	3.35	2.94	22	8.27	2.99	19	3.32	3.32	4	48.86	0.72
B	1	20	13.13	0.36	16	0.26	1.62	12	5.15	1.27	no markers retrieved		
	2	21	2.99	2.75	17	5.84	2.26	21	28.09	2.10	6	13.57	1.05
	3	20	7.83	2.68	14	19.59	2.17	16	39.68	1.90	13	31.78	1.60
C	1	15	6.89	1.16	16	5.48	1.37	9	9.59	0.99	3	19.80	1.03
	2	13	10.76	1.25	12	7.70	2.49	11	14.07	0.96	8	11.57	1.25
	3	18	5.00	1.57	16	5.63	2.37	17	14.68	2.03	10	53.61	0.74
	4	21	22.98	1.66	20	0.57	2.12	16	34.05	1.36	12	31.96	1.71
D	1	21	0.01	5.00	11	6.82	2.27	11	0.28	3.18	10	5.80	2.29
	2	19	1.96	3.68	13	28.51	1.22	11	8.68	2.98	3	9.72	1.41
	3	22	3.31	2.31	12	24.67	1.79	12	2.13	2.15	1	114.30	0

[a] N = number of clasts found. In 1979 at the beginning of the experiment $N = 25$ at each transect.
[b] $\bar{X}$ = mean value.
[c] CV = coefficient of variation.
[d] This value is the average annual movement over two years.

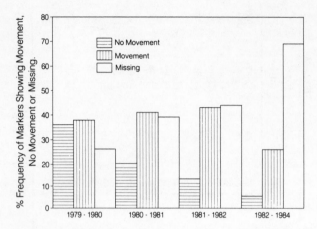

Figure 5.6 Relative frequency of markers showing no movement, movement, or missing over the four measurement intervals.

later reappearance, in some cases indicate that the mechanisms of debris transport involve extensive reordering (stirring) of channel sediments. Many of the markers relocated after movement showed abrasion (incipient rounding) of their painted surfaces and some were broken. Relocated markers were often found in tightly packed and imbricated channel bed deposits. These channel bed deposits, which included markers, were usually observed in pockets or bars rather than evenly distributed along the channel bed. Subsequent observations showed some of these pockets to be completely eroded to channel bed with only small, perched terracettes remaining along the channel margins. Such observations suggest that intermittent fluvial activity is an important agent for debris movement.

The mean values summarized in Table 5.2 are at least an order of magnitude greater than published values for debris shift, a process occurring in the same type of environment but on open debris-covered slopes or talus cones (Gardner, 1979; Caine, 1984). However, variability of transport distances within transects in measurement intervals as shown by the coefficient of variation are comparable to previously published measures of variability in debris shift (Gardner, 1982; Caine, 1984; Wallace, 1968). High within-transect variability is indicative of long-distance transport of individual or small groups of markers, while other markers either do not move or move very short distances. High coefficients of variation indicate that movement is not general within each transect.

Some inferences about mode of transport can be made by examining the patterns and magnitudes of movement through time. During the study period or until their eventual disappearance, 33 percent of the markers moved in only one measurement interval (e.g., 1980–81), 35 percent moved in two intervals (e.g., 1980–81 and 1981–82), 17 percent moved in three intervals, and 3 percent moved in all four intervals. The remainder did not move. Individually, the movements were classified by magnitude using a log 10 transformation. From these data frequency distributions of movement magnitude were constructed for each measurement interval as well as for the whole study period (Fig. 5.7). The log 10 transformation (see Caine, 1968) was used to encompass the wide range of movement magnitudes, from a few centimeters to well over 100 m. All distributions indicate that no movement was the most frequent observation followed by movements in the 0.1–0.99 m and 10–99 m classes. Approximately 5 percent of all movements were in the ⩾ 100 m category. These patterns suggest that the transport mechanisms produce generalized movement of channel sediment over relatively short 5-yr periods, reorder the sediment by means of mixing, and produce high average rates of movement but with considerable temporal and spatial variation within transects. The data do not suggest a significant difference in transport rates between the various study sites and transects. Moreover, there was no discernible relationship between movement distance and the size of the markers.

The quantitative results and qualitative observations together suggest a pattern of debris transport that is probably unique to steep bedrock channels. Channel profiles, which are static factors in transport, tend to be irregularly stepped with short, low-gradient reaches interspersed with longer, high-gradient reaches. Debris accumulations tend to be pocketed in the low-gradient reaches with little or no channel-bed debris in the intervening high-gradient reaches. The marked clasts were taken from these debris pockets where the transects were located as well. Movement data indicate a bimodal movement

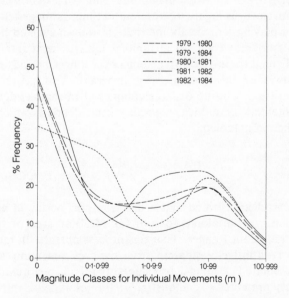

Figure 5.7 Movement magnitude–frequency curves.

frequency–magnitude distribution (Fig. 5.7), which suggests a tendency for markers to remain *in situ* or to move relatively long distances as individuals or small groups, giving movement the appearance of occurring in pulses. The stepped channel gradients would also favor this pattern. In the low-gradient debris pockets, movement must be initiated by an external force. Once moved beyond the threshold between low- and high-gradient reaches (usually less than 3–4 m), a clast would tend to move through the high-gradient reach under the influence of gravity and without further external stimulus. Such a clast may come to rest at the next low-gradient reach, becoming part of the debris pocket there and perhaps being buried. Alternatively, a clast with sufficient velocity and/or a continuing external stimulus may continue down channel. With the passage of time a distribution showing preference for no movement or substantial movement would emerge (Fig. 5.7). In addition, the disappearance and reappearance of markers can be explained by burial in debris pockets followed by reactivation and movement to lower pockets and redeposition there at or near the surface of debris.

The most plausible agent for initiating movement is water. Snow avalanches are another potential agent for movement. However, the fact that the gullies tend to accumulate snow preferentially through wind drifting and small, direct action surface avalanches producing a relatively stable and protective snowpack suggests that snow avalanches are less important in this regard. Most avalanches and results of avalanches that were observed indicate that snow slides over existing snow surfaces in the gullies. Water discharges in the channels are ephemeral or episodic being associated with snowmelt and/or rainfall. In addition, water discharge in snow-filled gullies is subnival. Further, not all discharges are of sufficient magnitude to mobilize debris in the channel. As a result, movement is highly intermittent in time, as suggested by the data.

Although displaying temporally intermittent movement and having a high probability for no movement in an annual measurement interval, marked clasts generally tend to move down channel over longer periods of time, such as the five years of this measurement program. This general movement mitigates against the build-up of large volumes of in-channel debris and thus reduces the potential for sudden, high-magnitude debris flows or torrents, at least in the channels studied.

Conclusions

This chapter has demonstrated that debris transport occurs in high-gradient, bedrock channels in the alpine zone of the Mt. Rae area of the Canadian Rocky Mountain Front Ranges. That this process operates in most mountain environments is intuitively apparent from the fact that many such bedrock channels act as conduits feeding debris slopes. In this chapter quantitative information has been presented to demonstrate that rates of debris movement in the bedrock channels are an order of magnitude greater than debris shift on debris slopes and several orders of magnitude greater than forms of slow mass wasting, such as solifluction, in similar environments. Again, such a finding might be expected given that the channels are conduits where sediment transport is confined and concentrated as opposed to being dispersed over a larger slope surface. Thus, in order to evaluate the regional importance and effectiveness of this concentrated form of erosion, it would have to be compared to the lower magnitude but spatially more ubiquitous forms of erosion. Such a comparison is beyond the scope of this paper, but it would be a worthy objective of future research and measurement.

A further conclusion of this chapter is that flows of water in the bedrock channels are a primary agent of sediment transport. The research has demonstrated that discharge, including subnival flow, is highly intermittent or ephemeral and is generated by snowmelt and/or occasional high-intensity rain. Although observations indicated some hydrological differences between sites with northerly and southerly exposures, these differences appear not to be reflected in debris-movement data. Discharge observations plus qualitative observations of clast abrasion, rounding, and fracture, packing and imbrication in freshly formed deposits, incision into channel bed sediment pockets, and considerable mixing and stirring of sediments in transport all support the contention that intermittent fluvial activity is a primary agent of debris transport.

Intermittent fluvial activity deserves more attention as an agent of erosion on high mountain slopes, taking its place with erosion by physical weathering and rockfall, snow avalanches, debris flows, and solifluction. Further, bedrock channels serve as conduits for sediment transport between areas of mountain-top and rock-face detritus and debris slopes at lower elevations. As conduits they are concentrations of energy that lead to progressive incision of the mountain slope. These findings support, with some quantitative evidence, observations made many years ago by Behre (1933) and Matthes (1930).

Acknowledgments

This research was funded by the Natural Sciences and Engineering Research Council of Canada, Grant A9152. The field assistance of Joe Desloges, Marnie Olson, David Sauchyn, Mary Sauchyn, Dan Smith, and William Barr is gratefully acknowledged.

References

Behre, C. H., Talus behavior above timberline in the Rocky Mountains, *Journal of Geology, 41*, 622–635, 1933.
Bones, J. G., Process and sediment size arrangement on high arctic talus, southwest Devon Island, NWT, Canada, *Arctic and Alpine Research, 5*, 29–40, 1973.

Caine, N., The log normal distribution and rates of soil movement: an example, *Revue Géomorphologie Dynamique, 18*, 1–7, 1968.

Caine, N., Elevational contrasts in contemporary geomorphic activity in the Colorado Front Range, *Studia Geomorphologica Carpatho-Balcanica, 18*, 5–31, 1984.

Church, M., R. F. Stock, and J. M. Ryder, Contemporary sedimentary environments on Baffin Island, NWT, Canada: debris slope accumulations, *Arctic and Alpine Research, 11*, 371–402, 1979.

Costa, J. E., and R. D. Jarrett, Debris flows in small mountain stream channels of Colorado and their hydrological implications, *Bulletin of the Association of Engineering Geologists, 18*, 309–322, 1981.

Day, T. J., The channel geometry of mountain streams, in *Mountain Geomorphology*, edited by O. Slaymaker and H. J. McPherson, pp. 141–150, Tantalus Press, Vancouver, Canada, 1972.

Desloges, J. R., and J. S. Gardner, Process and discharge estimation in ephemeral channels, Canadian Rocky Mountains, *Canadian Journal of Earth Sciences, 21*, 1050–1060, 1984.

Gardner, J. S., The movement of material on debris slopes in the Canadian Rocky Mountains, *Zeitschrift für Geomorphologie, 23*, 45–57, 1979.

Gardner, J. S., Alpine mass wasting in contemporary time: some examples from the Canadian Rocky Mountains, in *Space and Time in Geomorphology*, edited by C. E. Thorn, pp. 171–192, Allen and Unwin, London, 1982.

Gardner, J. S., Observations on erosion by wet snow avalanches, Mt. Rae area, Alberta, Canada, *Arctic and Alpine Research, 15*, 271–274, 1983*a*.

Gardner, J. S., Accretion rates on some debris slopes in the Mt. Rae area, Canadian Rocky Mountains, *Earth Surface Processes and Landforms, 8*, 347–356, 1983*b*.

Gardner, J. S., Rockfall frequency and distribution in the Highwood Pass area, Canadian Rocky Mountains, *Zeitschrift für Geomorphologie, 27*, 311–324, 1983*c*.

Gardner, J. S., D. J. Smith, and J. R. Desloges, *The Dynamic Geomorphology of the Mt. Rae Area: A High Mountain Region in Southwestern Alberta*, 237 pp., Department of Geography Publication Series No. 19, University of Waterloo, Canada, 1983.

Kellerhals, R., Hydraulic performance of steep natural channels, in *Mountain Geomorphology*, edited by O. Slaymaker and H. J. McPherson, pp. 131–140, Tantalus Press, Vancouver, Canada, 1972.

Matthes, F. E., Geological history of the Yosemite valley, *U.S. Geological Survey Professional Paper 160*, 108–109, 1930.

McPherson, H. J., Downstream changes in sediment character in a high energy mountain stream channel, *Arctic and Alpine Research, 3*, 65–79, 1971.

Mears, A. I., Flooding and sediment transport in a small alpine drainage basin in Colorado, *Geology, 7*, 53–57, 1979.

Pollock, D. M., and G. J. Gaye, One-day extreme rainfall statistics for the Prairie Provinces, *Atmospheric Environment Service Report CL1-5-73*, Environment Canada, 15 pp. 1973.

Rapp, A., Recent development of mountain slopes in Karkvagge and surroundings, northern Scandinavia, *Geografiska Annaler, 42*, 73–200, 1960.

Rapp, A., Slope erosion due to extreme rainfall, with examples from tropical and arctic mountains, *Nachrichten der Akademie der Wissenschaften in Göttingen*, 118–136, 1974.

Starkel, L., The role of extreme meteorological events in the shaping of mountain relief, *Geographica Polonica, 41*, 13–20, 1979.

Wallace, R. G., Type and rates of alpine mass movement, west edge of Boulder County, Colorado Front Range, Ph.D. Thesis, 200 pp., Ohio State University, Columbus, Ohio, 1968.

Whitehouse, I. E., and M. J. McSaveney, Diachronous talus surfaces in the Southern Alps, New Zealand and their implications to talus accumulation, *Arctic and Alpine Research, 15*, 53–64, 1983.

6
Sediment movement and storage on alpine slopes in the Colorado Rocky Mountains

T. Nelson Caine

Abstract

Contemporary rates of sediment movement on alpine slopes in three basins in the Colorado Rocky Mountains have been estimated for periods which range up to 20 years. In that time, hillslope processes on a basin-wide scale are dominated by mass transfers within the cliff–talus system, especially by work done in infrequent rockfalls involving more than 5 m^3 of debris. Almost all transfers of clastic sediment remain internal to the slopes which are therefore effectively decoupled from the stream channels below them. The only significant exception to this involves silt and clay, imported to the alpine system through the atmosphere, which is transported via the streams to alpine lakes. The closure of the slope systems appears to have been effective for most of the Holocene. Even on that time scale, slope processes have had little influence on landform generation.

Introduction

The concept of punctuated change during evolution (e.g., Gould, 1984) has an analog in geomorphology in the explanation of landscape generation. Evolutionary geomorphology is increasingly based on ideas that transient shocks, catastrophes, and crises, separated by intervening periods of recovery, are important in explaining landform development (Brunsden and Thornes, 1979; Thornes, 1983). This has obvious significance to the study of contemporary sediment transport if it is concerned with landform development rather than process mechanisms. Decades of measurement may define only gradual changes that may be quickly reversed by a single crisis. On hillslopes the importance of large-scale instability and landsliding implicitly fits this model as do modes of slope develoment that involve periods of sediment accumulation terminated by instabilities (e.g., Azimi and Desvarreux, 1974; Dietrich and Dunne, 1978).

The problem considered empirically here is that of evaluating clastic sediment movement and storage on alpine slopes in the Southern Rocky Mountains and the implications of these mechanisms for landscape development. Relatively long records of geomorphic processes are available for three alpine areas in Colorado and may be augmented by records of Holocene activity in the same areas. Disparities between these two records suggest the significance of intermittent transient events. Two sediment flux systems are considered here: that of coarse clastic material (greater than about 8 mm size) and that of finer clastic sediment (Caine, 1974). This separation is convenient for the two systems are not greatly interactive, though they are spatially coincident.

Field area

The alpine terrain of the Southern Rocky Mountains involves a mix of glaciated valleys and unglaciated interfluves. This has been described as the Southern Rocky Mountain type of mountain terrain by Barsch and Caine (1984). Within the alpine, four terrain units may be recognized: two forms of interfluve, one with a diamicton cover of uncertain age and origin (Madole, 1982) and one of bedrock aretes between glacial cirques; the valley floors (deglaciated more than 10 000 years ago: Carrara et al., 1984) of glaciated bedrock with few glacial deposits (most of these are below present tree-

Table 6.1 Catchment characteristics.

	Green Lakes	Williams Fork	Eldorado Lake
Morphology			
area (km^2)	2.12	0.98	1.17
elevation (m)	3550	3540	3813
orientation	East	East	North
relative relief (m/km^2)	257	367	292
mean slope	32°	15°	16°
channel length (km)	1.64	0.57	0.88
drainage density (km/km^2)	0.77	0.58	0.75
Surface cover (ha)			
bedrock	68.4	16.7	70.0
talus	41.3	5.5	5.0
blockfield	58.2	3.8	6.7
mesic tundra	38.2	37.9	24.5
willow scrub	5.0	24.4	0.6
wet tundra	2.5	2.2	1.8
bare soil	0.6	2.5	1.8
lakes and ponds	8.9	5.0	6.6
glacier	5.0	0.0	0.0

line); and the intervening valley walls consisting of a cliff–talus–rock glacier sequence (Caine, 1974).

Geomorphic energy within this environment is not high and is expended on resistant materials. Relative relief is rarely more than 500 m/km^2 in the alpine,

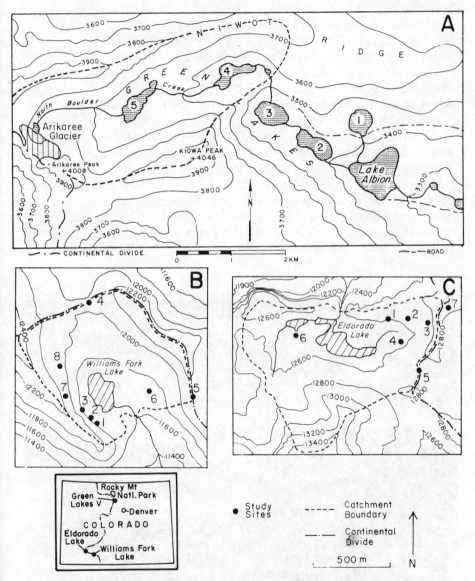

Figure 6.1 Location maps: (A) upper Green Lakes valley, Colorado Front Range (40° 03′ N, 105°37′ W), contour interval 100 m; (B) Williams Fork Lake basin, San Juan Mountains (37°37′ N, 107°09′ W), contour interval 200 ft; (C) Eldorado Lake basin, San Juan Mountains (37°43′ N, 107°33′ W), contour interval 200 ft.

and valley floor slopes only locally exceed $12°$. Despite the frequent occurrence of convective rainstorms in summer (Caine, 1976a), high stream flows are usually generated by snowmelt and rarely exceed $0.5 \text{ m}^3/\text{s km}^2$. Glacial energy is low at present, though it has been higher in recent centuries (Reheis, 1975).

Contemporary hillslope activity has been studied in three alpine valleys that have also been the sites of ecological studies (Table 6.1, Fig. 6.1). Holocene geomorphic activity has been studied in many ranges in the Colorado Rockies (e.g., Birkeland, 1973; Miller, 1973; Giardino et al., 1984), but contemporary activity has not usually been measured in these areas.

The upper Green Lakes valley (Fig. 6.2) appears typical of the high elevation environment of the east slope of the Front Range (Caine, 1982). It includes Niwot Ridge, the northern drainage divide of the basin, where work on the alpine system has been conducted for more than 30 years. The valley consists of a high alpine environment dominated by steep rock walls and talus slopes, a valley floor of glaciated bedrock and a number of permanent snow patches, including the Arikaree Glacier. The stream channels in the basin are generally steep and rocky. They are floored with cobbles and gravel and appear to be sufficiently well armored to prevent sediment movement even at times of peak flow. The drainage is channeled through two lakes that have a total area of 8.6 ha and a mean depth of 4.0 m. The outlet of Green Lake 4 at 3515 m defines the lower limit of the upper Green Lakes valley; below this point the

Figure 6.2 The upper Green Lakes valley. The view is to the west from Niwot Ridge with Green Lake 4 in the foreground.

Figure 6.3 The Williams Fork Lake basin from the southeast.

character of the valley and its vegetation cover change markedly (Caine, 1982).

The two basins in the San Juan Mountains (Fig. 6.1) are described in Caine (1976*b*). Each of them has an area of about 1 km^2 and contains a lake of about 5 ha (Table 6.1). The Williams Fork Lake basin (Fig. 6.3) is about 250 m lower in elevation than Eldorado Lake basin and is less alpine in appearance. The Williams Fork Lake basin is cut into Tertiary volcanic bedrock which, along with its lower elevation, probably accounts for its more extensive soil cover. The Eldorado Lake catchment is in Paleozoic quartzite, which is more similar than the Tertiary volcanics to the granodiorite and gneissic bedrock of the Green Lakes area.

The coarse debris system

As in many mountain areas, coarse sediment movement and storage tends to dominate the clastic sediment cascade in the alpine environments of the Southern Rockies (Caine, 1974). This dominance varies with spatial scale and is most evident at 1 km^2—that is, at the scale of cirques and valley heads (Barsch and Caine, 1984). The system consists of the bedrock cliffs, their associated taluses, and the rock glacier features of the talus foot. Movement of the coarse components of this debris is largely due to mass wasting, although debris flows and grainflows (Mears, 1979; Costa, 1984) from intense rainstorms frequently carry large-caliber material to the valley floors.

Conceptually, this system has often been treated as if it were physically closed at the lower end (Caine, 1969; Olyphant, 1983). Observations of alpine lake sediments support this assumption (Harbor, 1984). However, this system of debris production, transfer, and accumulation has been breached during glacials when coarse debris was transported out of the present alpine area. On the same time scale, large-scale instability in glacially steepened mountain walls may have had marked effects on coarse clastic deposits and slope form (Radbruch-Hall, 1978). Holocene variations in talus accumulation are also evident throughout the Southern Rocky Mountains.

Cliff retreat

In the Green Lakes valley, input to the talus system by rockfall from the cliffs may be estimated from a 15-year record of traps at the head of the talus below Kiowa Peak (Fig. 6.1), and from the rock debris found on the spring snow cover. The trap record (Table 6.2) suggests the addition of about 22.6 m^3 of debris to the head of the 41 ha of talus in the valley. It includes only relatively small rock fragments (maximum size of 10 cm). Larger fragments may have passed the traps and moved further downslope, but fewer and smaller clasts are recorded on traps there. Rockfall onto the spring snow cover shows similar characteristics (Table 6.2). In three years of record the volume of rock fragments on the snow cover over the talus tends to decrease with downslope distance (Fig. 6.2) and their size is much smaller than that of the underlying talus (compare clast sizes in Tables 6.2 and 6.4). These records suggest that

Table 6.2 Talus accumulation in upper Green Lakes valley[a].

Period	Area[b] (m^2)	Accumulation (cm^3/m^2)	Number of rock fragments	Mean size (ϕ)	Sorting coefficient (ϕ)
Arikaree Glacier					
1968	1500	0.861	89	-3.904	0.991
1969	800	0.744	40	-3.95	1.094
Arikaree, East Face					
1968	1900	4.026	58	-4.500	1.160
1969	1600	0.370	70	-3.929	0.888
1970	1300	0.193	28	-3.857	1.042
Kiowa, North Face					
1970	800	0.027	64	-2.750	0.707
1968–83	15	382.4	19	-6.605	0.788

[a] Observations are of rockfall accumulation on the snow cover, except for that of Kiowa (1968–83) which is the mean annual accumulation on the talus surface for that period.
[b] Total surface area from which fragments were collected.

debris accumulation on all the talus in the Green Lakes valley amounts to only 1.5 m^3/yr. Almost all of this debris is of such small caliber that it will be effectively "lost" into the present debris mantle, rather than contributing to the form of the slope.

In a 20-year period, which includes the years of recorded small fragment falls, at least three rockfalls involving more than 5 m^3 of debris have occurred. Each rockfall included clasts larger than 1 m which traversed the entire talus (Table 6.3). In total they involved more than 150 m^3 of rock and considerably more geomorphic work than occurs in talus deposition. They suggest no single mechanism of failure: the fall of July 1965 from Kiowa Peak was induced by an intense summer rainstorm whereas that of December 1983 was perhaps triggered by freeze-back in the rockwall. The fall from the east face of Arikaree Peak was not recorded until long after its occurrence.

Thus about 10 m^3 of rock are removed from the cliffs of Green Lakes valley each year, corresponding to a mean cliff retreat of about 0.02 mm/yr. More than 75 percent (by volume) of this material consists of large falls and is not added to the talus because of a long run-out.

In the two alpine catchments in the San Juan Mountains, the record of rockfall events spans only 4 years. In Williams Fork Lake basin, as in Green Lakes valley, it is dominated by large rockfalls. These began as small falls in 1971 and culminated with falls of more than 50 m^3 from the west side of the valley in May–August 1972 (Fig. 6.1). Individual falls were associated with periods of rain or with melting of a snow cornice at the head of the cliff. As in the Front Range, the larger fragments in each fall traveled across the talus to the valley floor. These rockfalls yield a cliff retreat rate of about 0.3 mm/yr, an order of magnitude higher than the Front Range estimate. Such a disparity is most readily explained by lithologic and structural differences.

The rockfall record from the quartzite cliffs of the Eldorado Lake basin

Table 6.3 Large rockfall events[a].

Period	Source	Sink	Volume (m^3)	Run-out distance (m)
Green Lakes valley				
1965	Kiowa peak	valley floor	50	80
1975–81	Arikaree peak	valley floor	8	220
1983	Kiowa peak	Green Lake 3	20	275
Williams Fork Lake				
1972	west cliff	lake shore	55	50
Eldorado lake				
	none recorded			

[a] Large rockfalls are those which involved more than 5 m^3 of rock.

shows no large events (Caine, 1976b). It amounts to only about 1.0 m^3/yr, is equivalent to a cliff retreat rate of 0.012 mm/yr, and is of the same order as the retreat rate in Green Lakes valley.

Talus movement

Clast movement at the talus surface has been estimated in all three catchments (Caine, 1976b, 1984) and shows the local variability found in other studies (e.g., Gardner, 1979, 1982). Over 14 years some clasts on the Kiowa Peak talus have not moved at all, while their neighbors have shifted more than 40 cm. A similar temporal variability is also evident in the movement of individual clasts in that short intervals of motion are interspersed with long periods of stability.

Movement rates of surface fragments on the Green Lakes talus slopes are generally less than 3 cm/yr. On open talus slopes (Kiowa and Arikaree), shift rates decrease in the downslope direction because clast size increases and slope gradient decreases slightly (Table 6.4). Where rockfalls, water flows, and snow avalanches are concentrated along couloirs through the cliffs (Couloir South and Couloir North), measured rates are one to two orders of magnitude higher than those on open slopes (Table 6.4). However, the rates are lower than those reported for more active alpine taluses (e.g., Rapp, 1960; Gardner, 1979, 1982; Luckman, 1971).

In Williams Fork Lake basin, talus shift tends to be faster than in Green Lakes valley. The 8-year record at Eldorado Lake shows only slightly higher rates than those of the Front Range (Table 6.5). In both areas in the San Juan

Table 6.4 Talus shift in upper Green Lakes valley.

Site	Position	Mean downslope shift (cm/yr)	S.D. (cm/yr)	Number of rock fragments	Mean size (ϕ)	Sorting coefficient (ϕ)
Kiowa	upper	0.99	0.54	26	−7.920	0.777
	middle	0.72	0.44	20	−9.060	1.080
	lower	0.47	0.94	30	−8.840	1.423
Arikaree	upper	3.10	2.72	26	−7.375	0.96
(East)	middle	2.82	3.05	45		
	lower	1.31	0.83	35	−7.423	1.071
Couloir	upper	7.2	22.6	11		
(South)	upper–middle	3.8	6.8	10		
	middle–lower	2.9	3.1	21		
	lower	52.2	156.6	53		
Couloir	upper	5.3	11.6	12		
(North)	middle	19.0	24.3	12		
	lower	10.3	38.6	31		

Data sources: Caine (1984) and Wallace (1968).

Table 6.5 Talus shift: San Juan Mountains.

Williams Fork				Eldorado Lake			
Distance from cliff (m)	Mean shift (cm/yr)	Mean size (ϕ)	Sorting coefficient (ϕ)	Distance from cliff (m)	Mean shift (cm/yr)	Mean size (ϕ)	Sorting coefficient (ϕ)
Site 1 (6-year record)				Site 4 (8-year record)			
2	126.0	−0.21	2.70	22	3.9	−3.76	0.68
30	88.3	−0.23	2.90	43	1.49	−3.78	0.66
60	8.4	−1.95	1.35	50	2.29	−4.42	0.52
70	6.2	−1.05	2.22	85	0.62	−3.47	1.62
95	3.2	−0.78	4.10				
110	0.3	−5.45	1.02				
Site 2 (6-year record)							
5	12.8	−0.19	2.91				
5	11.6	−0.81	2.50				
35	6.8	−1.34	2.66				
90	1.6	−0.97	3.25				

Mountains, shift rates decrease downslope, as on the open talus slopes of Green Lakes valley. The record also shows a marked seasonal pattern with most of the surface shift occurring in summer (Caine, 1976b).

Greater distances are involved in the transport of talus material by fluid flow of snow and water (White, 1981). In Green Lakes valley, wet snow avalanches are almost annual events on the east slopes of Kiowa Peak and Arikaree Peak. They are more effective agents of talus modification and sediment transport on the former peak where they occasionally transport large clasts onto the ice cover of Green Lake 3. On the east face of Arikaree Peak, wet snow avalanches seem to redistribute rockfall debris already on the snow cover (Fig. 6.4) rather than that on the talus beneath it.

The two basins in the San Juan Mountains appear to be largely uninfluenced by snow avalanches (Caine, 1976b). In each of them, less than 0.5 m^3 of clastic debris is incorporated into snow avalanches each year, which contrasts with the importance of avalanches below tree-line in the range (Carrara, 1979; Huber, 1982).

White (1981) has recorded the influence of debris flows on the form of the talus slopes of the Front Range. In the upper Green Lakes valley, no debris flows have been recorded in 15 years. However, trails on the north side of the valley have been cut by debris flows totaling more than 200 m^3 of debris in the past 60 years.

Apart from the event of September 1970 in Williams Fork Lake basin (Caine, 1976b), debris flow activity has not been observed directly in the two study areas in the San Juan Mountains. That event involved six flows that

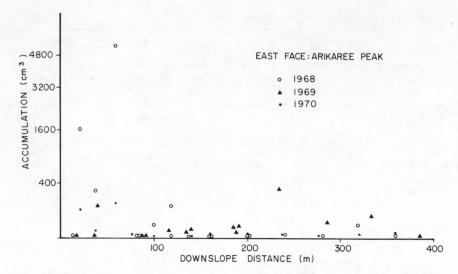

Figure 6.4 Debris accumulation on snow, Arikaree Peak. Accumulation is estimated by the volume of rock fragments on 100 m^2 areas on three transects on the east face of the peak. Downslope distance is the slope distance from the foot of the cliff. In 1968 and 1970 the snow cover on the transects was not affected by wet snow avalanches. In 1969 wet snow avalanches ran down the slope to 270 m from the cliff.

eroded debris from the higher parts of the talus and transported it beyond the foot of the slope. In all, 250 m^3 of debris were removed, sufficient to dominate empirical sediment budgets for the catchment (Caine, 1976b, c). More extensive surveys of debris flows in the San Juan Mountains (Clark, 1974; Sharpe, 1974) have suggested that the records from the two basins are typical of the range. Waste mantles derived from Tertiary volcanic bedrock, as in Williams Fork Lake basin, are more susceptible to such activity than those on Paleozoic metasediments like those of Eldorado Lake (Clark, 1974). Using stratigraphic evidence from debris flow fans, Sharpe (1974) concluded that events like that of 1970 have a return period of about 18 years in any 1 km^2.

Talus-foot transfers

Apart from large rockfalls, debris flows, and avalanches that travel to the valley floor, sediment transport away from the talus of continental alpine mountain ranges is effected usually by a variety of "rock glacier" features. Lobate rock glaciers (White, 1981) transfer debris from the talus, but the tongue-shaped rock glaciers of the valley floor are less clearly linked to the talus slopes, though they remain important in the budget of coarse clastic debris.

White's records of flow rates in the tongue-shaped rock glaciers of the Front Range (5 to 10 cm/yr) show them to move more slowly than rock glaciers in

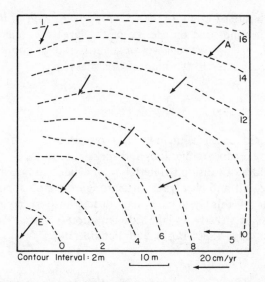

Figure 6.5 Movement in Kite Lake rock glacier, 1971–79. Extension of the strain net during the observation period: A to E, 76 cm ($\dot{\varepsilon} = 1.55 \times 10^{-3}$/yr); 1 to 5, -28 cm ($\dot{\varepsilon} = -0.52 \times 10^{-3}$/yr).

other ranges (White, 1971, 1976, 1981). The only tongue-shaped rock glacier in the upper Green Lakes valley is a small relict that is distal to the Arikaree Glacier. It has not been an active part of the sediment transport system since the early Holocene (Benedict, 1968; Carroll, 1974). The largest lobate rock glacier in the valley has been monitored by White (1981) and has a flow rate lower than those of tongue-shaped rock glaciers in the range. Surface velocities of 1 to 2 cm/yr are not greatly different from those of the talus further upslope. Other lobate forms in the Front Range have flow rates and debris discharges up to three times those of the Green Lake 5 case (White, 1981).

The rock glaciers of the San Juan Mountains have long been recognized as an important component of the alpine sediment flux system. Their movement is slightly greater than that in the Front Range examples. Eight years of record from the Kite Lake rock glacier (east of Eldorado Lake) show rates of 12 to 18 cm/yr at its center, with marked longitudinal extension (Fig. 6.5). A record of similar length from the Hurricane Basin rock glacier described by Brown

Table 6.6 Motion in the Hurricane Basin rock glacier, 1972–80.

Site	1	2	3	4	5	6	7[a]
Distance downslope (m)	0	100	130	160	185	242	355
Rate (cm/yr)	1.75	-1.0	4.88	4.63	3.63	5.25	46.38

[a] Site 7, just above the rock glacier snout, may reflect disturbance from mine operations on the snout.

(1925) suggests a rate of no more than 5 cm/yr over much of its length (Table 6.6). Burial of a mine road on the toe of the large rock glacier in Imogene Basin (37°57′N, 107°44′W) suggests an advance rate of about 2.5 cm/yr, but is not indicative of rates in the main debris mass.

The fine sediment system

The fine sediment cascade includes the movement of fine clastic waste (< 8 mm) by soil creep, solifluction, and surficial erosion, as well as the more catastrophic failure of hillslope material in landslides and debris flows. In the Colorado alpine, silt and clay transport through the atmosphere is an important component of this system. Eolian dust transport contributes to soil development and, indirectly, to lake sedimentation (Caine, 1974; Burns, 1980; Thorn and Darmody, 1980; Burns and Tonkin, 1982).

Creep processes
On Niwot Ridge (Fig. 6.1) rates of solifluction and frost creep in the upper 50 cm of the waste mantle are maximized in wet sites where they may reach 4.3 cm/yr at the soil surface (Benedict, 1970). Turf-banked lobes and terraces above 3350 m show the most rapid displacement rates, though they make up no more than 8 percent of the 10 km^2 area of alpine tundra on the ridge. The more well drained parts of Niwot Ridge show lower rates (< 0.1 cm/yr) that are probably typical of large parts of the Colorado alpine environment (Table 6.7). Similarly, low rates of mass wasting are evident in the approximately 46 ha of soil-covered area in upper Green Lakes valley. As on the interfluve, the greatest rates of soil movement are associated with wet soil (Table 6.7).

Table 6.7 Mass wasting rates in upper Green Lakes valley.

Site	Mean rate (cm/yr)	S.D. (cm/yr)	Sample size	Period
mesic alpine meadow[a]				
max.	0.249	0.140	15	1968–82
min.	0.058	0.057	23	1968–82
wet alpine meadow[a]				
max.	0.816	0.269	5	1968–82
min.	0.025	0.022	6	1968–82
solifluction lobes and terraces[b]				
max.	2.21	(0.4)	35	1964–67
min.	0.03	(0.1)	30	1964–67

[a] These data from Caine (1984) are based on the movement of stakes measured at the ground surface.

[b] These data from Benedict (1970) are displacements of stones at the surface. The standard deviations for these data have been estimated from the absolute range.

Table 6.8 Mass wasting rates: San Juan Mountains.

Environment[a]	Willliams Fork			Eldorado Lake		
	Mean rate[b] (cm/yr)	S.D. (cm/yr)	Number[c] of plots	Mean rate[b] (cm/yr)	S.D. (cm/yr)	Number[c] of plots
A	9.68	6.15	6	3.97	1.68	2
B	1.56	0.94	7	0.80	0.54	6
C	0.65	0.39	4	1.76	0.66	6
D	2.33	0.83	4	1.58	0.42	6
E	0.83	0.46	2			
F	1.81	0.76	2	0.81	0.58	4
G	2.06	0.78	2	0.53	0.35	6
H						
I				8.42	4.62	6

[a] Environments are defined by a cluster analysis of plot characteristics (Caine, 1976b): A—fine-textured talus; B—bare soil at low angle; C—snow-free mesic meadow at low angle; D—mesic meadow at varied angles; E—steep turf-covered slopes; F—broken turf upslope of snowdrifts; G—stable blockfield/talus; H—willow scrub; I—channel bank above intermittent stream (not in original classification).

[b] Rates are estimated from 10 cm and 20 cm length tilt-pegs.

[c] Plot size: 1×2 m.

Mass wasting transfers in the San Juan Mountains are summarized in Table 6.8. These results suggest slightly greater rates at Williams Fork Lake than at Eldorado Lake though these differences are usually not significant. When these fluxes are converted to surface velocities (approximately 1/10 the mass flux of Table 6.8), they do not differ greatly from those estimated for equivalent sites in Green Lakes valley (Table 6.7). The records from both areas show the influence of site characteristics such as slope gradient, soil texture, and vegetation (Caine, 1976b).

The Eldorado Lake data have been summarized into a synthetic sediment cascade that is noteworthy for its indication that net losses are slight (Caine, 1979). Although more than 50 m^3 of soil are moved on 240 m^2 of hillslope, only 400 cm^3 are removed annually. At the rate more than 100 000 years would be needed to remove the soil presently in transit. In contrast, the eroding channel bank (Segment 1 in Caine, 1979) shows a loss of more than 1200 cm^3/yr for each meter of channel length. However, the rest of the record suggests that solifluction and creep activity on these slopes involves only internal transfers, most of which are slight.

Surficial erosion
Estimates of soil erodibility in the alpine tundra suggest that when the vegetation is disturbed there is potential for rapid surficial erosion (Summer, 1982). In the Colorado Front Range, Bovis (1978) and Bovis and Thorn (1981) estimated surficial erosion rates from sediment trap records. Their results are corroborated by continuing work at the Martinelli Snowdrift (Table 6.9) and

Table 6.9 Soil loss rates: Colorado Front Range.

Site	Mean (g/m)	S.D. (g/m)	Number of sediment traps[a]	Period	Source
tundra meadow	17	49	15	3 summer months	Bovis and Thorn (1981)
dry tundra	38	30	14	3 summer months	Bovis and Thorn (1981)
Martinelli	1438	1904	25	3 summer months	Bovis and Thorn (1981)
Snowdrift	537	686	14	1981–82	Caine (1984)
	844	1345	16	1982–83	Caine (unpub.)
	337	355	15	1983–84	Caine (unpub.)

[a] Bovis and Thorn used sediment traps 10 cm and 100 cm long (along the contour), whereas Caine used ones 50 cm long.

Table 6.10 Surface tracer rates: San Juan Mountains[a].

	Williams Fork Lake (6-yr record)			Eldorado Lake (8-yr record)		
Environment[b]	Mean (cm/yr)	S.D. (cm/yr)	Number of plots	Mean (cm/yr)	S.D. (cm/yr)	Number of plots
A	21.56	20.71	5	6.52	5.12	8
B	3.86	2.92	13	3.25	2.98	10
C	0.37	0.22	4	0.67	0.17	6
D				0.39	0.16	6
E	5.05	4.00	4			
F	4.10	2.11	2	0.21	0.08	4
G	1.33	0.95	4	1.78	0.91	6
H	10.48	6.92	6			
I				12.37	5.47	4

[a] Measurements are based on marked particles of about 8 mm size.
[b] Environments are defined in Table 6.8.

by tracer studies in the San Juan Mountains (Table 6.10). All show little sediment movement in undisturbed tundra. In snowdrift sites where there is little vegetative cover, a complex response to the duration of snow cover is evident (Caine, 1976b). In areas where the cover is disturbed by burrowing animals (Stoecker, 1976; Thorn, 1978, 1982), sediment movement rates are increased by one or two orders of magnitude.

Eolian sediment transport

Eolian deposition of dust onto the soils and snow drifts of the Front Range has been documented by Burns (1980), Thorn and Darmody (1980), and Burns and Tonkin (1982), who suggest distant sources for this material. The

Table 6.11 Sediment on snow[a].

Distance[b] (m)	Weight[c] (g/m²)	Mean size (φ)	Sorting coefficient (φ)
5	28	4.98	3.03
15	18	5.57	2.95
25	17	5.33	3.07
35	18	5.95	1.68
45	20	4.83	3.33
55	27	5.42	2.90
65	20	4.85	2.57

[a] The snowdrift is on the east (downwind) side of the lobate rock glacier south of Green Lake 5.
[b] Measured from west (upslope) to east (downslope).
[c] Each sample was collected from a 2 m² area on 4 August 1970 by D. Nash. Samples contained about 15 percent organic material, including a high proportion of pine pollen.

observations of Nash (summarized in Table 6.11) show a bimodal texture and a reduction in total volumes and particle sizes with distance across snow drifts (Fig. 6.6). This suggests the selective transport of sand and granule fractions from bare soil surfaces adjacent to the drift site. If the silt and clay fraction of the wind-transported material is exotic, it suggests an annual input of about 14 g/m² to Green Lakes valley.

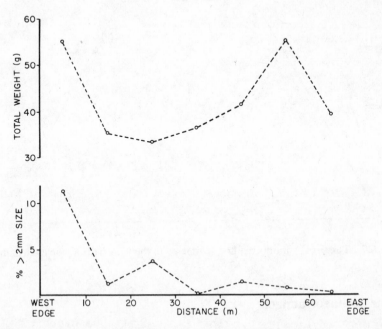

Figure 6.6 Sediment on a snowdrift at Green Lake 5. Collections from 2 m² quadrats on the snow surface were made by D. Nash on 4 August, 1970.

In the San Juan Mountains, no quantitative estimates of atmospheric inputs have been made. Observations of dust on the spring snow cover (e.g., in May 1972) suggest a significant dust accumulation. Stoecker (1976) showed the rapid degradation of gopher mounds by wind and suggested the local redistribution of this sediment. Frank et al. (1975) have also recorded the important contribution of burrowing animals to accelerated erosion by wind, rainsplash, and running water in similar terrain.

Fluvial channel transport

In contrast to the evidence of sediment transport on tundra slopes in the Southern Rocky Mountains, there is little evidence that this sediment moves into the stream systems. Streams are usually armored by coarse sediments or lost beneath blockfields and talus slopes. Records from upper Green Lakes valley (Harbor, 1984) suggest that, over the last 5000 years, lake sedimentation has amounted to no more than 0.15 mm/yr (~ 13.5 m^3/yr). These sediments are composed of silt and clay, corresponding in texture to dustfall onto the winter snow cover (Fig. 6.7). This suggests that the atmosphere, rather than the catchment bedrock, is the dominant source of the lake sediments (Caine, 1974), a conclusion which is supported by the mineralogy of dust trapped in Green Lakes valley (M.Z. Litaor, personal communication, 1985).

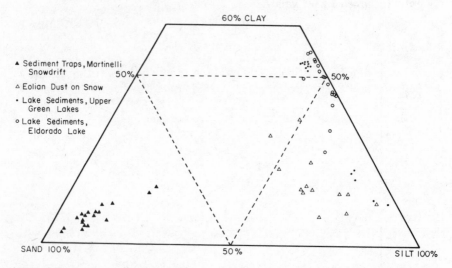

Figure 6.7 Textures of mobilized fine sediments in the Colorado alpine environment.

Clastic sediment budgets

Table 6.12 summarizes the estimates of coarse clastic sediment movement and expresses them as geomorphic work (Caine, 1976b). The two factors

Table 6.12   Coarse sediment budgets for Colorado alpine areas.

	Green Lakes		Williams Fork		Eldorado Lake	
	Volume (m^3/yr)	Work $(J \times 10^6/yr)$	Volume (m^3/yr)	Work $(J \times 10^6/yr)$	Volume (m^3/yr)	Work $(J \times 10^6/yr)$
rockfall	10	44.6	11.5	4.09	1 ·	2.55
talus accumulation	1.5	2.87	2.5	1.59	1	2.55
talus shift	206 500	14.43	27 500	16.40	25 000	4.89
debris flow	2.5	3.26	18.75	11.21		
rock glacier flow	373 000	3.59				

responsible for the variations in activity in Table 6.12 are the mass of material and the vertical distance of transport. In Green Lakes valley, rockfall is dominant because of the great distance of transport, especially in large rockfalls. Talus shift is less than this by a factor of 3, despite the great mass of debris involved. Besides these two mechanisms of debris movement, other mechanisms are relatively slight when estimated for the entire basin, although they may be significant on individual slopes. The disparity between present rates of talus accumulation and talus shift suggests that present-day transfers within the talus system are not steady state. This is indicated on a longer time scale by consideration of the total volume of talus in the valley $(\sim 0.8 \times 10^6 \, m^3)$, rather than just the fraction involved in shift. The talus in the valley must have accumulated in about 12 000 years but extrapolation of present rates of accumulation would yield less than $0.04 \times 10^6 \, m^3$ in that time.

The records from the two basins in the San Juan Mountains show patterns similar to that of Green Lakes valley (Table 6.12). In these smaller basins, there is proportionately less talus ($\sim 0.2 \times 10^6 \, m^3$ at Williams Fork Lake basin and $\sim 0.25 \times 10^6 \, m^3$ at Eldorado Lake) but, even so, that extrapolation of present accumulation rates would not account for talus volumes. At Williams Fork Lake it would account for $0.07 \times 10^6 \, m^3$ (~ 30 percent of that in the basin) and at Eldorado Lake for $0.024 \times 10^6 \, m^3$ (about 10 percent).

The order of magnitude difference between present rates of talus accumulation and a mean rate for postglacial time is supported by talus ages. In the Front Range (Harbor, 1984) and in the San Juan Mountains (Andrews et al., 1976), active talus is usually restricted to the upper parts of the slopes. In contrast, these authors ascribe a mid-Holocene age to most of the lower talus slopes.

Estimates of movement in the rock glaciers of both the Front Range and the San Juan Mountains are generally compatible with talus shift rates further upslope. Present flow rates are usually capable of accounting for rock glacier emplacement in the Holocene. Rates of about 5 cm/yr maintained for 10 000 years could account for the present position of most rock glaciers and this is supported by a tentative date of 10 300 B.P. for debris at the snout of the

Paradise Basin rock glacier (Andrews et al., 1976). However, such a conclusion does not contradict the idea that some rock glaciers were initiated by catastrophic slope failure (Whalley, 1974; Johnson, 1984).

Table 6.13 summarizes activity in the fine sediment cascade of the three catchments. In the upper Green Lakes and Eldorado Lake basins, these estimates are dominated by the large volume of material involved in soil creep and solifluction. Surficial erosion is low in the Green Lakes valley, reflecting the small area of disturbed soil surfaces there, and is much greater in the San Juan basins. In the Williams Fork Lake basin, the work performed by surface erosion, because of greater travel distances, exceeds that due to soil creep. The differences in surficial erosion between the three catchments (Table 6.13) reflect their climatic and vegetational environments and land use. This process affects the alpine tundra with a soil cover rather than the high alpine environment and has been accelerated by disturbance from mining and grazing in the San Juan basins.

The fine sediment system on alpine hillslopes like the coarse sediment system is effectively decoupled from the stream channels. This is clearly reflected in the contrast between the fine sediments in motion on the hillslopes and those deposited in alpine lakes. As in other alpine lakes, most of the lake sediments in the three basins consist of clayey silts, texturally similar to contemporary eolian deposits (Fig. 6.7) (Thorn and Darmody, 1980). The correspondence between the volume of fine sediment deposited annually in the lakes of the upper Green Lakes valley (13.5 m^3/yr) and that of silt and clay deposited as dust on the unvegetated area (~ 1.4 km^2) of the basin (13.12 m^3/yr) supports this conclusion.

The estimates of surficial sediment flux on the bed of the Martinelli Snowdrift (Thorn, 1976; Table 6.9) is further evidence to suggest a break in the sediment transport system between the slopes and the stream channels. High rates of sediment mobilization (in the range of 4000 to 17 500 kg/yr) contrast with the removal rates for particulates through the channel draining

Table 6.13 Fine sediment budgets for Colorado alpine areas.

	Green Lakes		Williams Fork		Eldorado Lake	
	Volume (m^3/yr)	Work (J $\times$ 10^6/yr)	Volume (m^3/yr)	Work (J $\times$ 10^6/yr)	Volume (m^3/yr)	Work (J $\times$ 10^6/yr)
soil creep and solifluction	95 440	0.90	154 000	3.39	49 000	0.25
surficial wasting	240	0.03	1 600	4.54	1 315	0.05
lake sedimentation	13.5	1.00	n.d.		n.d.	
suspended sediment transport	2.47	0.315				
eolian transport	20.0	—				

the snowdrift (~2000 kg/yr, including organic debris). This suggests a sediment delivery ratio of 10 to 50 percent from a drainage area of about 8 ha.

Thus the alpine hillslope systems of the southern Rocky Mountains seem to be closed to the export of locally derived clastic sediment. In contrast, eolian dust appears to be transported through them, albeit with a long residence time (e.g., organic debris associated with silt and clay on ablation surfaces of Arikaree Glacier gave an age of 2355 ± 90 B.P.). At the present time, much of this mobile sediment is trapped in the high elevation lakes or in the tundra soils.

On a longer time scale, sediment export occurs through glaciation, although the survivial of ancient clastic sediments within the alpine zone (Madole, 1982) shows that this is not fully effective. During glacial periods the clastic debris accumulated in the preceding interglacial will be incorporated into glacial deposits. Below 3250 m elevation on the north side of the North Boulder Creek catchment, there are about 400 ha of Pinedale till and moraine. If this is assigned a mean thickness of 2 m, it represents a volume of 8.2×10^6 m^3, almost all of which was derived from the 7.1 km^2 area of Green Lakes valley above 3250 m. Since deglaciation, almost 2.0×10^6 m^3 of clastic debris has accumulated in the upper Green Lakes valley (2.1 km^2), equivalent to about 6.6×10^6 m^3 for the entire basin. The correspondence between this volume and that of the Pinedale glacial deposits suggests that the effect of glaciation in this valley has been that of redistributing previously produced sediments rather than eroding bedrock. Unfortunately, the two basins in the San Juan Mountains were parts of larger glacial systems during the Pinedale (Carrara et al., 1984), and so a similar evaluation cannot be made for them.

The estimates in Tables 6.12 and 6.13 can be compared to the geomorphic work done in solute transport from the three basins. As in other studies (e.g., Barsch and Caine, 1984), the work of solute transport is equivalent to or greater than that of most other processes. In the upper green Lakes valley, it amounts to 10.4×10^6 J/yr compared to 3.24×10^6 J/yr from the Williams Fork Lake basin and 1.49×10^6 J/yr from the Eldorado Lake basin (Caine, 1976b, c). This indicates that, as in lowland areas, much of the present activity of mountain geomorphic systems is associated with geochemical processes. Further, these are the only processes capable of exporting large volumes of material from systems that are effectively closed to clastic debris export under present environmental conditions.

Finally, it is noteworthy that, with respect to most components in the sediment budget, the three Colorado catchments show considerably lower rates of activity than those estimated for mountain basins elsewhere (Barsch and Caine, 1984). The coarse debris cascade in the upper Rhine basin (Jackli, 1956) and in Karkevagge (Rapp, 1960) show much higher levels of geomorphic work by rockfall, talus shift, debris flow, and rock glacier flow. The difference frequently amounts to an order of magnitude, as in the results from other studies of parts of the debris cascade in mountains (e.g., Gardner, 1979, 1982). Similar contrasts are evident in the fine sediment cascade, where they are especially marked in the estimates of lake sedimentation rates. In the much larger, more complex catchment of the upper Rhine, Jackli's (1956) estimate of lake sedimentation is four orders of magnitude higher than that for the upper Green Lakes valley after correction for the difference in catchment area (Barsch and Caine, 1984). Such contrasts emphasize the stability of the geomorphic environment of Colorado's alpine areas in recent millenia.

Conclusion

The empirical data assembled here clearly support the view that the hillslope systems of the high alpine environments of the Southern Rocky Mountains are essentially closed to the export of clastic sediments and have been in that state for much of the Holocene. At present, the hillslope sediment budgets are dominated by internal transfers of coarse debris within the cliff–talus system. Even where fine sediment is involved in movement, it is not transported out of the slope system. An exception to this closure of the hillslope sediment systems is provided by the silt and clay fractions, transported into and redistributed within the alpine area by eolian processes. This material continues to move, probably very slowly, into the fluvial system and eventually into the alpine lake sediments, although some of it is filtered out by the soil and vegetation covered slopes.

Contemporary records of sediment transport, even those of 20 years duration, have little relevance to the evolution of the alpine landscape. Sediment fluxes estimated from them fail by orders of magnitude to account for the volumes of Holocene terrestrial sediments. Such a disparity suggests that intervals of more active sediment production and transport have occurred since deglaciation of the alpine valleys at the end of the Pinedale. Even during these intervals, the hillslope systems remained closed at their lower ends.

Acknowledgments

Work in the Green Lakes valley has been supported by the National Science Foundation under Grant DEB-80-12095 to P. J. Webber and by the Council for Research and Creative Work, University of Colorado. The Bureau of Reclamation, U.S. Department of Interior supported research in the San Juan Mountains under Contract 14-06-D-7052. I am also indebted to many members of the Institute of Arctic and Alpine Research, University of Colorado, in particular to those whose work I have drawn on here.

References

Andrews, J. T., L. D. Williams, and P. E. Carrara, Ecological overview, Part II. The geomorphological overview, in *Ecological Impacts of Snowpack Augmentation in the San Juan*

Mountains of Colorado, edited by H. W. Steinhoff and J. D. Ives, pp. 87–104, Colorado State University, Fort Collins, 1976.

Azimi, C., and P. Desvarreux, A study of one special type of mudflow in the French Alps, *Quarterly Journal of Engineering Geology, 7*, 329–338, 1974.

Barsch, D., and N. Caine, The nature of mountain geomorphology, *Mountain Research and Development, 4*, 287–298, 1984.

Benedict, J. B., Recent glacial history of an alpine area in the Colorado Front Range, U.S.A. II. Dating the glacial deposits, *Journal of Glaciology, 7*, 77–87, 1968.

Benedict, J. B., Downslope soil movement in a Colorado alpine region. Rates, processes and climatic significance, *Arctic and Alpine Research, 2*, 165–226, 1970.

Birkeland, P. W., Use of relative age dating methods in a stratigraphic study of rock glacier deposits, Mt Sopris, Colorado, *Arctic and Alpine Research 5*, 401–416, 1973.

Bovis, M. J., Soil loss in the Colorado Front Range: Sampling design and areal variation, *Zeitschrift für Geomorphologie Supplementband 29*, 10–21, 1978.

Bovis, M. J., and C. E. Thorn, Soil loss variation within a Colorado alpine area, *Earth Surface Processes and Landforms, 6*, 151–163, 1981.

Brown, W. H., A probable fossil glacier, *Journal of Geology, 33*, 464–466, 1925.

Brunsden, D., and J. B. Thornes, Landscape sensitivity and change, *Transactions of the Institute of British Geographers, 4*, 462–483, 1979.

Burns, S. F., Alpine soil distribution and development, Indian Peaks, Colorado Front Range, Ph.D. thesis, 235 pp., University of Colorado, Boulder, 1980.

Burns, S. F., and P. J. Tonkin, Soil-geomorphic models and the spatial distribution and development of alpine soils, in *Space and Time in Geomorphology*, edited by C. E. Thorn, pp. 25–43, George Allen and Unwin, London, 1982.

Caine, N., A model for alpine talus slope development by slush avalanching, *Journal of Geology, 77*, 92–100, 1969.

Caine, N., The geomorphic processes of the alpine environment, in *Arctic and Alpine Environments*, edited by J. D. Ives and R. G. Barry, pp. 721–748, Methuen, London, 1974.

Caine, N., Summer rainstorms in an alpine environment and their influence on soil erosion, San Juan Mountains, Colorado, *Arctic and Alpine Research, 8*, 183–196, 1976*a*.

Caine, N., The influence of snow and increased snowfall on contemporary geomorphic processes in alpine areas, in *Ecological Impacts of Snowpack Augmentation in the San Juan Mountains*, edited by H. W. Steinhoff and J. D. Ives, pp. 145–200, Colorado State University, Fort Collins, 1976*b*.

Caine, N., A uniform measure of subaerial erosion, *Bulletin of the Geological Society of America, 87*, 137–140, 1976*c*.

Caine, N., The problem of spatial scale in the study of contemporary geomorphic activity on mountain slopes, *Studia Geomorphological Carpatho-Balcanica, 13*, 5–22, 1979.

Caine, N., Water and sediment flows in the Green Lakes Valley, Colorado Front Range, in *Ecological Studies in the Colorado Alpine*, edited by J. C. Halfpenny, pp. 13–22, INSTAAR Occasional Paper 37, 1982.

Caine, N., Elevational contrasts in contemporary geomorphic activity in the Colorado Front Range, *Studia Geomorphologica Carpatho-Balcanica, 16*, 5–31, 1984.

Carrara, P. E., The determination of snow avalanche frequency through tree-ring analysis and historical records at Ophir, Colorado, *Bulletin of the Geological Society of America, 90*, 773–780, 1979.

Carrara, P. E., W. N. Mode, M. Rubin, and S. W. Robinson, Deglaciation and postglacial timberline in the San Juan Mountains, Colorado, *Quaternary Research, 21*, 42–55, 1984.

Carroll, T., Relative age dating techniques and late Quaternary chronology, Arikaree Cirque, Colorado, *Geology, 2*, 321–325, 1974.

Clark, J. A., Mudflows in the San Juan Mountains, Colorado: controls and work, M.A. thesis, 102 pp. University of Colorado, Boulder, 1974.

Costa, J. E., Physical geomorphology of debris flows, in *Developments and Applications of Geomorphology*, edited by J. E. Costa and P. J. Fleisher, pp. 268–317, Springer Verlag, Berlin, 1984.

Dietrich, W. E., and T. Dunne, Sediment budget for a small catchment in mountainous terrain, *Zeitschrift für Geomorphologie Supplementband, 29*, 191–206, 1978.

Frank, E. C., M. E. Brown, and J. R. Thompson, Hydrology of Black Mesa watersheds, western Colorado, 11 pp., *U.S. Forest Service General Technical Report RM-13*, 1975.

Gardner, J. S., The movement of material on debris slopes in the Canadian Rockies. *Zeitschrift für Geomorphologie Supplementband, 23*, 45–57, 1979.

Gardner, J. S., Alpine mass wasting in contemporary time: some examples from the Canadian Rocky Mountains, in *Space and Time in Geomorphology*, edited by C. E. Thorn, pp. 171–192, George Allen and Unwin, London, 1982.

Giardino, J. R., J. F. Shroder, and M. P. Lawson, Tree ring analysis of movement of a rock–glacier complex on Mount Mestas, Colorado, U.S.A., *Arctic and Alpine Research, 16*, 299–309, 1984.

Gould, S. J., Toward the vindication of punctuational change, in *Catastrophes and Earth History*, edited by W. A. Berggren and J. A. Van Couvering, pp. 9–34, Princeton University Press, Princeton, NJ., 1984.

Harbor, J. M., Terrestrial and lacustrine evidence for Holocene climatic/geomorphic change in the Blue Lake and Green Lakes valleys of the Colorado Front Range, M.A. thesis, 205 pp., University of Colorado, Boulder, 1984.

Huber, T., The geomorphology of subalpine snow avalanche runout zones: San Juan Mountains, Colorado, *Earth Surface Processes and Landforms, 7*, 109–116, 1982.

Jackli, H., Gegenswartsgeologie des bundnerischen Rheingebeites—ein Beitrag zur exogen Dynamik Alpiner Gebirgslandschaften, *Beitrage zur Geologie der Schweiz, Geotechnische Series No. 36*, 126 pp., 1956

Johnson, P. G., Rock glacier formation by high-magnitude low-frequency slope processes in the southwest Yukon, *Annals of the Association of American Geographers 74*, 408–419, 1984.

Luckman, B. H., The role of snow avalanches in the evolution of alpine talus slopes, in *Slopes: Form and Process*, edited by D. Brunsden, pp. 93–110, Institute of British Geographers Special Publication 3, 1971.

Madole, R. F., Possible origins of till-like deposits near the summits of the Front Range in north-central Colorado, *U.S. Geological Survey Professional Paper 1243*, 1982.

Mears, A. I., Flooding and sediment transport in a small alpine drainage basin in Colorado, *Geology, 7*, 53–57, 1979.

Miller, C. D., Chronology of Neoglacial deposits in the northern Sawatch Range, Colorado, *Arctic and Alpine Research, 5*, 385–400, 1973.

Olyphant, G. A., Analysis of the factors controlling cliff burial by talus within Blanca Massif, southern Colorado, U.S.A., *Arctic and Alpine Research, 15*, 53–64, 1983.

Radbruch-Hall, D. H., Gravitational creep of rock masses on slopes, in *Rockslides and Avalanches, I. Natural Phenomena*, edited by B. Voight, pp. 607–657, Elsevier, New York, 1978.

Rapp, A., Recent developments of mountain slopes in Karkevagge and surroundings, northern Scandinavia, *Geografiska Annaler, 42*, 71–200, 1960.

Reheis, M. J., Source, transportation and deposition of debris on Arapaho Glacier, Front Range, Colorado, U.S.A., *Journal of Glaciology, 14*, 407–420, 1975.

Sharpe, D. R., Mudflows in the San Juan Mountains, Colorado: flow constraints, frequency, and erosional effectiveness, M.Sc. thesis, 126 pp., University of Colorado, Boulder, 1974.

Stoecker, R. E., Pocket gopher distribution in relation to snow in alpine tundra, in *Ecological Impacts of Snowpack Augmentation in the San Juan Mountains of Colorado*, edited by H. W. Steinhoff and J. D. Ives, pp. 281–288, Colorado State University, Fort Collins, 1976.

Summer, R. M., Field and laboratory studies on alpine soil erodibility, southern Rocky Mountains, Colorado, *Earth Surface Processes and Landforms, 7*, 253–266, 1982.

Thorn, C. E., Quantitative evaluation of nivation in the Colorado Front Range, *Bulletin of the Geological Society of America, 87*, 1169–1178, 1976.

Thorn, C. E., A preliminary assessment of the geomorphic role of pocket gophers in the alpine zone of the Colorado Front Range, *Geografiska Annaler, 60A*, 181–187, 1978.

Thorn, C. E., Gopher disturbance: its variability by Braun-Blanquet vegetation units in the Niwot Ridge alpine tundra zone, Colorado Front Range, U.S.A., *Arctic and Alpine Research, 14*, 45–51, 1982.

Thorn, C. E., and R. G. Darmody, Contemporary eolian sediments in the alpine zone, Colorado Front Range, *Physical Geography, 1*, 162–171, 1980.

Thornes, J. B., Evolutionary geomorphology, *Geography 68*, 225–235, 1983.

Whalley, W. B., Rock glaciers and their formation as part of a glacier debris-transport system, *Geographical Papers No. 24*, 60 pp., Department of Geography, University of Reading, 1974.

White, S. E., Rock glacier studies in the Colorado Front Range, 1961 to 1968, *Arctic and Alpine Research, 3*, 43–64, 1971.

White, S. E., Rock glaciers and block fields: review and new data, *Quaternary Research, 6*, 77–97, 1976.

White, S. E., Alpine mass movement forms (noncatastrophic): classification, description and significance, *Arctic and Alpine Research, 13*, 127–137, 1981.

7
Solute movement on hillslopes in the alpine environment of the Colorado Front Range

John C. Dixon

Abstract

The pattern of solute movement on two hillslopes developed on glacial moraines in the alpine and subalpine zones of the Colorado Front Range was investigated. Grain size distributions of the <2 mm size fraction provide evidence of the lateral movement of fine grained particles from higher topographic positions to lower topographic positions on hillslopes. Silica, alkalis, and alkali earths also show evidence of removal from higher topographic positions on hillslopes, and transportation to lower topographic positions by throughflow. Soil pH trends down the toposequences reflect the pattern of movement of metal cations on the hillslopes. Accumulation of moisture and associated solutes at the bottom of hillslopes results in enhanced biotite weathering. Secondary clay minerals associated with biotite weathering are smectite where calcium and magnesium are abundant and vermiculite where potassium is removed. Vertical movement of solutes within individual soil profiles is reflected in increases in the relative abundance of silt and clay at depth. Soil pH trends with depth poorly reflect vertical solute movement within the catena soils.

Introduction

The importance of solution as a denudational process is widely recognized. In fact the amount of material removed from hillslopes by this process may exceed that of all other erosional processes combined (Carson and Kirkby, 1972, p. 233, 237; Young, 1972, p. 47; Selby, 1982, p. 113). In the alpine environment the importance of solution within the entire alpine geomorphic process suite was first quantitatively recognized by Rapp (1960). Rapp found that of all transporting processes acting on slopes in Kärkevagge and surroundings of northern Scandinavia, transportation of dissolved salts exceeded all others. Subsequent studies by Reynolds and Johnson (1972) and Gardner (1984) have drawn similar conclusions to those of Rapp.

Most studies of the solutional denudation of the Earth's land surface have concentrated on the dissolved loads of major streams. Preoccupation with the ionic concentration of stream waters has stemmed from the belief that solutes transported across and through hillslopes ultimately find their way to river channels. Little attention, however, has been given to the way in which solutes are transported on individual hillslopes. Carson and Kirkby (1972, p. 234) suggest that not all solutes transported across hillslopes necessarily reach major drainage channels. Rather, some material may be deposited in lower topographic positions as a result of the downslope movement of solutes in throughflow. In addition, solutes are moved from soil A horizons to soil B horizons as a result of leaching.

This study examines the pattern of movement of solutes on two hillslopes developed on glacial moraines in the alpine and subalpine zones of the Colorado Front Range. Solute movement patterns are interpreted from the distribution of silts and clays, metal oxide molar ratios, and secondary clay minerals in soils developed in three different topographic positions in two soil catenas. Soil physical and chemical properties are used as surrogate measures of hillslope solute movement because these properties primarily represent the products of the long-term pattern of water movement through hillslopes. It has been suggested that soil chemistry in particular reflects the pattern of movement of throughflow water, with higher topographic positions on the hillslope being areas of solute removal, toeslope positions being areas of accumulation, and midslope positions being areas of transportation and removal. Throughflow is considered to be the dominant hillslope process responsible for the observed soil property differences. However, other hillslope processes may also be contributing to these differences (Birkeland, 1984, p. 239).

Study area

The two hillslopes examined in this study have developed on glacial moraines deposited by mountain glaciers during the Late Quaternary. Both of the catenas are developed on glacial moraines that were deposited approximately 10 000 B.P. (Benedict, 1973; Birkeland and Shroba, 1974; Davis et al., 1984). One of the catenas has developed on glacial deposits in Caribou Lake cirque and the second is developed on similar deposits in Arapaho cirque (Fig. 7.1). Caribou Lake cirque is a NNE-facing cirque on the western side of the Continental Divide approximately 40 km west of Boulder, Colorado. The steep cirque walls rise to 4000 m at the Continental Divide to the east. Satanta Peak (3651 m) is the dominant peak of the cirque.

Arapaho cirque is a NE facing cirque on the eastern side of the Continental Divide approximately 35 km west of Boulder. It is surrounded by

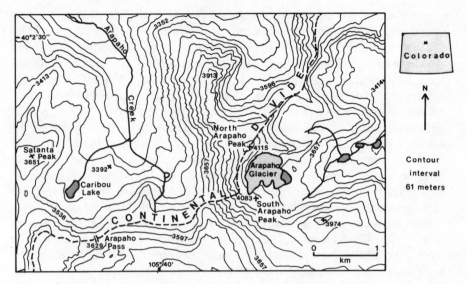

Figure 7.1 Location of study areas.

steep walls which rise 200 m above the cirque floor. Two peaks dominate the area: North Arapaho Peak (4115 m) and South Arapaho Peak (4083 m). The cirque is occupied by a small Ural-type glacier, which results from the accumulation of snow drifted by prevailing northwesterly winds (Outcult and MacPhail, 1965).

The study area lies in the Precambrian crystalline complex of the Colorado Front Range. Metamorphosed sedimentary and volcanic (?) rocks of high metamorphic grade are widely distributed between plutons of igneous rock (Pearson and Johnson, 1980). Several stocks and many sills and dikes have intruded the Precambrian rocks. These are composite bodies consisting of monzonite, syenite, and gabbro (Smith, 1938; Wahlstrom, 1940; Cree, 1948).

Arapaho and Caribou Lake cirques are both chiefly cut into Precambrian biotite gneiss (Pearson and Johnson, 1980). On the floors of the cirques are extensive glacial and periglacial deposits which span the last 10 000 years (Benedict, 1973; Burke and Birkeland, 1983).

Climate of the Colorado Front Range is summarized by Barry (1972, 1973) from a 20-year record at a 3650 m station on the eastern side of the Continental Divide. Annual precipitation is approximately 102 cm, which occurs as snow and summer rain. Precipitation is derived from three sources. In winter, most of the snow is from Pacific maritime air masses, spring precipitation is from the Gulf of Mexico and is characterized by wet snow. Summer precipitation is in the form of heavy afternoon thunderstorms that result from local convective activity (Greenland, 1978). Orographic effects give rise to higher precipitation on the western side of the Continental Divide than on the eastern side (Hjermstad, 1970).

Mean annual air temperature in the study area is $-3.8°C$. Monthly mean temperatures range from $-13.2°C$ in January to $8.3°C$ in July (Barry, 1973). Temperatures are, in general, a few degrees cooler on the western side of the Divide than on the eastern side (Hansen et al., 1978). The region is characterized by extremely high winds with mean speeds of 10.3 m/s (Barry, 1972).

Detailed studies of the alpine vegetation of the study area have been conducted by Komarkova (1976) and Komarkova and Webber (1978). The two cirques discussed in this study show differences in vegetation patterns. In Caribou Lake cirque, subalpine forest extends far up into the cirque. The forest is dominated by spruce (*Picea engelmanni*) and subalpine fir (*Abies lasiocarpa*). To the south and southeast of Caribou Lake the cirque is dominated by alpine herbaceous plants, notably *Kobresia myosuroides, Carex rupestris*, and *Deschampsia caespitosa*, with isolated stands of krummholz spruce. In moist areas around Caribou Lake and ponds of standing water, scrub willow (*Salix*) occurs. The crest and midslope positions on the moraine where the catena was sampled are vegetated by *Carex rupestris, Dryas octopetala*, and *Trifolium dasyphyllum*. Vegetation cover at the lowest topographic position in the catena is dominated by *Kobresia myosuroides, Carex elynoides*, and *Acomastylus rosii* (Burns and Tonkin, 1982).

Arapaho cirque lies above treeline and the vegetation is dominated by the herbaceous plants discussed above. At the higher elevations within the cirque, vegetation cover is sparse with only occasional herbaceous plants and lichens. At lower elevations, stands of krummholz spruce and fir are found in favorable sites. Extensive thickets of scrub willow occur around lakes and along stream courses. The crest and midslope positions on the moraine where the catena was sampled are vegetated by *Carex rupestris, Dryas octopetala, Trifolium dasyphyllum*, and *Acomastylus rosii* (Burns, personal communication). Vegetation cover at the lower topographic position in the catena is dominated by *Kobresia myosuroides, Carex elynoides*, and *Acomastylus rosii* (Burns and Tonkin, 1982).

Soils of the study area have been mapped and described in detail by Burns (1980) and Burns and Tonkin (1982). The two cirques are dominated by Cryic and Pergelic Entisols, Inceptisols and some Mollisols. Dry alpine sites are dominated by Dystric Cryochepts and Pergelic Cryumbrepts while moist sites are characterized by Cryaquolls and Cryaquepts. Soils in the crest and midslope positions on the moraines where the catenas were sampled are Dystric Cryochrepts. The soils in the toeslope position of the two catenas are Pergelic Cryumbrepts (Burns and Tonkin, 1982).

Parent materials of the soils from both catenas have been derived from the glacial erosion of biotite gneiss present in the Precambrian crystalline complex of the Colorado Front Range. The biotite gneiss is composed of biotite, quartz, and oligoclase–andesine, with variable amounts of sillimanite, microcline, garnet, and cordierite, and small amounts of muscovite and green spinel (Pearson and Johnson, 1980).

Methods

The two catenas discussed in this paper were excavated on glacial moraines of similar age, covered by the same type of vegetation and composed of similar parent material in similar climatic settings. Control of these soil-forming factors ensures that chemical and mineralogical changes in the soil profiles of the catenas reflect the influence of topography and associated solute movements (Jenny, 1941). Soil pits were dug in crestslope, midslope, and toeslope positions down one side of each moraine (Figs. 7.2 and 7.3). Crestslope soils are characterized by dry moisture regimes. These are generally sites where snow fails to accumulate during winter because of wind drift. The midslope soils are also relatively dry but do have more snow cover in the winter than crestslope soils. Toeslope soils are commonly wet to saturated as a result of water movement from higher on the slope.

Soil texture was determined using the methods of Day (1965). The <2 mm size fraction was separated from coarser size fractions by dry sieving. Organic matter was removed using a combination of decantation and hydrogen peroxide oxidation. The sand fraction was separated by sieving through a 63 μm mesh screen and then dried and weighed. The remaining silt-clay fraction was dispersed in sodium pyrophosphate, washed into 1000 ml cylinders and filled with water to 1000 ml. Silt and clay size fractions were

Figure 7.2 Arapaho cirque catena showing location of crestslope (C), midslope (M), and toeslope (T) soil profiles.

Figure 7.3 Caribou Lake cirque catena showing location of crestslope (C), midslope (M), and toeslope (T) soil profiles.

drawn off with a pipette at appropriate depths and times. These draws were then dried and weighed, and relative percentages of each were determined.

Whole soil geochemistry of the <2 mm size fraction was determined by atomic absorption spectrophotometry (AA). The data obtained from the AA analyses were initially calculated as metal oxide weight percent, then recalculated as metal oxide molar ratios by dividing weight percent values by the molecular weight of each oxide. Soil pH was determined from a 1 : 1 soil–water paste using a glass electrode.

Molar ratios were used because the stoichiometric proportions of elemental oxides are more representative of absolute changes due to weathering than are weight percent values (Colman, 1982). In general, as weathering progresses the value of molar ratios decreases (Gerrard, 1981, p. 160; Birkeland, 1984, p. 81). The molar ratios selected for this study were silica : alumina ($SiO_2 : Al_2O_3$), alkali : alumina ($K_2O + Na_2O : Al_2O_3$), and alkaline earths : alumina ($CaO + MgO : Al_2O_3$) (Gerrard, 1981). These indices approach measures of absolute chemical change to the extent that Al_2O_3 remains immobile (Colman, 1982).

Al_2O_3 was selected as the "stable" metal oxide against which to compare the mobility of other elemental oxides, as it represents the least mobile of all of the elements examined in this study. In the pH range of 4.5–9.5, Al^{3+} is

Table 7.1 Grain size and pH distributions in the soils of the Arapaho cirque catena.

Sample number	Topographic position	Depth (cm)	Horizon	Percent sand	Percent silt	Percent clay	pH
AC11	crestslope	0–6	A1	70.6	15.3	14.1	5.5
AC13		12–22	A3	61.3	19.3	19.4	4.3
AC15		32–42	IIB2	70.4	15.3	14.3	4.9
AC31	midslope	0–6	A1	76.3	11.9	12.2	5.3
AC32		6–16	A3	68.7	18.8	12.5	4.4
AC33		16–22	IIB1	64.9	21.2	13.9	4.0
AC35		32–42	IIB2	57.9	24.0	18.1	4.0
AC91	toeslope	0–4	A1	74.4	15.3	10.3	4.4
AC92		6–14	A3	71.4	17.3	11.3	4.4
AC93		14–24	IIB2	67.1	19.3	13.6	4.4
AC95		34–44	IICox	60.2	25.8	14.0	5.2

immobile (Loughnan, 1969, p. 52). Examination of the pH values of the soils in both catenas (Tables 7.1 and 7.2) shows that they approach the upper limit of aluminum mobility. For this reason aluminum in the catena soils may be slightly mobile, and therefore the molar ratios may only be minimal measures of the mobility of silica, alkalis, and alkaline earths (Birkeland, 1984, p. 80). Further, aluminum mobility is strongly influenced by organic matter content.

Table 7.2 Grain size and pH distributions in the soils of the Caribou Lake cirque catena.

Sample number	Topographic position	Depth (cm)	Horizon	Percent sand	Percent silt	Percent clay	pH
CL11	crestslope	0–8	A1	33.0	33.7	33.3	4.3
CL12		8–18	A3	64.9	24.1	11.1	4.0
CL13		18–28	IIB2	67.6	20.2	12.2	3.9
CL15		38–48	IICox	73.1	16.8	10.1	4.4
CL41	midslope	0–4	A1	71.3	16.2	12.5	4.5
CL42		4–15	IIB2	54.7	27.7	17.6	4.0
CL43		15–25	IIB3	56.0	28.3	15.7	4.4
CL45		35–45	IICox	66.7	22.0	11.1	4.6
CL71	toeslope	0–4	A1	34.9	37.8	27.3	4.0
CL72		4–8	IIB1	31.9	42.0	26.1	3.9
CL73		8–20	IIB2	25.3	46.0	28.7	4.0
CL75		30–40	IICox	67.7	11.9	20.4	4.6

While organic matter content was not determined for the soils in the catena, comparative data from similar soils in the study area show that organic matter contents are generally less than 5 percent (Burns, 1980). As a result, aluminum mobility probably remains quite low, a situation supported by the relative mobility sequence calculated for this study.

The clay size ($< 2 \mu m$) fraction of the soils from the study area was separated by sedimentation. The clay was then concentrated by centrifuging at 1600 rpm for one hour. Following decantation of excess water, the remaining clay slurry was applied to heated ceramic plates with an eyedropper and allowed to dry. This method produced well oriented clay mounts. Clay samples were sodium saturated because of deflocculation in sodium pyrophosphate. No other chemical pretreatments of the clays were conducted.

X-ray analysis was performed on the samples after each of the following treatments: (1) air drying, (2) glycolation in an ethylene glycol atmosphere at $60°C$ for 12 h; and (3) heating at $550°C$ for 2 h. The samples were X-rayed on a Norelco diffractometer using Cu Kα radiation with nickel filter, at a scanning rate of one degree per minute.

Clay minerals were identified by observing changes in the basal reflections of treated and untreated samples. Methods of identification were based on those of Grim (1968), Birkeland (1969), Birkeland and Janda (1971), Mahaney (1974), and Shroba (1977) and are described below.

Biotite is indicated by a sharp, narrow (0 0 1) peak at 10.5 Å while weathered biotite displays a broad asymmetric peak between 10.16 Å and 10.28 Å. Glycolation and heat treatment do not change the position or intensity of the biotite peaks. In some instances, random interstratification of biotite with vermiculite or chlorite may account for peaks greater than 10.1 Å (Birkeland and Janda, 1971; Shroba, 1977).

Vermiculite can be tentatively identified by a poorly defined (0 0 1) peak between 13.5 Å and 14.5 Å. This peak expands upon glycolation to about 15 Å and produces a broad to sharp peak that collapses to about 10 Å after heating to $550°C$ for 2 h (Birkeland and Janda, 1971; Isherwood, 1975).

Hydrobiotite displays either a sharp (0 0 2) peak between 11.63 Å and 11.79 Å or a broad (0 0 2) peak between 11.19 Å and 12.45 Å. Peak positions remain unchanged by glycolation, but decrease to 10 Å after heating to $550°C$ for 2 h (Shroba, 1977).

Smectite is identified by a broad (0 0 1) peak between 13.39 Å and 15.53 Å. This peak expands to about 17 Å upon glycolation and subsequently collapses to 10 Å after heating to $550°C$.

Kaolinite displays a sharp, first-order (0 0 1) peak between 7.14 Å and 7.25 Å, and a second-order (0 0 2) peak between 3.58 Å and 3.59 Å (Shroba, 1977). These peaks are unaffected by glycolation but disappear after heating to $550°C$ for 2 h (Grim, 1968, p. 87).

Chlorite is identified by a sharp, symmetric (0 0 1) peak between 13.31 Å and 14.28 Å that is unaffected by glycolation and heating at $550°C$ for

2 h. Second-order (0 0 2) reflections at 7 Å and (0 0 4) reflections at 3.5 Å disappear or undergo partial collapse after heating (Mahaney, 1974; Dixon, 1977).

Non-clay minerals in the < 2 μm fraction of the soils include quartz, which has two sharp (0 0 1) peaks, a minor peak at 4.23 Å and a major peak between 3.3 Å and 3.42 Å. Plagioclase is identified by a sharp (0 0 2) peak at approximately 3.2 Å. Potassium feldspar (microcline) is identified by a small (0 0 2) peak between 3.18 Å and 3.29 Å. The peaks of these non-clay minerals remain unaffected by glycolation or heating at $550°C$ (Isherwood, 1975; Markos, 1977).

The B2 horizon is used for the purpose of comparison of soil physical and chemical characteristics down the catenas. This horizon was chosen because it is the most intensely weathered (Colman, 1982) and, therefore, should most strongly reflect the pattern of long-term solute movement. Whole soil analysis was selected as the basis for this study, rather than the analysis of soil solutions. Soils reflect the accumulated long-term history of land surface processes, whereas soil solutions reflect only short-term events.

Natural water samples were collected from streams, lakes and seeps in the two cirques during the early summer. At this time of the year the hydrologic regime receives dissolved solids from spring snowmelt and from groundwater flow, and total solute concentrations are low (Caine, 1984). Ionic concentrations were determined by atomic absorption spectrophotometry. Each sample was acidified in 1 percent HNO_3 to a pH of approximately 2. The samples were then aspired into the spectrometer flame for analysis.

Results

Soil texture trends

The distribution of the < 2 mm particle size fraction both between and within the soil profiles of the catenas suggests that water moves laterally and vertically on hillslopes and carries silt and clay with it (Tables 7.1 and 7.2). In Caribou Lake cirque, relative silt and clay abundances show a progressive increase in soil B2 horizons from crestslope to toeslope positions in the catena as a result of throughflow on hillslopes from higher topographic positions to lower topographic positions. In the Arapaho cirque catena, however, similar trends are not as strongly developed. There is an increase in relative silt and clay contents in the B2 horizons of the midslope profile compared to the crestslope position. Abundances then decrease in the toeslope profiles compared to the midslope profile.

Vertical movement of water within the profiles is reflected in the generally higher relative abundances of silt and clay in lower soil horizons compared to surface horizons (Tables 7.1 and 7.2). Only in the crestslope soil of the Caribou Lake cirque catena is such a trend not evident. In this profile a silt- and clay-rich eolian horizon apparently exists and masks any evidence of translocation. Evidence for clay translocation, rather than *in situ* weathering, is found in the presence of clay films on ped faces, the occurrence of silt cutans, and the development of clay cutans observed on mineral grain edges and along channels in soil thin sections (Dixon, 1983).

Soil chemistry trends

Chemical data indicate that there are substantial differences between the soils in the crest-, mid-, and toeslope positions of the two catenas. These differences are interpreted to be a reflection of the pattern of water movement through the hillslopes and associated transportation of dissolved cations.

In both of the catenas examined the $SiO_2 : Al_2O_3$ ratios of the soil B2 horizons show a progressive increase from the crestslope to the toeslope positions (Tables 7.3 and 7.4). This trend suggests that silica is being transported from higher topographic positions to lower topographic positions in the catena, where it accumulates. The increase in silica may be the result of a variety of processes, such as transportation in solution. Alternatively, the increase may simply be a reflection of the downslope movement of aluminosilicate minerals in slopewash. If the increase in silica is the result of the transportation of aluminosilicate minerals downslope by overland flow, accompanying the increase would be a thickening of the solum. Such a thickening was not observed; thus, transport of the silica in solution is the most likely mechanism. The silica has probably been derived from the dissolution of quartz and aluminosilicate minerals, a source supported by scanning electron micrograph observations (Dixon, 1983).

Table 7.3 Oxide molar ratios of Arapaho cirque soils.

Sample number	Topographic position	Depth (cm)	Horizon	$\dfrac{SiO_2}{Al_2O_3}$	$\dfrac{K_2O + Na_2O}{Al_2O_3}$	$\dfrac{CaO + MgO}{Al_2O_3}$
AC11	crestslope	0–6	A1	8.31	0.52	0.57
AC13		12–22	A3	7.93	0.46	0.45
AC15		32–42	IIB2	7.63	0.41	0.43
AC31	midslope	0–6	A1	8.79	0.55	0.54
AC32		6–16	A3	8.80	0.54	0.49
AC33		16–22	IIBI	8.20	0.48	0.48
AC35		32–42	IIB2	8.16	0.28	0.36
AC91	toeslope	0–4	A1	8.78	0.50	0.48
AC92		6–14	A3	8.74	0.50	0.45
AC93		14–24	IIB2	8.38	0.45	0.46
AC95		34–44	IICox	8.12	0.46	0.35

Table 7.4 Oxide molar ratios of Caribou Lake cirque soils.

Sample number	Topographic position	Depth (cm)	Horizon	$\dfrac{SiO_2}{Al_2O_3}$	$\dfrac{K_2O + Na_2O}{Al_2O_3}$	$\dfrac{CaO + MgO}{Al_2O_3}$
CL11	crestslope	0–8	A1	7.33	0.39	0.51
CL12		8–18	A3	5.39	0.35	0.20
CL13		18–28	IIB2	5.10	0.35	0.22
CL15		38–48	IICox	5.29	0.36	0.20
CL41	midslope	0–4	A1	8.32	0.43	0.27
CL42		4–15	IIB2	6.05	0.32	0.26
CL43		15–25	IIB3	5.55	0.35	0.31
CL45		35–45	IICox	6.40	0.41	0.34
CL71	toeslope	0–4	A1	8.65	0.35	0.29
CL72		4–8	IIB1	7.91	0.36	0.26
CL73		8–20	IIB2	7.60	0.34	0.27
CL75		30–40	IICox	6.94	0.42	0.20

A comparison of the $SiO_2 : Al_2O_3$ ratios in the soil A horizons with those of the soil B horizons shows that in the latter the values of the ratios are lower. The higher values in the A horizons are probably due to eolian addition. No evidence exists therefore for net vertical movement of silica within the soil profile (Tables 7.3 and 7.4).

The alkali : alumina ratio ($K_2O + Na_2O : Al_2O_3$) in both catenas is lowest in the midslope position and shows an increase in the toeslope position (Tables 7.3 and 7.4). This trend suggests that alkalis are removed from the midslope position, transported downslope, and reprecipitated. However, only in the Arapaho cirque toposequence is the highest alkali : alumina ratio observed in the toeslope position. In the Caribou Lake cirque toposequence, the highest ratio occurs in the crestslope profile. This crestslope ratio, however, is probably not significantly greater than that in the toeslope position.

The lowest value of the alkali : alumina ratio in the midslope position of the catena is interpreted to be the result of the greatest removal of alkalis on the hillslope. This section of the hillslope probably experiences greater alkali loss because of throughflow from higher on the hillslope.

Higher alkali : alumina ratios in the crestslope position of the catenas compared to the midslope positions suggest that throughflow and associated solute movement is lower than in the midslope position. This observation may result in part from the preferential addition of eolian sediments to the crestslope, a process that is both widespread and spatially discontinuous in the Colorado alpine (Thorn and Darmody, 1980). Alternatively, the surface may be geomorphically more stable.

Accumulation of alkalis in the toeslope position of the catenas is suggested by the higher alkali : alumina ratios compared to the midslope value. Alkalis removed from the midslope and crestslope positions are apparently transported downslope in solution by throughflow and precipitated at the foot of the hillslope. Potassium and sodium moved on the hillslopes have probably been derived from the dissolution of biotite and feldspars, respectively. Evidence of this process is observed in feldspar and biotite etching in scanning electron micrographs (Dixon, 1983).

Alkali : alumina ratios in A horizons are higher than those of B horizons in the Arapaho cirque. In the toeslope profile of the Caribou Lake cirque there is no significant increase in this ratio in the IIB1 horizon compared to the A horizon. Again there is little evidence for the vertical movement of alkalis within the soil profiles (Tables 7.3 and 7.4). Concentration of alkalis in the A horizon may be the result of eolian addition or biological recycling.

The pattern of distribution of alkaline earth : alumina ($CaO + MgO : Al_2O_3$) ratios is similar to that of the alkali : alumina ratios (Tables 7.3 and 7.4). In the Arapaho cirque catena (Table 7.3) the highest alkaline earth : alumina ratio in the B2 horizons occurs in the toeslope profile, with lower ratios in the midslope and crestslope positions. The lower ratios in the midslope and crestslope profiles suggest alkaline earth removal from these higher topographic positions. Alkaline earth removal from the crestslope profile, however, is small. Alkaline earths removed from higher topographic positions are apparently transported down the hillslope by throughflow and reprecipitated in the toeslope position. The source of the alkaline earths is probably the weathering of feldspars and biotite in the soil B2 horizons of profiles in higher topographic positions.

In the Caribou Lake cirque catena the pattern of movement of alkaline earths on the hillslope is not as pronounced (Table 7.4). Alkaline earth : alumina ratios in the B2 horizons of the midslope and toeslope profiles are similar. However, the ratios in both profiles are higher than that of the crestslope profile. In this catena there may be some movement of alkaline earths on the hillslope, but it appears to be minimal. Alkaline earth : alumina ratios in any of the soil profiles show little evidence of vertical movement of solutes. Again, higher concentrations in the A horizon may be the result of either eolian additions or biological recycling.

Soil pH data also reflect the pattern of movement of solutes on hillslopes. In the Arapaho cirque catena the pH of the B2 horizon is greatest in the crestslope profile, least in the midslope profile, and intermediate between the two in the toeslope profile (Table 7.1). The slightly higher value in the toeslope profile compared to the midslope profile supports the suggestion that cations are being removed from the midslope to the toeslope position in the catena. The higher pH is probably the result of the transportation of bases to the toeslope position. The highest pH in the crestslope position is probably best explained by the addition of base-rich eolian sediment.

In the Caribou Lake cirque catena, soil pH is lowest in the B2 horizon of the crestslope profile and increases very slightly in the topographically lower profiles (Table 7.2). There is no difference in the pH of the B2 horizons of the midslope and toeslope profiles. This trend suggests that some cations are being transported from the crestslope position of the catena to the lower topographic positions. The slightly higher pH of the topographically lower profiles is probably the result of alkaline earth transportation.

Soil pH poorly reflects the pattern of vertical solute movement because of the presence of an eolian horizon at the top of most profiles. The eolian horizon is reflected in the higher pH. The B2 horizons of the profiles generally possess the lowest pH values.

Clay mineral trends

The nature, abundance, and extent of alteration of minerals in the clay size fraction of the soils in the catenas also illustrate the pattern of water and associated solute movement on hillslopes.

Quartz, plagioclase, feldspar, biotite, and chlorite are the primary silicate minerals present within the clay fraction of the soils. The dominant secondary clay minerals are smectite, vermiculite, and hydrobiotite. Quartz, feldspar, and chlorite show no evidence of transformation to secondary clay minerals, but biotite is extensively altered in numerous profiles. The extent and nature of biotite alteration varies according to hillslope position.

Biotite weathering is enhanced in the toeslope position of both catenas (Figs. 7.4 and 7.5). Extensive alteration is reflected in lower, broader, and less symmetric biotite X-ray peaks than those found in other locations in the catena. Enhanced biotite alteration in the toeslope position is interpreted to be primarily the result of the concentration of moisture in this topographic setting. In higher topographic positions, biotite peaks are higher, sharper, and more symmetric in shape, suggesting substantially less weathering.

The weathering of biotite has resulted in the formation of a variety of secondary clay mineral products, which show some variability with topographic setting and between cirques (Figs. 7.4 and 7.5). In Arapaho cirque the principal alteration product is smectite (Fig. 7.4). Relative abundances of smectite in the soils of the catena do not show strong differences down the hillslope. Biotite alteration also increases with depth in the solum. The formation of smectite from biotite is probably the result of the presence of bases moved from higher topographic positions in the catena.

In Caribou Lake cirque, biotite alters to different secondary clay products depending on topographic position. In the toeslope profile, biotite has altered almost exclusively to vermiculite (Fig. 7.5). The vermiculite peaks are relatively sharp and symmetric suggesting that this mineral is well crystallized. Vermiculite is also the dominant secondary clay mineral in the midslope position. However, the X-ray peaks are lower, broader, and less symmetric, which suggests that the vermiculite is less crystallized. The crestslope profile is dominated by smectite with substantially smaller amounts of vermiculite.

The predominance of vermiculite in the midslope and toeslope topographic positions may be a reflection of lower amounts of potassium in these topographic positions compared to that in the crestslope position, a trend

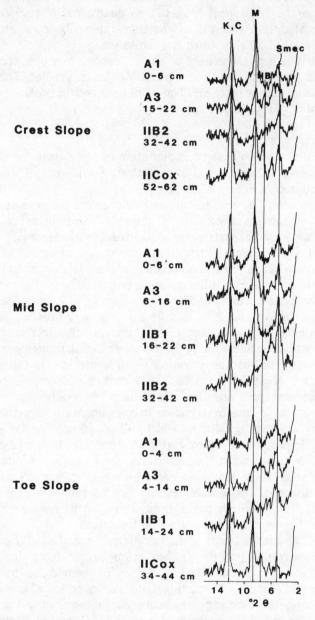

Figure 7.4 X-ray diffraction traces of glycolated clays from the Arapaho cirque catena: K,C = kaolinite/chlorite, M = biotite mica, HB = hydrobiotite, V = vermiculite, C = chlorite, Smec = smectite.

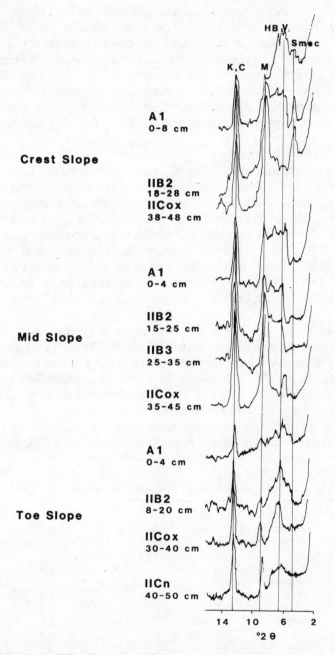

Figure 7.5 X-ray diffraction traces of glycolated clays from the Caribou Lake cirque catena: K,C = kaolinite/chlorite, M = biotite mica, HB = hydrobiotite, V = vermiculite, C = chlorite, Smec = smectite.

reflected in the alkali : alumina ratios discussed earlier in this paper. Vermiculite formation is inhibited where potassium is abundant (Douglas, 1977). The predominance of vermiculite may also be a result of vegetational influences. Caribou Lake cirque is at present partially vegetated by spruce and subalpine fir, and it is likely that during the early Holocene this vegetation type was more extensive. Shroba (1984) suggested that vermiculite and hydrobiotite are key clay minerals in determining the former extent of spruce/fir forests, which appear to promote the rapid production of these clay minerals.

Natural water geochemistry

The cationic concentrations in natural waters are shown in Table 7.5 for Arapaho cirque and Table 7.6 for Caribou Lake cirque. Ionic concentrations of water in both cirques are low compared to those found in non-alpine environments (Hem, 1970). Similar low concentrations, however, have been reported in other studies of alpine and subalpine hydrochemistry (Zeman and Slaymaker, 1975; Patterson, 1980). The concentration of cations in natural waters of both cirques is greater than that of incoming precipitation (Lewis and Grant, 1979). The greater concentration of cations in the natural water presumably results from chemical weathering of the surficial deposits in the cirques.

A comparision of the concentration of cations in natural waters with that of the bedrock fragments in the surficial deposits (Dixon, 1983) permits the relative mobility of elements to be determined. For the two cirques in this study the cation mobility sequence is Ca > Mg > Na > K. When additional cation data including Fe, Al, and Si are considered (Patterson, 1980), the mobility sequence is Ca ≫ Mg ≃ Na ≫ Si > K ≫ Fe ≃ Al.

The complete cation mobility sequence is about the same as that reported from the White Mountains of California (Marchand, 1974). The sequence is

Table 7.5 Chemistry of natural waters from Arapaho cirque[a].

Sample number	Source	Mg^{2+} (mg/l)	Ca^{2+} (mg/l)	K^+ (mg/l)	Na^+ (mg/l)
AC 1–3[b]	lake	0.13	0.95	0.2	0.1
AC 7–9	seep	0.33	3.50	1.4	0.1
AC10–12	snowmelt	0.31	1.92	0.6	0.2
AC13	lake	0.24	1.68	0.5	0.1
AC14	stream	0.24	1.59	0.5	0.1
AC15	lake	0.63	3.30	0.7	1.2
AC16	lake	0.33	2.00	0.6	0.2
AC17	lake	0.36	2.30	1.0	0.3
AC18	stream	0.23	1.42	0.5	0.2
AC19	stream	0.23	1.45	0.5	0.2

[a] The range of accuracy is ± 5 percent of reported concentration.
[b] Samples 1–3, 7–9, 10–12 are averages of three samples.

Table 7.6 Chemistry of natural waters from Caribou Lake cirque[a]

Sample number	Source	Mg^{2+} (mg/l)	Ca^{2+} (mg/l)	K$^+$ (mg/l)	Na$^+$ (mg/l)
CL5	snowmelt	0.06	0.4	0.1	< 0.1
CL6	lake	0.70	2.2	0.4	0.3
CL7	seep	1.76	3.4	0.4	0.6
CL8	lake	1.38	3.7	0.4	0.7
CL9	stream	1.32	3.6	0.4	0.7

[a] The range of accuracy is ± 5 percent of the reported concentration.

also similar to that reported from more temperate environments (Loughan, 1969, p. 52). In the alpine environment of Colorado, however, the mobility of silicon appears to be enhanced.

Discussion and conclusions

Trends in the pattern of distribution of fine grained particles, oxide molar ratios, and secondary clay minerals in catenas provide insights into the pattern of solute movements on hillslopes in alpine and subalpine environments. Silt and clay size particles are moved laterally through the hillslopes as well as vertically within individual soil profiles. Increases in the relative abundance of silt and clay in the B2 horizons of soils in the catena toeslope positions compared to higher topographic positions suggest that in addition to dissolved ions, throughflow waters also carry fine particles. The development of silt and clay bulges at depth within the soil profiles, together with field and microscope evidence, also shows vertical translocation of fine particle sizes within the soil profiles.

The distribution of metal oxide molar ratios in the B2 horizons of soils in different topographic settings on glacial moraines also reflects the long-term pattern of solute movement on alpine hillslopes. In general, solutes derived from the solutional weathering of aluminosilicate minerals in higher topographic positions, together with those in incoming precipitation, are transported to toeslope positions in the catenas. Transport is probably achieved by throughflow moving downslope along the soil A/B horizon boundary. At this boundary there is a marked textural change with finer grain sizes increasing markedly in abundance in the B horizon (Tables 7.1 and 7.2).

Trends observed in metal oxide molar ratios could simply result from different rates of solute removal in different topographic settings. However, a comparison of the molar ratios in soil B2 horizons with those of parent material C horizons provides evidence of absolute change. The B2 horizon of soils in the midslope position of the Caribou Lake cirque shows a decrease in metal oxide molar ratios compared to the parent material C horizons. This observation suggests that metal cations are being removed from this topographic position. In the crestslope soil, removal of cations is less apparent. These is a very slight and probably insignificant decrease in the alkali : alumina ratio, whereas the alkaline earth : alumina ratio shows a slight increase. This observation suggests that the crestslope positions are relatively stable geomorphically. Although comparative data for Arapaho cirque (Table 7.3) are not available because of sampling contraints, the trend of decreasing ratios does suggest a loss of cations in the B2 horizon in these profiles. In the toeslope profiles of both cirques there is evidence of the absolute accumulation of cations from higher on the hillslopes. In both cirques the $SiO_2 : Al_2O_3$ ratio in the soil B2 horizon is greater than that of the C horizon, confirming an absolute accumulation of silicon relative to aluminum (Tables 7.3 and 7.4). Alkaline earth : alumina ($CaO + MgO : Al_2O_3$) ratios also show an absolute increase in the B2 horizons of toeslope soils compared to the parent materials, supporting the conclusion of downslope solute movement (Tables 7.3 and 7.4). The alkali : alumina ($K_2O + Na_2O : Al_2O_3$) ratios, however, do not show evidence of absolute accumulation of these cations in the toeslope positions. The failure of alkalis to accumulate in the toeslope soil profiles is not presently understood.

Available soil chemical data do not support grain size distribution evidence of vertical leaching within the soils of the catena. The vertical movement of solutes is believed to be poorly reflected because of eolian additions to the alpine landscape (Thorn and Darmody, 1980). It appears from the chemical data available that the rate of addition of metals via eolian processes is clearly greater than that of vertical translocation or removal of metal cations in solution. The source of eolian sediment has been suggested by a number of authors (Retzer, 1976; Willard, 1979) to be the Four Corners region of the southwestern United States. Some of it is also undoubtedly derived from reworking of local surficial debris.

The concentration of moisture in the toeslope position in the catenas is probably responsible for the more intense weathering of biotite observed in this position. The concentration of bases at the foot of the hillslope exerts a geochemical control on the nature of the secondary clay minerals formed. Where potassium is preferentially removed, such as in Caribou Lake cirque, vermiculite is the dominant clay mineral formed. Alternatively, this may be a vegetational influence. Smectite forms when calcium and magnesium are present in available solutes.

Secondary clay formation associated with the weathering of biotite occurs in all topographic positions of the catenas. However, the greatest intensity of secondary clay formation is to be found in the toeslope position (Figs. 7.4 and 7.5). Enhanced secondary clay mineral formation in the toeslope position is possible because of the availability of metal cations for substitution into the biotite crystal lattice. Secondary clay mineral formation, both on hillslopes and at the base of the hillslopes, means that at least some of the solutes derived from mineral dissolution fail to reach drainage channels and, therefore, fail to be accounted for when stream ionic concentrations are extrapolated to measures of landscape denudation.

Natural water geochemistry of the two cirques shows that chemical weathering of surficial deposits is clearly occurring and is responsible for the addition of metal cations to natural waters. However, the water chemistry is not clearly related to the results of soil studies. The cation mobility sequence calculated for the two cirques shows that calcium and magnesium are the two most mobile cations. Yet these two elements show evidence of accumulation in soils in catena toeslopes. On the other hand, potassium and sodium, which show no evidence of accumulation in the soils, are less mobile than calcium and magnesium. The accumulation of silica observed in catena toeslopes does show some consistency with its position in the mobility sequence. However, the overall lack of correspondence between soil catena chemistry and natural water geochemistry is striking and may reflect significant eolian input or weathering processes not recognized at present.

Acknowledgments

This research was supported in part by the Department of Geological Sciences, University of Colorado and National Science Foundation grant EAR-8023718. Geochemical analyses were conducted by Skyline Labs, Inc., Golden, Colorado, and X-ray facilities were kindly provided by the Department of Geological Sciences, University of Colorado, Boulder. The author thanks Greg Mitchell, University of Arkansas for drafting the figures and Nel Caine, University of Colorado, for commenting on an earlier draft of the chapter. The comments of two anonymous reviewers also contributed to the improvement of the chapter; their time and effort are greatly appreciated.

References

Barry, R. G., Climatic environment of the east slope of the Colorado Front Range, *University of Colorado, Institute of Arctic and Alpine Research Occasional Paper 3*, 1972.
Barry, R. G., A climatological transect along the east slope of the Front Range, Colorado, *Arctic Alpine Research, 5*, 80–110, 1973.
Benedict, J. B., Chronology of cirque glaciation, Colorado Front Range, *Quaternary Research, 3*, 584–599, 1973.
Birkeland, P. W., Quaternary paleoclimatic implications of soil clay mineral distribution in a Sierra Nevada–Great Basin transect, *Journal of Geology, 77*, 289–302, 1969.
Birkeland, P. W., *Soils and Geomorphology*, 372 pp., Oxford University Press, Oxford, 1984.
Birkeland, P. W., and R. J. Janda, Clay mineralogy of soils developed from Quaternary deposits of the eastern Sierra Nevada, California, *Bulletin of the Geological Society of America, 82* 2495–2512, 1971.
Birkeland, P. W., and R. R. Shroba, The status of the concept of Quaternary soil-forming intervals in the western United States, in *Quaternary Environments,* edited by W. C. Mahaney, pp. 541–576, York University Series in Geography, Geographical Monographs 5, Toronto, 1974.

Burke, R. M., and P. W. Birkeland, Holocene glaciation in the mountain ranges of the western United States, in *Late Quaternary Environments of the United States, Vol. 2, The Holocene,* edited by H. E. Wright, Jr., pp. 3–11. University of Minnesota Press, Minneapolis, 1983.

Burns, S. F., Alpine soil distribution and development, Indian Peaks, Colorado Front Range, Ph.D. dissertation, 360 pp., University of Colorado, 1980.

Burns, S. F., and P. J. Tonkin, Soil geomorphic models and the spatial distribution of alpine soils, in *Space and Time in Geomorphology,* edited by C. E. Thorn, pp. 25–43, George Allen and Unwin, London, 1982.

Caine, T. N., Elevational contrasts in contemporary geomorphic acitivity in the Colorado Front Range, *Studia Geomorphologica Carpatho-Balcanica, 18,* 5–30, 1984.

Carson, M. A., and M. J. Kirkby, *Hillslope Form and Process,* 475 pp., Cambridge University Press, Cambridge, 1972.

Colman, S. M., Chemical weathering of basalts and andesites: evidence from weathering rinds, *U.S. Geological Survey Professional Paper 1246,* 51 pp., 1982.

Cree, A., Tertiary intrusives of the Hessie–Tolland area, Boulder and Gilpin Counties, Colorado, Ph.D. dissertation, 44 pp., University of Colorado, Boulder, 1948.

Davis, P. T., S. F. Burns, and T. N. Caine, Holocene deposits in Arapaho cirque, Front Range, Field Trip Guide 10, *American Quaternary Association, Eighth Biennial Meeting,* 24 pp., Boulder, 1984.

Day, P. R., Particle fractionation, and particle size analysis, in *Methods of Soil Analysis, Part I,* edited by C. A. Black, pp. 545–567, American Society of Agronomy Monograph 9, Madison, 1965.

Dixon, J. B., Kaolinite and sepentinite group minerals, in *Minerals in Soil Environments,* edited by J. B. Dixon and S. B. Weed, pp. 357–403, Soil Science Society of America, Madison, 1977.

Dixon, J. C., Chemical weathering of late Quaternary cirque deposits in the Colorado Front Range, Ph.D. dissertation, 174 pp., University of Colorado, Boulder, 1983.

Douglas, L. A., Vermiculites, in *Minerals in Soil Environments,* edited by J. B. Dixon and S. B. Weed, pp. 357–403, Soil Science Society of America, Madison, 1977.

Gardner, J. S., Sediment storage and transfer in an alpine basin, Canadian Rockies (Abstract), *Association of American Geographers Annual Meeting Program Abstracts,* 144, 1984.

Gerrard, A. J., *Soils and Landforms: An Integration of Geomorphology and Pedology,* 219 pp., George Allen and Unwin, London, 1981.

Greenland, D., Spatial distribution of radiation in the Colorado Front Range, *Climatological Bulletin 24,* 1–14, 1978.

Grim, R. E., *Clay mineralogy,* 598 pp., McGraw-Hill, New York, 1968.

Hansen, W. R., J. Chronic and J. Matelock, Climatography of the Front Range urban corridor and vicinity, *U.S. Geological Survey Professional Paper 1019,* 59 pp., 1978.

Hem, J. T., Study and interpretation of the chemical characteristics of natural water, *U.S. Geological Survey Water Supply Paper 1473,* 363 pp., 1970.

Hjermstad, L. M., The influence of meteorological parameters on the distribution of precipitation across the central Colorado mountains, *Atmospheric Science Paper 163,* 78 pp., 1970.

Isherwood, D., Soil geochemistry and rock weathering in an arctic environment, Ph.D. dissertation, 154 pp., University of Colorado, Boulder, 1975.

Jenny, H., *Factors of Soil Formation,* 281 pp., McGraw-Hill, New York, 1941.

Komarkova, V., Alpine vegetation of the Indian Peaks area, Front Range, Colorado Rocky Mountains, Ph.D. dissertation, 655 pp., University of Colorado, Boulder, 1976.

Komarkova, V., and P. J. Webber, An alpine vegetation map of Niwot Ridge, Colorado, *Arctic and Alpine Research, 7,* 1–29, 1978.

Lewis, W. M., and M. C. Grant, Changes in the output of ions from a watershed as a result of the acidification of precipitation, *Ecology, 60,* 1093–1097, 1979.

Loughnan, F. C., *Chemical weathering of the silicate minerals,* 154 pp., Elsevier, Amsterdam, 1969.

Mahaney, W. C., Soil stratigraphy and genesis of Neoglacial deposits in the *Quaternary Environments,* edited by W. C. Mahaney, pp. 197–240, York University Series in Geography, Geographical Monographs 5, Toronto, 1974.

Marchard, D. E., Chemical weathering, soil development and geochemical fractionation in a part of the White Mountains, Mono and Inyo Counties, California, *U.S. Geological Survey Professional Paper 352-J,* 379–424, 1974.

Markos, G., Geochemical alteration of plagioclase and biotite in glacial and periglacial deposits, Ph.D. dissertation 236 pp., University of Colorado, Boulder, 1977.

Outcult, S. I., and D. MacPhail, A survey of Neoglaciation in the Front Range of Colorado, *University of Colorado Studies Series in Earth Sciences 4,* 124 pp., 1965.

Patterson, C. G., Geochemistry of Boulder Creek, Boulder, Jefferson, and Gilpin Counties, M.S. thesis, 231 pp., University of Colorado, Boulder, 1980.

Pearson, R. C., and G. Johnson, Mineral resources of the Indian Peaks study area, Boulder and Grand Counties, Colorado, *U.S. Geological Survey Bulletin 1462,* 109 pp., 1980.

Rapp., A., Recent development of mountain slopes in Kärkevagge and surroundings, northern Scandinavia, *Geografiska Annaler, 42A,* 65–201, 1960.

Retzer, J. L., Alpine soils, in *Arctic and Alpine Environments,* edited by J. D. Ives and R. G. Barry, pp. 771–802, 1969.

Reynolds, R. C., and N. M. Johnson, Chemical weathering in the temperate glacial environment of the northern Cascade Mountains, *Geochimica et Cosmochimica Acta, 36,* 537–554, 1972.

Selby, M. J. *Hillslope Materials and Processes,* 264 pp., Oxford University Press, Oxford, 1982.

Shroba, R. R., Soil development on Quaternary tills, rock glacier deposits and taluses, southern and central Rocky Mountains, Ph.D. dissertation, 424 pp., University of Colorado, Boulder, 1977.

Shroba, R. R., Pedologic evidence for Holocene treeline 100 meters above its present upper limit in the Colorado Rocky Mountains (Abstract), *American Quaternary Association, Eighth Biennial Meeting,* 116, 1984.

Smith, W., Geology of the Caribou stock in the Front Range, *American Journal of Science 36,* 8–26, 1938.

Thorn, C. E., and Darmody, R. G., Contemporary eolian sediments in the alpine zone, Colorado Front Range, *Physical Geography 1,* 162–171, 1980.

Wahlstrom, E. E., Audubon–Albion stock, Boulder County, Colorado, *Bulletin of the Geological Society of America, 51,* 1781–1820, 1940.

Willard, B. E., Plant sociology of alpine tundra, Trail Ridge, Rocky Mountain National Park, Colorado, *Quarterly of the Colorado School of Mines 64,* 1–119, 1979.

Young, A., *Slopes,* Oliver and Boyd, Edinburgh, 1972.

Zeman, L. J., and Slaymaker, H. O., Hydrochemical analysis to discriminate variable runoff source areas in an alpine basin, *Arctic and Alpine Research, 7,* 341–351, 1975.

8
Hillslope hydrology models for forecasting in ungauged watersheds

M. G. Anderson and S. Howes

Abstract

In the context of hydrological forecasting from ungauged catchments there are conflicting requirements with respect to minimizing data inputs (which restricts physically based modeling of hillslope runoff processes) and simultaneously developing a soundly based, portable operational model. This paper shows that empirical preprocessing of data and the inclusion of selected physically based model elements within an essentially empirical model provide a sound operational approach. Selected examples of the model performance are presented in application to catchments in Texas and Arkansas.

Introduction

Improvements in hydrologic forecasting for ungauged catchments will only be realized by selective application of elements of physically based models of hillslope processes whose parameters relate directly to measurable, physical basin characteristics. Forecasting in this context estimates the magnitude and timing of occurrence of a flood event. Advances in this application will not be achieved merely by further refinement or elaboration of empirical models, accompanied by progression further up the "dead-end street of parameter optimization" (Klemes, 1982, p. 102).

Efforts in hydrology can very broadly be divided into two approaches: those directed toward scientific advancement and research, and those toward practical goals and application (Amorocho and Hart, 1964; Amorocho, 1979; Beven and O'Connell, 1982). Mathematical modeling is associated with both approaches. However, the role and hence the characteristics of the models are very different. Scientific and research objectives are achieved by application of physically based, distributed mathematical models. These complex, rigorous models require detailed and accurate spatially distributed data. The models are calibrated and contain a large number of parameters that are not necessarily either independent or physically based. They are used as research tools to improve understanding and for explanation of hydrologic processes (Freeze, 1971; Stephenson and Freeze, 1974; Freeze, 1974). Models designed to meet practical application goals are limited to simpler structures due to fundamental data limitations and operational logistics. They comprise statistical, empirical, or conceptual models in which exact replication of the prototype system is not critical. These models are designed for forecasting and prediction and provide no insight into the internal hydrologic mechanisms.

It is proposed here to illustrate the potential of certain elements of physically based models in a very important application of mathematical hydrology models, that of forecasting in ungauged catchments. For this, use of mathematical models is desirable, but is heavily constrained by the limited quantity and quality of data and, more particularly, by an absence of historical streamflow data for calibration. It is first necessary to review briefly the manner in which ungauged catchment modeling has been approached and to draw attention to the limitations of such attempts.

Traditionally, ungauged catchment models have been empirical, such as the synthetic unit hydrograph (Nash, 1960). Verma and Advani (1973) reviewed a number of empirical relations. More recently, conceptual models such as USDAHL (Holton et al., 1975) and the Stanford Watershed model (Crawford and Linsley, 1966) have been developed and widely applied. These models contain calibrated parameters, and in order to provide values for them, two solutions have been proposed. The model may be calibrated on a nearby gauged catchment, which is assumed to be hydrologically similar (Crow et al., 1978). Alternatively, the model may be calibrated on a number of basins, and the parameter values correlated to measurable basin characteristics (Ross, 1970; James, 1972; Jarboe and Haan, 1974; Douglas, 1974).

Certain criticisms can be made of this calibration approach. First, use of parameters calibrated outside the basin introduces error into forecasts. Both methods outlined above involve extrapolation of the model outside the range for which it has been calibrated. In answer to this criticism, models such as HYSIM (Manley, 1977) have been developed. This is a conceptual model, but all its parameters, except three, are physically based (three groundwater parameters are derived from adjacent catchments). A second criticism concerns the application of empirical and conceptual models: use of these is to admit to a reductionist philosophy.

Recent attempts have consequently been made to demonstrate the potential of physically based and distributed models for application to ungauged situations. Engman and Rogowski (1974) proposed a relatively simple physically based model founded on the concept of variable contributing area in which all model parameters can be related to basin characteristics. Results of the application of TOPMODEL (Beven and Kirkby, 1976, 1979; Beven et al., 1984) suggest that this model is also suitable for application to ungauged watersheds.

Although there are examples in the literature of the application of physically based models to ungauged watersheds, certain reasons may be identified why this has been resisted by many. First, we must consider the requirements of such detailed and complex models. It is necessary to assemble detailed and accurate spatially distributed catchment data and historical flow records for calibration. Subsurface hillslope processes are described by nonlinear, partial differential equations. Approximate solutions to these for particular boundary conditions are achieved by application of numerical methods, such as finite difference or finite element. For stable and convergent solutions, access to extensive computer resources is required. Operation of such models demands detailed knowledge, experience and familiarity by the user. Clearly, such models cannot be described as portable, and they are not easily transferable to alternative watersheds because of data constraints and the prohibitive costs of data collection. Model application may at best be confined to an instrumented watershed or, at worst, to a single hillslope element. Nor is such a model easily transferable to a new user who may not have access to either suitable computing facilities or support personnel with the required degree of hydrologic and mathematical expertise. Beven (1975) showed that for these reasons a detailed physically based distributed model is inappropriate for ungauged applications.

Second, many have contributed to the debate as to the degree of realism that should be sought in a model in order to provide the prescribed levels of accuracy. A number of studies (Kirkby, 1975; Naef, 1981; James, 1982) have demonstrated the superiority of results derived from simpler models when compared to more complex models. This has led to a general feeling of satisfaction with empirical or conceptual models for practical applications. Dooge (1972, p. 172) captured this feeling when, in the context of water resource development, he commented that "a model is something to be used rather than something to be believed," and that "individual models must always be regarded as tools which are designed to be useful for a particular purpose, rather than dogmas which support an ideology." Amorocho (1979, p. 92) also summarized another widely held view that has tended to prevent the replacement of simpler, empirical models: "The range of trade-offs between model complexity and model coarseness is narrowed by the natural behavior of catchments. Hydrologic systems may be heavily dampened."

Third, there is concern that the magnitude of error involved in physically based, distributed models precludes the attainment of reliable forecasts. Several possible sources of error have been identified. Error in the theoretical structure of the model is attributable to gaps in our understanding of hydrologic processes. Error in the data includes both measurement error and the spatial variability of soil properties and precipitation. Error may also be introduced by the choice of mathematical solution and the assumptions made to render the solution tractable.

Although these three points may be a fair assessment of the reasons why very detailed, complex fully distributed models are confined to a role in scientific enquiry, it is suggested that model elements, or generalizations derived from these models, could be considered as useful inputs to ungauged modeling application. Kirkby et al. (1976) observed that immense catchment variability and the complexity of hydrologic processes, only a very limited variety of flood hydrograph form is displayed. This observation suggests that a great deal of averaging occurs and that only a few key variables are of importance. It futher suggests that it may be feasible to sacrifice the fully distributed nature of many hillslope hydrology models while still retaining acceptable resolution. This notion is doubly attractive in the context of simultaneously presenting the possibility of reducing the input data needs for ungauged catchment applications.

This possibility forms the basis of this chapter, and the following proposals are made in respect of the model formulation that is subsequently undertaken.

1. Any model designed for flood forecasting in the ungauged catchment should contain parameters that relate directly to measurable basin characteristics.
2. Physically based models provide the means for allowing this.
3. There are certain recent developments concerning the relationship between basic soil textural information and soil hydrologic properties that provide potential for allowing the ungauged application of certain subsurface hydrology models (Anderson et al., 1985).
4. Modeling in ungauged applications must admit to all possible sources of error. An estimate of this must be incorporated and its effects propagated through the model. This will produce probability distributions of outputs. In an applications context the "honest presentation is to convey hydrologic estimates stochastically as probability distributions rather than single quantities" (James, 1982, p. 297).

The basic model used is that of HYMO (Williams and Hann, 1972, 1973). This model is an empirically based model using the well established curve number (CN) routine to predict rainfall excess. We will argue that a more realistic infiltration model can be used in place of the CN routine to predict hillslope runoff contributing to streamflow. In the above discussion and the subsequent analysis, we take the term *ungauged catchment* to define a circumstance in which the only available data are those of topographic and soil maps together with precipitation and antecedent moisture conditions.

Model formulation

The original Williams and Hann model
HYMO is a flood hydrograph forecast model that is suitable in terms of its data requirements for application to the unguaged catchment. It is distributed only in the limited sense that modeling of the hydrologic response of a larger

catchment involves a subdivision of the total area into smaller subcatchment units. These are assumed to be internally homogeneous with respect to hydraulic and hydrologic characteristics. A rainfall hyetograph for each subcatchment is transformed into the outflow flood hydrograph, which is then routed through the next subcatchment and added to the flood hydrograph produced by that area. Solution thus begins at the upstream portion of the catchment and proceeds downstream.

Generation of the runoff hydrograph for each subcatchment is a three-stage procedure. First, the unit hydrograph is derived from a dimensionless hydrograph that requires details of subcatchment area, height difference, and channel length. (Successful application of this method recommends that subcatchment area be not greater than 65 km^2 (25 mile^2).) Second, incremental runoff is determined using the CN method (USDA, 1972), an empirical relationship given by

$$Q = (P - 0.2S)^2/(P + 0.8S) \tag{1}$$

where Q = storm runoff (mm), P = storm rainfall (mm), $S = (25\,400/CN) - 254$, potential maximum storage (mm), and CN = runoff curve number.

There are two methods for deriving the value of the watershed CN. For the ungauged application, data concerning the hydrologic soil group, land use, agricultural practices, and hydrologic condition can be derived from soil survey maps. This information together with USDA tables (USDA, 1972) is used to estimate the CN. Where variability of these characteristics cannot be ignored, a single watershed CN is derived by weighting each CN according to the proportion of the watershed area in which it occurs. The CN thus derived represents an average antecedent moisture condition (AMC II) for the watershed. For application of this method to any particular storm event, this CN may be further adjusted according to the previous 5-day rainfall totals to either a drier (AMC I) or wetter (AMC III) condition. Alternatively, for the gauged application, where rainfall and the corresponding runoff data are available, an optimum CN can be calculated according to the following equation established by Hawkins (1979):

$$CN = \frac{100}{1 + 0.5[P + 2Q - (4Q^2 + 5PQ)^{1/2}]} \tag{2}$$

Finally, this direct runoff is convolved with the unit hydrograph to produce the flood hydrograph for the outflow of the subcatchment.

Flood routing through the channel network is achieved using the variable storage coefficient method (Williams, 1969) whereas routing through reservoirs is accomplished by the storage indication method. HYMO also determines sediment yield by application of the universal soil loss equation. However, none of these three hydrologic procedures are the subject of interest here. This paper is concerned with the use of the CN method for derivation of direct runoff. The inclusion of this hydrologic procedure in HYMO may be attractive because of its simplicity, ease of use, and conservative data requirements; but it is proposed that predictions may be achieved at the cost of reduced accuracy. Indeed, Bales and Betson (1982) drew attention to the regularity with which the use of ungauged estimates of CN (AMC II) underestimated runoff volumes from a total of 585 storms in 36 catchments. These CN were substantially lower than the optimum, calculated values.

A number of possible sources of inaccuracy on the CN model may be suggested.

1. Runoff predictions, and hence the flood hydrograph prediction produced by HYMO, are highly sensitive to CN. Hawkins (1975) demonstrated that over a considerable range of rainfall totals, runoff predictions are more sensitive to errors in CN than to errors of a similar magnitude in precipitation. This sensitivity is greatest for low runoff, low precipitation conditions. Smith (1976) examined the sensitivity of HYMO to the CN value and demonstrated that a 10 percent change in CN produces a 55 percent in peak discharge. Consequently, an accurate estimate of the ungauged catchment CN is critical. However, this is constrained by two factors: first, the inadequate and very simple method by which antecedent moisture conditions are incorporated into the model despite their significant influence on runoff, and second, the dependence of CN not only on basin and antecedent conditions but also on storm size (Hawkins, 1973).

2. The model does not incorporate time as an independent variable. The effects of rainfall intensity are not taken into account despite their significance in the generation of runoff. Indeed, Morel-Seytoux and Verdin (1981) suggest that once surface ponding occurs, the CN method will continue to predict runoff provided that there is some runoff, but regardless of its intensity and relationship to soil conditions.

3. Certain theoretical problems with the CN method have been identified. Morel-Seytoux and Verdin (1981) question the theoretical basis of the model. Both they and Hjelmfelt (1980) demonstrate the infiltration bahavior implied by the model to be in direct disagreement with physical infiltration theory, especially when applied to storms which are not of uniform intensity. Under these conditions highly discontinuous infiltration occurs. Misleading runoff predictions are therefore made.

In response to such criticisms, certain improvements to the CN method have been proposed. Hawkins (1973) related the CN to storm characteristics. Williams and LaSeur (1976) and Hawkins (1978) proposed alternative methods for a more realistic adjustment of the CN to antecedent moisture conditions. However, a more interesting and innovative extension to the method was provided by Morel-Seytoux and Verdin (1981). They advocated the inherent superiority of physically based infiltration models and developed such a model whose parameters are then related to the CN.

Model development

In this paper the CN model is completely replaced by a physically based infiltration model of the type documented by Hillel (1977). This will now be described.

The physical law governing the flow of water through a rigid, homogeneous, isotropic and isothermal porous medium is describable by a nonlinear Fokker–Planck equation. This is derived from Darcy's law and the principle of continuity. The form of this equation, simplified for vertical flow only, is given by

$$\frac{\partial \theta}{\partial t} = \frac{\partial}{\partial z}(K(\theta)) - \frac{\partial K(\theta)}{\partial z} \qquad (3)$$

where θ = volumetric moisture content, K = hydraulic conductivity, z = vertical distance (positive downwards), and t = time.

Values of unsaturated hydraulic conductivity vary with soil moisture content, and it is unlikely that these data will be available for the ungauged catchment. They are therefore numerically derived from the suction–moisture curve using the following relationship established by Marshall (1958) and Millington and Quirk (1959), and later developed by Campbell (1974) and Jackson (1972):

$$K_i = K_s \left(\frac{\theta_i}{\theta_s}\right)^p \frac{\sum\limits_{j=1}^{m}[(2j+1-2i)\psi_j^{-2}]}{\sum\limits_{j=1}^{m}[(2j-1)\psi_j^{-2}]} \qquad (4)$$

where K_i = hydraulic conductivity at corresponding moisture content θ_i, K_s = saturated hydraulic conductivity, θ_s = saturated soil moisture content, ψ = suction head, m = number of equal-sized increments of moisture content, and p = a constant. (Jackson (1972) determined that a value of unity for p allows a more accurate determination of the K_i over a greater range of soils.)

Equation (3) is a nonlinear partial differential equation to which exact solutions are available only for specific initial and boundary conditions. Following Hillel (1977), the equations are converted into explicit finite difference equations and solutions are defined at discrete points in space and time.

This solution is computationally manageable but is usually only conditionally stable. A check on stability is thus performed throughout the simulation to identify when errors become large. The time and/or the space grid is then reduced. The profile to be modeled may be divided into three layers, each with different hydrologic properties. Each layer is divided into cells. Flow between the midpoint of each cell is simulated under both saturated and unsaturated conditions. Detention capacity, expressed as an equivalent depth of water on the soil surface, has to be exceeded by rainfall excess before runoff begins. When precipitation ceases this store is depleted by infiltration and evaporation. The evaporation rate e is derived from the following single isothermal equation:

$$e = \frac{e_{max} \sin(2\pi t)}{86\,400} \tag{5}$$

where t = time in seconds from 0600 (sunrise), and e_{max} = maximum midday evaporation rate. (Between 1800 and 0600 e is set to one hundredth of e_{max}.)

Detention capacity is the only model parameter that is not a measurable characteristic. It is not physically based but represents the net effect of vegetation interception, litter interception, and surface detention. Its value also reflects the antecedent moisture conditions of vegetation and litter. The model allows dynamic changes in its structure. It allows water tables and perched water tables to develop and fluctuate throughout a storm. The data requirements of the infiltration model are indicated in Table 8.1. Certain of the soil hydrologic parameters may not be commonly available for the ungauged catchment, but it is suggested that a series of charts and regression equations developed by Brakensiek and Rawls (1983) may prove very useful in deriving these parameters and allowing the routine use of the infiltration model for the ungauged catchment.

These charts were developed from simulations based on approximately 5000 soil data sets in the United States and represent average soil conditions prior to a particular agronomic practice. Anderson et al. (1985) demonstrated that suction moisture curves derived from this method exhibit similar characteristics to experimentally derived curves. The suction moisture curve, saturated hydraulic conductivity, and saturated moisture content can all be derived from percent clay, sand, and organic matter.

A further extension of the infiltration model is also proposed. It is important that the model takes into consideration the high degree of variability of soil characteristics exhibited by the catchment (Russo and Breshler, 1981). This variability includes both actual field variability and experimental variation or error. Such variability leads to a lack of confidence in a deterministic model; thus a probabilistic approach is adopted. The five soil hydrologic properties required by the model are assumed to be independent random variables. Each can be described by a suitable probability density function, derived from the literature. Rogowski (1972), Nielson et al. (1973), Coelho (1974), Baker (1978), and Russo and Breshler (1981) provided evidence for log-normally distributed hydraulic conductivity. Other soil hydrologic properties are shown to display normal distributions (Nielson et al., 1973; Rogowski, 1972; Russo and Breshler, 1981). Detention capacity was assumed to be normal also. Random values for each of the input parameters are derived from their respective probability density functions for a given mean and standard deviation. As neither the normal nor log-normal distribution is bounded at the tails, there is a small probability of randomly generated values assuming

Table 8.1 The infiltration model data requirements.

---

1. Soil profile moisture characteristics
   For each layer:
   soil water content at saturation
   saturated hydraulic conductivity
   suction–moisture curve (a maximum of 20 observations)
   For each cell:
   initial soil water content

2. Soil profile dimensions
   total number of cells in column
   number of cells in layer 1
   number of cells in layer 2
   thickness of each cell

3. Surface conditions
   detention capacity (in meters)
   maximum evaporation during the day

4. Precipitation
   rainfall data time increment
   rainfall data for each time increment
   rainfall start time
   rainfall stop time

5. Program controls
   iteration time for simulation
   simulation start time
   simulation stop time
   number of profiles for the catchment area

---

negative or unrealistic values. Checks are therefore performed on generated values to ensure physical consistency.

To derive a forecast for any one storm the infiltration model is executed repeatedly for the same initial and boundary conditions, but with varying combinations of randomly generated soil water properties. This produces a distribution of hydrologic responses from which the probability of a certain response may be evaluated.

Figure 8.1 indicates very simply the manner in which both the deterministic and stochastic infiltration model have been inserted into the runoff hydrograph procedure of HYMO. Both infiltration models were developed in Fortran 77 on a mainframe and fully incorporated into HYMO. The newly modified HYMO has been ported onto a micro, the HP 9186, which contains a 68000 microprocessor. The program occupies 65 kbytes, and the compiled code 90 kbytes of memory. Execution time on the micro is approximately double that on the mainframe for any given storm, but the model still operates at acceptable speeds and further optimization of code is possible.

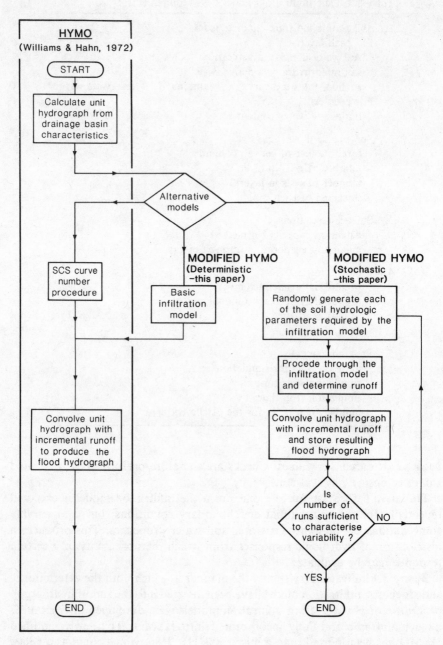

Figure 8.1 Three alternative procedures for deriving the flood hydrograph for a subcatchment.

Any one subcatchment may be represented by more than one soil column. In order to combine the relative contributions of runoff provided by each of the soil types, the complete storm is applied to each of the soil columns, and the incremental runoff produced by each is weighted according to the percentage area of the catchment occupied by that particular soil type. These relative contributions are then summed to produce the total runoff volume derived from the subcatchment.

It is important to note, therefore, that in the proposed model configuration the more classical hillslope distributed model of runoff generation has effectively been collapsed to a simple infiltration model, being distributed only in the sense that different soil types within the catchment are treated as distinct modeling units.

Validation and verification

Validation and verification provide a methodology for assessing the level of confidence associated with information derived from a mathematical model. Hydrologic models are rarely fully evaluated (Dooge, 1972, 1981; Amorocho, 1973, 1979), and Miller et al., (1976) warned that failure to do so can lead ultimately to a lack of faith in modeling. One possible reason for this lack of model evaluation is that nobody really has a clear idea as to how the process should be carried out other than that the procedure must be objective and carried out in the context of the model's proposed application. The process of validation and verification is here considered to be a three-stage activity (Fig. 8.2).

The first stage, design validation, aims to establish the model's face validity by an assessment of the model's assumptions. It is considered that the infiltration model developed above provides an improved basis for derivation of rainfall excess. It conforms not only to basic scientific principles but also to data availability in the context of its proposed application, the ungauged catchment.

The second stage, output validation, consists of a series of simple tests which confirm that the computer program actually does carry out those logical processes expected of it; that is, that the computer program is consistent with the mathematical model and that infiltration behaves rationally. Details of these tests are given in Anderson and Howes (1984). As part of output validation, and especially within the context of the ungauged application, it is necessary to explore the sensitivity of the modified HYMO to error in soil hydrologic properties. This was achieved by application of a stochastic sensitivity analysis. The sensitivity analysis was designed to illustrate for a variety of storms the modified flood hydrograph model's sensitivity to error in the five soil hydrologic properties (Table 8.1). The following major points are derived from the analysis; a more detailed consideration is given by Anderson and Howes (1984).

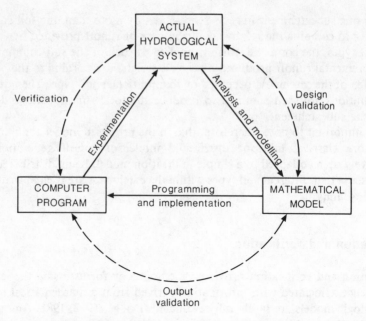

Figure 8.2   The three-stage process of model validation and verification.

First, variability of the flood hydrograph is positively related to the magnitude of error in the input parameters, but it is also strongly related to storm characteristics. The magnitude of output variability increases as storm intensity decreases and storm duration increases. High-intensity, short-duration storms can therefore be identified as conditions where sensitivity to data errors is minimal. Second, for a range of storm conditions, the model is most sensitive to error in saturated hydraulic conductivity, then to saturated volumetric moisture content and the suction moisture curve, then to initial moisture content, and it is least sensitive to error in detention capacity. These results are encouraging because (1) the most sensitive parameters are derived from the Brakensiek and Rawls method, (2) choice of the least sensitive parameters is more subjective, and (3) an operational evaluation of their values has not yet been provided. Finally, when the magnitude of variation in input parameters increases, significantly lower runoff volume and hence peak discharge is predicted from the case where no variation in the parameters occurs.

Verification, the third stage of model assessment, aims to establish a measure of the extent to which the model and the program implementing it represent an accurate representation of reality. This is achieved by a comparison of model predictions with the measured hydrograph. For this purpose, data for the North Creek catchment, Texas, and for the Sixmile Creek catchment, Arkansas, are used (Fig. 8.3). It should be stressed, however, that verification does continue with further applications of the model. These are reported in detail by Anderson and Howes (1984).

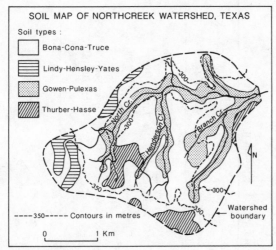

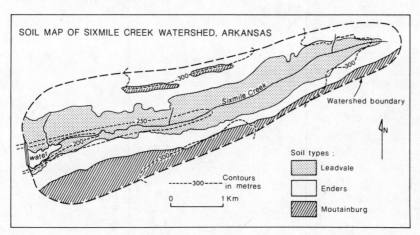

Figure 8.3 Location of catchments.

Table 8.2 Catchment and storm characteristics for model verification.

Catchment	Area (km^2)	Difference in height (m)	Channel length (km)	Land use	Storm dates	Precipitation totals (mm)	Storm duration (h)	CN
								Calibrated[a]
North Creek, Texas	61.6	108.0	5.3	rangeland and pastureland	May 6, 1969	45.2	8.75	87
					July 27, 1962	76.7	9.0	70
					September 18, 1965	107.2	1.3	55
								Ungauged
Sixmile Creek, Arkansas	11.0	79.0	8.3	rangeland	November 3–4, 1959	101.6	8.5	62
					March 20, 1955	69.6	8.0	91
					June 25, 1958	108.5	14.0	62

[a] Equation (2).

Information concerning the landuse and soil texture characteristics of both catchments necessary for application of the modified model were derived from the soils map and accompanying description alone. The hydrologic characteristics of each soil type were estimated from the charts compiled by Brakensiek and Rawls. The exact percent clay and percent sand information is not available and, therefore, the suction–moisture curve, saturated moisture content, and saturated hydraulic conductivity values were set equal to the centroid position of each soil texture group. The organic matter was assumed to be 0.5 percent. The initial relative saturation of the soil was estimated from the rainfall information for the five-day period previous to each storm. For most of the storms a very high initial relative saturation was required to generate sufficient runoff. For this reason, detention capacity was assumed to be zero. Further catchment and storm information is provided in Table 8.2.

Verification of HYMO incorporating the deterministic infiltration model

Figures 8.4 and 8.5 illustrate that for a range of storms applied to the North Creek and the Sixmile Creek catchments, the modified model together with the input data supplied by the Brakensiek and Rawls information provides a hydrograph more closely approximating the measured hydrograph than does the original model. These improvements are considered to be especially significant for North Creek, as the CN values utilized in these comparisons are optimized (equation (2)) rather than ungauged estimates. In contrast, no fine tuning of input parameters for the modified deterministic model has been performed. The CN values used for Sixmile Creek are ungauged estimates.

To assess the accuracy of predictions for Sixmile Creek, the details of the absolute error (measured discharge minus predicted) are plotted for each storm (Fig. 8.5). A positive error indicates that the model is underpredicting, and a negative error that it is overpredicting. There are two possible sources of model errors: random and systematic. In the applications illustrated in Figure 8.5 the errors appear to be systematic. The errors derived from both models for any storm follow a similar temporal pattern but differ in their magnitude, the modified model displaying the lower absolute error in all cases apart from the two exceptions identified above. This implies that there remains a systematic source of error in the modified model that has not been removed by replacement of the runoff component. This source of error can be attributed to the unit hydrograph method.

Certain goodness-of-fit criteria can also be evaluated to aid an assessment of the relative performance of the models. Those used in this paper are given in the appendix. Figure 8.6 shows clearly for Sixmile Creek that the modified HYMO model more closely predicts the measured hydrographs than does the original model. The hydrographs predicted by the modified model have the lower error standard derivation, time to peak discharge error, and peak discharge error, and a value of the McCuen and Snyder statistic closer to unity.

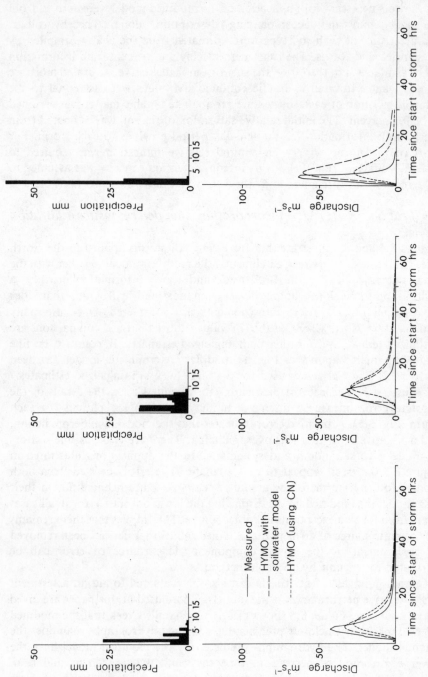

Figure 8.4 Comparison of measured hydrograph to those predicted by both models for three storms, North Creek catchment.

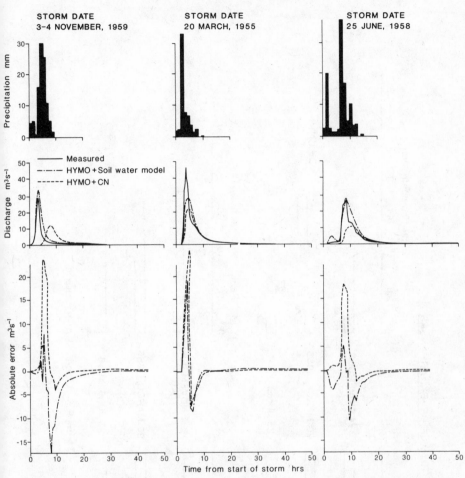

Figure 8.5 Comparison of measured hydrograph to those predicted by both models for three storms, Sixmile Creek catchment.

Table 8.3 illustrates that the predictions provided by the modified model are not very sensitive to the exact positioning adopted on the soil texture triangle for derivation of hydrologic characteristics. This is an encouraging characteristic, as the specific percent clay and percent sand information may not always be available. Under these circumstances the centroid position can be safely assumed. Indeed, this sensitivity is not as great as, for example, the sensitivity of the original model to an error in the estimate of the CN of ± 10 (Table 8.4).

The response of predictions for Sixmile Creek to two levels of variation (standard deviation) of soil hydrologic parameters can now be considered. The lower level of standard deviation of saturated moisture content and the suction–moisture curve were derived from Rawls et al. (1982, Table 2), and the upper level of standard deviation was obtained by doubling the figure for

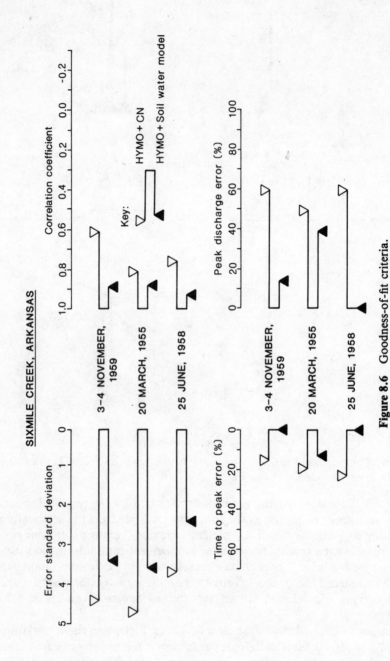

Figure 8.6 Goodness-of-fit criteria.

Table 8.3 Sensitivity of modified HYMO's predictions on the flood hydrograph for North Creek, Texas, to soil hydrologic parameters derived from the Brakensiek and Rawls method.

Storm date	Percent peak discharge error[a]	Error standard deviation[a]
May 6, 1969		
centroid	32	179
highest percent clay, lowest percent sand	33	180
July 27, 1962		
centroid	21	219
highest percent clay, lowest percent sand	24	213
September 9, 1965		
centroid	79	2102
highest percent clay, lowest percent sand	73	2020

[a] Refer to the appendix.

Table 8.4 Sensitivity of original HYMO to value CN, North Creek, Texas.

Storm date	Percent peak discharge error[a]	Error standard deviation[a]
May 6, 1969		
CN = 97	29	465
CN = 87[b]	37	283
CN = 77	72	405
July 27, 1962		
CN = 80	39	545
CN = 70[b]	17	364
CN = 60	54	425
September 9, 1965		
CN = 65	19	905
CN = 55[b]	57	708
CN = 45	84	729

[a] Refer to the appendix.
[b] Calibrated value.

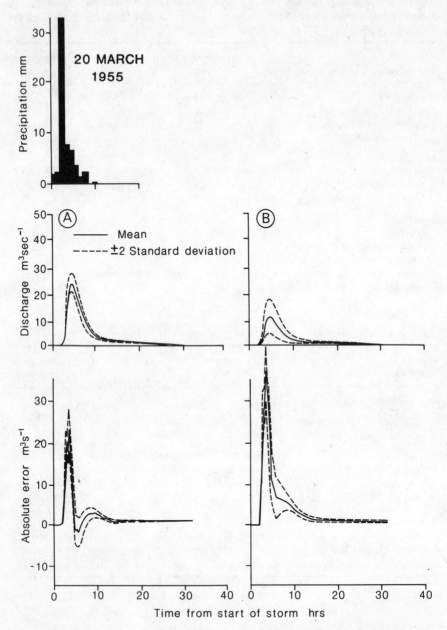

Figure 8.7 Mean and ± 2 standard deviations derived from application of stochastic infiltration model to storm of March 20, 1955, Sixmile Creek: (A) lower level of input parameter variation; and (B) upper level of variation.

Table 8.5 Comparison of hydrograph predictions derived from the deterministic and stochastic infiltration model for the storm of March 20, 1955, Sixmile Creek.

	Measured	Modified HYMO with *no* variation	Modified HYMO with	
			Low level variation[a]	Higher level variation[a]
peak discharge (m^3/s)	46	28.20	22.50 (1.70)	9.90 (3.10)
time to peak discharge (h)	4	4.50	4.30 (0.134)	4.40 (0.126)
sum of errors		3016.00	9193.00 (1368.00)	17552.00 (2326.00)
McCuen and Snyder index		0.889	0.917 (0.013)	0.913 (0.02)
error standard deviation		140.00	162.00 (12.00)	268.00 (30.00)
peak discharge error (%)		39.00	52.00 (4.00)	79.00 (7.00)

[a] Mean and standard deviation in parentheses.

the lower level. Warrick and Nielson (1980) stated that the coefficient of variation of saturated hydraulic conductivity varies from 86 to 190 percent. Accordingly, standard deviations corresponding to coefficients of variation of 90 percent for the lower level and 190 percent for the upper level were utilized. The standard deviation of initial moisture content was assumed to correspond to a coefficient of variation of 10 to 15 percent for the lower level and doubled for the upper level. No variation of detention capacity, which has been assumed to be zero, was considered. The mean values for the soil parameters obtained for application of the deterministic model were used.

Figure 8.7 depicts for Sixmile Creek the mean and two standard deviations of the hydrograph and prediction error associated with the two levels of variability of soil parameters for the storm of March 20, 1955. For this storm, 30 runs of the model were completed. As would be expected, Figure 8.7 indicates that as the magnitude of variation of input parameters increases so does the variation in the predicted hydrologic response. Associated with this, the mean simulated peak discharge decreases, and the absolute error increases. The predictions provided by the deterministic model (Fig. 8.5) are superior to those from the stochastic one and appear to be further substantiated by the findings illustrated in Table 8.5.

Discussion

This chapter demonstrates the improvements to hydrological forecasts for ungauged catchments that are to be gained from modeling one of the more crucial elements of hillslope processes, infiltration, with a physically based model. The model developed here still remains operational and meets the requirements of this application in the following respects:

1. The model is suitable for routine practical application. It is consistent with the quantity and quality of data commonly available for the ungauged catchment. The necessary catchment data can be derived from soil and topographic maps, and thus no fieldwork is necessary. In terms of computer resources, the model is not costly to operate. It will execute on a Hewlett–Packard 9816, a powerful 32-bit microcomputer, and there are further plans for its transfer onto a more basic 8-bit machine, the Apple II. This model is also easy to apply. Potential users can operate the model with a minimum of external aid.
2. The model parameters have physical interpretation. No calibration is necessary.
3. The model appears to provide better predictions than the original HYMO model for a variety of catchment types and scales.
4. It is possible to include the effects of spatial variability and error into the model's forecasts in a stochastic manner.
5. The nature of errors in predictions suggests that further improvements could be derived by replacement of the unit hydrograph method. The possibility of its replacement by a physically based overland flow model could be considered.
6. Elements of distributed physically based models of hillslope hydrology can be used in an extremely parsimonious fashion to develop flood forecasting models of the type discussed here.

The basis of the work reported here was the development of a portable operational flood forecasting model for ungauged catchments. The articulation of this objective was seen to be most readily achieved by the coupling of an empirical model (HYMO) to a physically based submodel of infiltration driven primarily by preprocessed empirical data.

In general terms and given the overriding objective, hillslope process data minimization, model portability, and a continuous forecasting capability cannot be simultaneously satisfied by a common (say physically based) modeling methodology. This approach may be seen in traditional terms as philosophically inconsistent, but this inconsistency does not *de facto* negate the realization of the objective as measured by the empirical validation approaches we have illustrated. Indeed, we would argue that the inclusion of certain aspects of physically based or distributed models into ungauged catchment models should be part of a new research effort. The model produced by Engman and Rogowski (1974) and TOPMODEL were both developed to satisfy the conflicting objectives of (1) reflecting the current status of knowledge of hydrological processes, and (2) attempting to retain operational feasibility for selected applications. However, some degree of realism must be forfeited, as outlined in this chapter. This may on occasions prove to be of a nature relating to methodological consistency as well as a simple enforced increase in abstraction. New objectives may in this way establish new definitions of consistency.

Acknowledgments

This work was funded in part by the U.S. Army Corps of Engineers through its European Research Office. The authors are particularly grateful to W. P. James of Texas A&M University and Dr. E. T. Engman of the U.S. Department of Agriculture, Beltsville, Maryland, for providing certain of the data used in the analysis.

Appendix

1. Error standard deviation is given by

$$\text{ESD} = \left(\frac{\sum_{i=1}^{n} [q_m(i) - q_s(i)]^2}{n} \right)^{\frac{1}{2}}$$

For a perfect fit, this index approaches zero.

2. McCuen and Snyder (1975) index is given by

$$r = \frac{1}{n} \sum_{i=1}^{n} \left[\left(\frac{q_m(i) - \bar{q}_m}{\sigma q_m} \right) \left(\frac{q_s(i) - \bar{q}_s}{\sigma q_s} \right) \right]$$

For a perfect fit, this index approaches unity.

3. Time to peak discharge error is given by

$$te = \left(\frac{tQ_m - tQ_s}{tQ_m} \right) \times 100$$

4. Peak discharge error is given by

$$eq = \left(\frac{Q_m - Q_s}{Q_m} \right) \times 100$$

where n = number of pairs of coordinates, q_m = measured discharge, q_s = simulated discharge, Q_m = peak measured discharge, Q_s = peak simulated discharge, $\bar{q}_m$ = mean measured discharge for storm event, $\bar{q}_s$ = mean simulated discharge for storm event, σq_m = standard deviation of measured discharge for storm event, σq_s = standard deviation of simulated discharge for storm event, tQ_m = time to peak measured discharge, and tQ_s = time to peak simulated discharge.

References

Amorocho, J., Quantifying ignorance—the problems of uncertainties in hydrology, *Transactions of the American Geophysical Union, U.S. National Committee for the International Hydrological Decade Bulletin, 7*, 931–933, 1973.

Amorocho, J., Spatially distributed variables in hydrologic modelling, in *The Mathematics of Hydrology and Water Resources*, edited by E. H. Lloyd, T. O'Donnell and J. C. Wilkinson, pp. 87–94, Academic, London, 1979.

Amorocho, J., and W. E. Hart, A critique of current methods in hydrologic systems identi-fication, *Transactions of the American Geophysical Union, 45*, 307–321, 1964.

Anderson, M. G., and S. Howes, Streamflow modelling, *Final Technical Report, DAJA 37-81-C-0221*, 130 pp., U.S. Army European Research Office, 1984.

Anderson, M. G., S. Howes, P. E. Kneale, and J. M. Shen, On soil retention curves and hydrological forecasting in ungauged catchments, *Nordic Hydrology, 16*, 11–32, 1985.

Baker, F. G., Variability of hydraulic conductivity within and between 9 Wisconsin soil series, *Water Resources Research, 14*, 103–108, 1978.

Bales, J., and R. P. Betson, The curve number as a hydrologic index, in *Rainfall–Runoff Relationships*, edited by V. P. Singh, pp. 371–386, Water Resources Publications, Littleton, Colo., 1982.

Beven, K. J., A deterministic spatially distributed model of catchment hydrology, Ph.D. dissertation, 265 pp., University of East Anglia, 1975.

Beven, K. J., and M. J. Kirkby, Towards a simple physically-based variable contributing model of catchment hydrology, *Working Paper 154*, 11 pp., Geography Department, University of Leeds, 1976.

Beven, K. J., and M. J. Kirkby, A physically-based variable contributing area model of basin hydrology, *Hydrological Sciences Bulletin, 24*, 43–69, 1979.

Beven, K. J., M. J. Kirkby, N. Schofield, and A. F. Tagg, Testing a physically-based flood forecast model (TOPMODEL) for 3 U.K. catchments, *Journal of Hydrology, 69*, 119–143, 1984.

Beven, K. J., and P. E. O'Connell, On the role of physically-based distributed modelling in hydrology, *Report No. 81*, 66 pp., Institute of Hydrology, Wallingford, 1982.

Brakensiek, D. L., and W. J. Rawls, Use of infiltration procedures for estimating runoff, paper presented to *Soil Conservation Service Workshop*, Tempe, Ariz., 1983.

Campbell, G. S., A simple method for determining unsaturated conductivity from moisture retention data, *Soil Science, 117*, 311–314, 1974.

Coelho, M. A., Spatial variability of water related soil physical parameters, Ph.D. dissertation, 110 pp., University of Arizona, Tucson, 1974.

Crawford, N. H., and R. K. Linsley, Digital simulation in hydrology: Stanford Watershed Model IV, *Technical Report, 39*, Department of Civil Engineering, Stanford University, 1966.

Crow, F. R., T. Ghermazien, and R. L. Bengtson, Transferability of the USDAHL hydrology model to ungauged grassland catchment, *Technical Paper 78–2066*, 13 pp., American Society Agricultural Engineers, 1978.

Dooge, J. C. I., Mathematical models of hydrologic systems, in *Proceedings of the International Symposium on Mathematical Modelling Techniques in Water Resource Systems, 1*, edited by A. K. Biswas, pp. 171–189, Environment Canada, Ottawa, 1972.

Dooge, J. C. I., General report on model structure and classification, in *Logistics and Benefits of using Mathematical Models of Hydrologic and Water Resource Systems*, edited by A. K. Askew, F. Grego, and J. Kindler, pp. 1–21, Pergamon, Oxford, 1981.

Douglas, J. R., Conceptual modelling in hydrology, *Report No. 24*, 60 pp., Institute of Hydrology, Wallingford, 1974.

Engman, E. T., and A. S. Rogowski, A partial area model for stormflow synthesis, *Water Resources Research, 10,* 464–472, 1974.

Freeze, R. A., Three-dimensional transient saturated–unsaturated flow in a groundwater basin, *Water Resources Research, 7,* 347–366, 1971.

Freeze, R. A., Streamflow generation, *Review of Geophysics and Space Physics, 12,* 627–647, 1974.

Hawkins, R. H., Improved prediction of storm runoff in mountain watersheds, *Journal of Irrigation and Drainage Division, Proceedings of the American Society of Civil Engineers, 106,* 519–523, 1973.

Hawkins, R. H., The importance of accurate curve numbers in the estimation of storm runoff, *Water Resources Bulletin, 11,* 887–891, 1975.

Hawkins, R. H., Runoff curve numbers with varying site moisture, *Journal of Irrigation and Drainage Division, Proceedings of the American Society of Civil Engineers, 104,* 389–398, 1978.

Hawkins, R. H., Runoff curve numbers from partial area watersheds, *Journal of Irrigation and Drainage Division, Proceedings of the American Society of Civil Engineers, 105,* 375–390, 1979.

Hillel, D., *Computer Simulation of Soil Water Dynamics,* 214 pp., I.D.R.C., Ottawa 1977.

Holton, H. N., G. J. Stiltner, W. N. Henson, and N. C. Lopez, USDAHL-74 revised model of watershed hydrology, *USDA–ARS Technical Bulletin 1518,* United States Department of Agriculture, Washington, D.C., 1975.

Hjelmfelt, A. T. Jr., Curve number procedure as infiltration method, *Journal of Hydraulics Division, Proceedings of the American Society of Civil Engineers, 106,* 1107–1111, 1980.

Jackson, R. D., On the calculation of hydraulic conductivity, *Journal of the Soil Science Society of America, 36,* 380–382, 1972.

James, L. D., Hydrologic modelling parameter estimation, and watershed characteristics, *Journal of Hydrology, 17,* 283–307, 1972.

James, L. D., Precipitation–runoff modelling: future developments, in *Applied Models of Catchment Hydrology,* edited by V. P. Singh, pp. 291–312, Water Resources Publications, Littleton, Colo., 1982.

Jarboe, J. E., and C. T. Hann, Calibrating a water yield model for small ungauged watersheds, *Water Resources Research, 10,* 256–262, 1974.

Kirkby, M. J., Hydrograph modelling strategies, in *Processes in Physical and Human Geography,* edited by R. F. Peel, M. D. I. Chisholm, and P. Haggett, pp. 69–90, Heinemann, London, 1975.

Kirkby, M. J., J. Callan, D. Weyman, and J. Wood, Measurement and modelling of dynamic contributing areas in very small catchments, *Working Paper 167,* Geography Department, University of Leeds, 31 pp., 1976.

Klemes, V., Empirical and causal models in hydrology, *Studies in Geophysics: Scientific Basis of Water Resource Management,* National Academy Press, Washington, pp. 95–104, 1982.

McCuen, R. H., and W. M. Snyder, A proposed index for comparing hydrographs, *Water Resources Research, 11,* 1021–1024, 1975.

Manley, R. E., Simulation of flows in ungauged basins, *Hydrological Sciences Bulletin, 23* 85–101, 1977.

Marshall, T. J., A relation between permeability and size distribution of pores, *Soil Science, 9,* 1–8, 1958.

Millington, R. J., and J. P. Quirk, Permeability of porous media, *Nature, 183,* 387–388, 1959.

Miller, D. R., G. Butter, and L. Bramall, Validation of ecological system models, *Journal of Environmental Management, 4,* 383–401, 1976.

Morel-Seytoux, H. J., and J. P. Verdin, Extension of the SCS rainfall–runoff methodology for ungauged watersheds, *Final Report DOT-FH-PO-9-3-0015*, U.S. Department of Transportation, Washington, D.C., 1981.

Naef, F., Can we model the rainfall–runoff process today? *Hydrological Sciences Bulletin, 26*, 281–289, 1981.

Nash, J. E., A unit hydrograph study with particular reference to British catchments, *Proceedings of the Institute of Civil Engineers, 17*, 249–282, 1960.

Nielson, D. R., J. W. Biggar, and K. T. Erh, Spatial variability of field measured soil water properties, *Hilgardia, 42*, 215–259, 1973.

Rawls, W. J., D. L. Brakensiek, and K. E. Saxton, Estimation of soil water properties, *Transactions of the American Society of Agricultural Engineers, 25*, 1316–1320, 1982.

Rogowski, A. S., Watershed physics: soil variability criteria, *Water Resources Research, 8*, 1015–1023, 1972.

Ross, G. A., The Stanford Watershed Model: the correlation of parameter values selected by a computerized procedure with measurable physical characteristics of the watershed, *Research Report 35*, Water Resources Institute, University of Kentucky, Lexington, 1970.

Russo, D., and E. Bresler, Effect of field variability in soil hydraulic properties on solutions of unsaturated water and salt flows, *Journal of the Soil Science Society of America, 45*, 675–681, 1981.

Smith, V. E., The application of HYMO to study areas of S. W. Wyoming for surface runoff and soil loss estimates, *Water Research Series, No. 60*, Water Resources Institute, University of Wyoming, Laramie, 1976.

Stephenson, G. R., and R. A. Freeze, Mathematical simulation of sub-surface flow contributing to snow melt runoff, Reynolds Creek, Idaho, *Water Resources Research, 10*, 284–298, 1974.

United States Department of Agriculture, *Soil Conservation Service National Engineering Handbook, Hydrology, Section 4*, 1972.

Verma, R. D., and R. M. Advani, Flood estimation methods for ungauged catchments—a review, *International Symposium on River Mechanics, 2*, 253–264, 1973.

Warrick, A. W., and D. R. Nielson, Spatial variability of soil physical properties in the field, in *Application of Soil Physics*, edited by D. Hillel, pp. 319–355, Academic, New York, 1980.

Williams, J. R., Flood routing with variable travel time or variable storage coefficients, *Transactions of the American Society of Agricultural Engineers, 12*, 100–103, 1969.

Williams, J. R., and R. W. Hann, Jr., HYMO: a problem-orientated computer language for building hydrologic models, *Water Resources Research, 8*, 79–85, 1972.

Williams, J. R., and R. W. Hann, Jr., HYMO: a problem-orientated computer language for hydrologic modelling—users' manual, *Report ARS-S-9*, Agricultural Research Service, Southern Region, 1973.

Williams, J. R., and W. V. LaSeur, Water yield model using the SCS curve numbers, *Journal of Hydraulics Division, Proceedings of the American Society of Civil Engineers, 102*, 1241–1253, 1976.

9
Hillslope runoff processes and flood frequency characteristics

Keith Beven

Abstract

A model of hillslope hydrology is described that attempts to simulate a variety of different runoff production mechanisms in a way that takes account of both hillslope form and the spatial variability of soil characteristics. The model has been designed to reflect the essence of physically based theory but is computationally very simple, so that it can be used to explore the effects of hillslope runoff production on flood frequency characteristics by means of Monte Carlo simulation. Storm period calculations are carried out using an hourly time step, and interstorm period calculations are performed using analytical equations that include the effects of unsaturated zone drainage and evapotranspiration losses. A number of 100-year simulations are described representing runoff production on a single catchment topography with a variety of soil and climatic conditions.

Introduction

A number of recent papers have been concerned with the prediction of the flood frequency characteristics of catchments. These papers have differed in the way in which they have estimated runoff production during flood producing events. Hebson and Wood (1982) used a constant partial contributing area; Cordova and Rodriguez-Iturbe (1983) used the SCS curve number technique and Diaz-Granados et al. (1984) used the Philip infiltration equation. This study makes further use of the model of Beven (1985*a*, *b*), which is an attempt to simulate a number of runoff production mechanisms on the basis of simplified physical theory. This model incorporates the effects of both hillslope topographic form and the spatial characteristics of soil properties on predicted runoff production.

Hillslope hydrology and flood runoff production

The analysis of process response times of Kirkby (1976) suggests that except in very large catchments, the form and peak of flood hydrographs will be primarily controlled by rate of runoff production on hillslopes, rather than overland or channel flow routing. Correctly predicting hillslope runoff production rates therefore becomes a central problem in predicting flood frequency characteristics for all situations in which runoff coefficients are significantly less than 1. A review of the field evidence (e.g., Dunne, 1978) reveals a pattern of complexity of runoff production processes. This complexity is not adequately described by resort to the classic models of Horton "infiltration excess" runoff, Dunne "saturation excess" runoff or Hewlett "subsurface stormflow." We should rather envisage a continuum of processes by which water can reach a stream channel during a flood event.

We should recognize that any single water particle may take both surface and subsurface flow paths on its way to the stream dependent on rainfall intensities, soil characteristics and antecedent conditions. We should note that in many locations, studies using oxygen and hydrogen isotopes as natural tracers suggest that the major part of some storm hydrographs, including some surface runoff, may be made up of "pre-event" water displaced by the incoming rainfall (e.g., Sklash and Farvolden, 1979). Field studies and modeling studies suggest that spatial variability of soil characteristics may be important in controlling runoff production (e.g., Smith and Hebbert, 1979; Luxmoore and Sharma, 1980; Freeze, 1980).

It has also been suggested that continuous large voids in the soil may be important in channeling flow into the soil and possibly into the soil matrix in a way that bypasses the bulk soil matrix (Beven and Germann, 1982). Variations in effective rainfall intensity at the soil surface (for example, the locally increased intensities associated with stemflow) may be important in inducing and maintaining such channeling flows. Pipe flows may be important in transmitting excess water rapidly down hillslopes (e.g., Gilman and Newson, 1980). Small scale variations in soil depth and bedrock surface topography may also have an important effect on subsurface flow rates, water table heights, and consequently runoff production.

What emerges from this brief summary of current understanding of hillslope hydrology is that complexity of runoff production processes arises from the interaction of climatically controlled inputs (that may exhibit important spatial and temporal variations) with characteristically heterogeneous hillslopes. Heterogeneity arises from three-dimensional variability in topographic form, soil hydraulic characteristics, soil depths, vegetation controls on surface fluxes, and antecedent moisture conditions; variations in time may also be important. However, most hydrologists have not been taught to think in ways that reflect this heterogeneity, and this is readily apparent in the models that have been used to simulate hillslope hydrology. The model used in this chapter goes some way to incorporating both different runoff production mechanisms and the effects of variability in hillslope characteristics in a relatively simple structure.

Theory: background

The model structure builds upon that of TOPMODEL, variants of which are described in the papers of Beven and Kirkby (1979), Beven and Wood (1983), and Beven et al. (1984). It is assumed that at any point i on a hillslope, downslope saturated subsurface flow rate q_i may be described by

$$q_i = K_i \tan \beta \, e^{-S_i/m} \tag{1}$$

where $\tan \beta$ is the local slope angle, K_i is a soil transmissivity, S_i is the local deficit of readily drained soil moisture, and m is a parameter dependent on the rate of change of hydraulic conductivity with depth. Beven (1985b) shows how equation (1) can be related to a similar exponential decline in hydraulic conductivity with depth:

$$K_s = K_0 \, e^{fz} \tag{2}$$

where K_0 is the hydraulic conductivity at the soil surface and z is depth into the soil. Beven (1984) gives data to suggest that equation (2) is an adequate fit to the hydraulic conductivity data for a variety of soils and that K_0 may vary between 0.01 and 100 m/h and f between 1 and 12 m^{-1}. In soils, for which (2) holds and conductivity at depth is small, $K_i \simeq K_0/f$, and $m \simeq -f/\Delta\theta$ where $\Delta\theta$ is a volumetric moisture deficit below saturation.

Prediction of saturation excess surface runoff

Under steady-state conditions due to an input flux r,

$$q_i = a_i r \tag{3}$$

where a_i is the area of hillslope per unit contour length draining through point i. The value of a_i will be relatively high for convergent hillslopes and low for divergent hillslopes. Using (1) and (3),

$$S_i = - m \ln\left(\frac{ar}{K_i \tan \beta}\right) \tag{4}$$

Integrating over the hillslope or catchment area A we can obtain an expression for mean storage deficit $\bar{S}$ as

$$\bar{S} = \frac{1}{A} \int_A - m \ln\left(\frac{ar}{K_i \tan \beta}\right) \tag{5}$$

or using (4) in (5) to eliminate $\ln(r)$, we have

$$\bar{S} = S_i - m\gamma + m \ln\left(\frac{a}{K_i \tan \beta}\right)_i \tag{6}$$

where

$$\gamma = \frac{1}{A} \int_A \ln\left(\frac{a}{K_i \tan \beta}\right)_i$$

Since (6) no longer involves the input rate r it can be used, given knowledge of $\bar{S}$, to predict the *pattern* of local storage deficits all over the hillslope or catchment area. During storm conditions, continuous accounting of $\bar{S}$ allows the change in the pattern of storage deficits over time to be predicted subject to the assumption that the steady state relationship (6) holds in the transient case. In particular, the area for which $S_i \leqslant 0$ can be predicted—that is, the area in which the soil is completely saturated (saturation from below). Any rainfall falling on this area is assumed to reach a stream channel as saturation excess surface runoff.

In addition, there will also be areas for which S_i is greater than 0 but smaller than the rainfall input during a time step. In this case the rainfall that is in excess of the storage deficit will also add to saturation excess surface runoff (saturation from above).

This theory implies that all points with the same value of $a/(K_i \tan \beta)$ are hydrologically similar. Clearly, high values of a and small values of K_i and $\tan \beta$ will serve to increase the likelihood of surface saturation. Thus hillslope forms that are convergent hollows with low hydraulic conductivities will be the first to generate saturation excess runoff in a catchment area.

Prediction of infiltration excess surface runoff

For soils in which (1) and (2) are adequate to describe the lateral subsurface discharge and hydraulic conductivity profile, respectively, infiltration excess surface runoff may be predicted in a consistent way by making use of the infiltration equation reported in Beven (1984) based on (2) and the Green–Ampt assumptions. In this model, time to ponding t_p is given by the equality

$$r(t_p) = \frac{K_o f}{\Delta\theta} \frac{I_p + C}{e^{f I_p / \Delta\theta} - 1} \tag{7}$$

where r is the rainfall intensity, $I_p = \int_{t_p} r(t)\,dt$ is the volume of water infiltrated to time t_p, $\Delta\theta = \theta_s - \theta_i$ is an initial moisture deficit assumed constant with depth, and C is the storage suction factor of Morel-Seytoux and Khanji (1974).

After ponding the total volume of infiltration at any time t is given by the implicit relationship

$$t - t_p = \frac{\Delta\theta}{K_0 f}\left[\ln(I+C) - \frac{1}{e^{-fc/\Delta\theta}}\left(\ln(I+C)\right.\right.$$
$$\left.\left. + \sum_{n=1}^{\infty}\frac{[-f(I+C)/\Delta\theta]^n}{n!n}\right) - \text{constant}\right] \quad (8)$$

where

$$\text{constant} = \ln(I_p + C) - \frac{1}{e^{-fc/\Delta\theta}}\left(\ln(I_p + C) + \sum_{n=1}^{\infty}\frac{[-f(I_p+C)/\Delta\theta]^n}{n!n}\right)$$

Prediction of subsurface storm runoff and drainage

At any time, outflows from the saturated zone into the channel q_b will be given by an integration over the channel length of the local q_i values, so that

$$q_b = \frac{1}{A}\int_L K_i \tan\beta e^{-Si/m} \quad (9)$$

where L is the channel length and the integration is taken as applying to both banks. Using (6) to substitute for S_i in (9), this simplifies to

$$q_b = q_0 e^{-\bar{S}/m}$$

where $q_0 = e^{-\gamma}$ is a constant.

Two points should be noted here. First, the steady-state assumptions on which this simplified theory is based imply that (1) applies both to flows within the soil and to the return flow contribution to surface flow owing to any supply of water from upslope in excess of the saturated flow capacity of the soil profile. In that (1) is exponential, this may not introduce undue error. Second, the channel length in (9) is assumed fixed. It is known that extensions of the channel network during storm periods may be important factors in storm response. In this study, an hourly time step is used during storm periods, and it is assumed that if overland flow is generated anywhere on a hillslope, it reaches a channel within one time step. While this should cope with normal extensions of the channel network in humid areas, the measured overland flow velocities reported in Beven et al. (1984) would suggest a slower surface flow response than the procedure used here under conditions where unchanneled surface flows provide a significant proportion of the storm hydrograph.

Prediction of interstorm outflows

The interstorm period calculations require additional equations to define drainage from the unsaturated zone and for evapotranspiration losses. Since it was intended to use the model for long time period calculations, forms were chosen that allowed analytical solutions during interstorm periods as follows:

unsaturated zone drainage

$$q_v = \alpha K_0 e^{-S_i/m}$$

potential evapotranspiration

$$E_p = \bar{E}_p [1 + \sin(0.0174\, t - b)]$$

and actual evapotranspiration

$$E_a = \begin{cases} E_p & S_{RZ} < S_{RZ}^* \\ 0 & S_{RZ} \geqslant S_{RZ}^* \end{cases}$$

where α is the vertical hydraulic gradient at the water table (assumed constant and equal to 1.00 throughout), t is time in days, $\bar{E}_p$ is mean daily potential evapotranspiration rate, b is a coefficient approximately equal to $\pi/2$, S_{RZ} is a root zone storage deficit (in addition to the local drainage deficit S_i), and S_{RZ}^* is a maximum root zone storage deficit.

Details of the analytical calculations for the interstorm periods are given in Beven (1985b).

Flood frequency predictions

In long-term predictions for the purpose of calculating flood frequency curves, the model can be used either with a long-term observed rainfall record (in a similar way to the use of actual extreme rainstorms by Cordova and Rodriguez-Iturbe (1982)) or with synthetic rainstorms based on specified distributions for storm intensities, durations and interarrival times (in a similar way to Diaz-Granades et al. (1984)). In both cases the use of a full sequence of rainstorms in the present study allows variations in antecedent conditions prior to events to be taken into account.

Beven (1985a) has shown how such a model can be used to predict the flood frequency characteristics of an actual catchment (in that case with some knowledge of the distributions of antecedent conditions from measurements). Beven (1985*b*) extended this work to investigate the variations in flood frequency characteristics due to different topographies and soil characteristics within a single climatic regime. This paper now investigates variations due to different climatic regimes for a particular topography and two sets of spatially variable soil characteristics. The topography chosen is based upon a simple collection of hillslopes forming an actual 11 km^2 catchment (Bottoms Beck, Lancashire, UK). The topography and derived $\ln(a/\tan\beta)$ distribution function for this catchment are shown in Figure 9.1.

The climatic data on which the inputs to the model for the different climatic regimes were based is summarized in Table 9.1. It was assumed that the distributions of intensities, durations, and interarrival times were all exponential and mutually independent, and that the probability of occurrence of a rainstorm was equally likely throughout the year. As far as was known, the data of Eagleson and Tellers (1982) were based on all rainstorms, and in simulating random rainstorms here, no threshold values were used for rainstorm intensities, durations or interstorm periods (as used, for example, in Beven (1985*b*)). This led to some more extreme storm rainfalls being

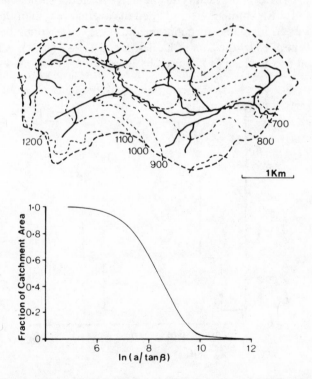

Figure 9.1 Topography and derived distribution function of $\ln(a/\tan\beta)$ used in all simulations.

Table 9.1 Parameters of exponential distribution of rainfall intensity $\bar{i}$, duration $\bar{t}_d$, interarrival times $\bar{t}_a$, and mean hourly potential evapotranspiration $\bar{e}_p$.

Location	$\bar{i}$ (mm/h)	$\bar{t}_d$ (h)	$\bar{t}_a$ (h)	$\bar{e}_p$ (mm/h $\times 10^4$)
upland Wales[a]	2.77	6.55	67.58	0.625
Massachusetts[b]	1.318	7.68	72.00	0.625
California[b]	1.031	33.6	524.4[c]	1.125
Georgia[b]	2.05	7.68	91.68	1.125
Kansas[b]	3.94	3.60	142.04[c]	1.417

[a] Beven (1985b).
[b] Eagleson and Tellers (1982).
[c] See comments about length of rainfall season in text.

simulated for the "upland Wales" regime compared with the simulations of Beven (1985b) who used the same mean values (including the threshold value). In addition, Eagleson and Tellers (1982) give data for the length of rainy season for the different stations. In the case of the "California" and "Kansas" regimes these are significantly less than 1 year (212 and 322 days on average, respectively). There is no facility in the current model to handle seasonality explicitly, so for the simulations presented the average length of the interstorm periods has been increased to maintain the same average number of storms each year as implied by the data of Eagleson and Tellers (1982). A single storm profile based on the mean profile used by Beven (1985a) was used throughout.

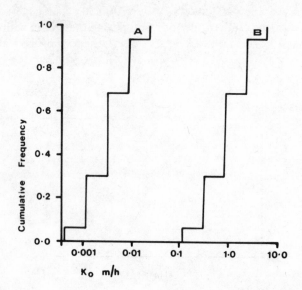

Figure 9.2 Distributions of K_0 used in simulations: (A) $\bar{K}_0 = 0.003$ m/h, and (B) $\bar{K}_0 = 0.3$ m/h.

Table 9.2 Model parameters used in the simulations.

Parameter	Value
$m \simeq -f/\Delta\theta$	0.15 m
$\bar{K}_0$	0.3 m/h
	0.003 m/h
$\sigma_{\ln(K_0)}$	1.0
S_{RZ}^*	0.1 m
C	0.04 m
initial root zone deficit	0.02 m
initial $\bar{S}$	0.02 m

Two soil hydraulic conductivity K_0 distributions were used with means of 0.3 m/h and 0.003 m/h, both having log-normal distributions with a standard deviation of 1 log unit. The distributions are represented in discrete form for the model calculations, as shown in Figure 9.2. The other parametric data in the model are summarized in Table 9.2.

Results

The results of the ten 100-year simulations are summarized in Tables 9.3, 9.4, and 9.5.

Depending on the number of storms simulated during the 100-year periods, the jobs took between 17 and 45 min to run on a GEC 32 bit virtual memory minicomputer. Table 9.3 shows the simulated average water balance components over the 100-year period. Comparing these data with those given by Eagleson and Tellers (1982) suggests that, in the case of the California and Kansas regimes, discharge volumes are overestimated using the combination

Table 9.3 Simulated average water balance components for the 100-year simulations.

Location	$\bar{K}_0$ (m/h)	Mean annual rainfall (mm)	Mean annual discharge (mm)	Mean annual evapotranspiration (mm)
upland Wales	0.3	2075.6	1538.7	535.6
	0.03	2075.6	1530.8	543.2
Massachusetts	0.3	1078.4	567.9	509.5
	0.03	1078.4	553.8	523.2
California	0.3	576.3	226.4	350.4
	0.03	576.3	226.8	349.4
Georgia	0.3	1353.9	597.5	755.4
	0.03	1353.9	592.0	760.6
Kansas	0.3	836.3	231.5	604.5
	0.03	836.3	238.5	597.0

Table 9.4 Number of storms, mean storm rainfall, and mean annual flood for 100-year simulations.

Location	$\bar{K}_0$ (m/h)	Number of storms (100 yr)	Mean storm rainfall, all storms (mm)	Mean annual flood (mm/h)
upland Wales	0.3	11837	17.54(29.4)[a]	20.81(6.71)
	0.03	11837	17.54(29.4)	23.50(6.9)
Massachusetts	0.3	11013	9.79(16.3)	7.57(2.78)
	0.03	11013	9.79(16.3)	9.95(3.07)
California	0.3	1564	36.85(68.2)	3.32(3.79)
	0.03	1564	36.85(68.2)	4.79(4.35)
Georgia	0.3	8842	15.31(25.31)	10.61(3.91)
	0.03	8842	15.31(25.31)	13.39(4.17)
Kansas	0.3	5953	14.05(24.4)	13.43(9.59)
	0.03	5953	14.05(24.4)	17.86(10.06)

[a] Standard deviations in parentheses.

of parameter values used in these simulations. In the other cases the simulated volumes of discharge and evapotranspiration are close to observed values. In all cases but one, the lower K_0 values give lower discharge values. The exception is the Kansas regime, where higher rainfall intensities lead to a greater contribution of infiltration excess overland flow to the total discharge.

Table 9.5 Summary data on mechanisms of runoff production for 100-year simulations. Mean values over all storms given with standard deviations in parentheses.

Location	$\bar{K}_0$ (m/h)	Maximum saturation excess, contributing area (%)	Maximum infiltration excess, contributing area (%)	Surface runoff at peak (%)	Surface runoff volume (%)
upland Wales	0.3	47.2(39.8)	0.0(0.0)	56.9(47.7)	41.9(40.0)
	0.03	58.1(48.3)	5.5(14.7)	60.3(48.7)	57.0(47.1)
Massachusetts	0.3	27.6(34.9)	0.0(0.0)	37.4(46.6)	25.5(35.9)
	0.03	37.9(47.2)	1.6(6.7)	41.4(49.1)	38.4(46.4)
California	0.3	13.8(28.7)	0.0(0.0)	20.2(39.2)	13.0(28.2)
	0.03	18.2(36.2)	6.3(14.9)	33.7(47.0)	27.7(41.4)
Georgia	0.3	23.8(34.9)	0.0(0.0)	31.8(45.4)	22.5(35.4)
	0.03	31.2(44.9)	4.0(11.8)	39.8(48.7)	36.4(45.8)
Kansas	0.3	11.08(25.6)	0.0(0.0)	16.96(37.1)	11.69(27.9)
	0.03	15.5(34.5)	4.8(14.6)	26.9(44.3)	24.1(40.7)

Table 9.4 shows that there was a seven-fold difference in the number of storms simulated for the different regimes, with a similar range in the simulated mean annual floods. All the simulated distributions of mean annual floods were of extreme value 1 or Gumbel type, as illustrated in Figure 9.3. Growth curves for the different regimes varied markedly, being steepest for California, which has the smallest number of events per year and the lowest mean annual flood, but the highest mean storm rainfall of the locations considered. Table 9.1 shows that the high storm rainfalls for this case result from the storm durations; the mean intensity is in fact the lowest of the regimes considered.

Table 9.5 summarizes the simulated behavior for the different regimes in terms of four indexes of runoff production: the maximum saturation excess contributing area for each storm, the maximum infiltration excess contributing area, the proportion of surface runoff at the hydrograph peak, and the proportion of surface runoff during each storm period. These figures show that the surface response for all these cases is dominated by saturation excess surface runoff. The higher conductivity soils produce no infiltration excess runoff at all, and even with the lower conductivities the infiltration excess volumes on average are small. It is interesting to note that the low intensity, long duration rainfalls of California produce higher average infiltration excess contributing areas than the high intensity, short duration rains of Kansas.

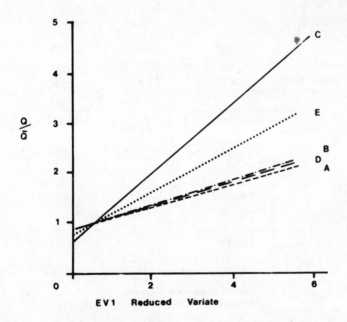

Figure 9.3 Normalized flood frequency curves for the simulations with $\bar{K}_0 =$ 0.003 m/h: (A) upland Wales, (B) Massachusetts, (C) California, (D) Georgia, and (E) Kansas.

With the exception of upland Wales, all the regimes produce the bulk of the simulated discharges by subsurface flow, which on average provides the major proportion even of the peak discharges. It is worth noting that these proportions are averaged over all simulated storms, and that for the flood flows the proportion of surface flow would be expected to be much higher (see, for example, Beven (1985b)). This is revealed by histograms of the values of these indices over all storms, examples of which are shown in Figure 9.4, 9.5, and 9.6. All the distributions show a major peak at zero contributing area or surface flow, whereas in many cases the secondary peak associated with higher flows is close to 1.0.

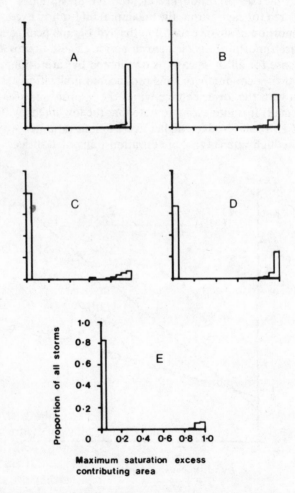

Figure 9.4 Histograms of values of maximum saturation excess contributing area for all simulated storms: (A) upland Wales, (B) Massachusetts, (C) California, (D) Georgia, and, (E) Kansas. $\bar{K}_0 = 0.003$ m/h for all cases.

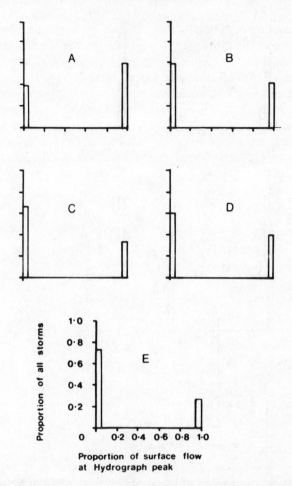

Figure 9.5 Histograms of values of maximum infiltration excess contributing area for all simulated storms: (A) upland Wales, (B) Massachusetts, (C) California, (D) Georgia, and (E) Kansas. $\bar{K}_0 = 0.003$ m/h for all cases.

Discussion and conclusions

It must be emphasized that the intention of this chapter has been to introduce a methodology that will allow consideration of hillslope hydrology, and the way in which it is affected by variability in topography, soils, and climate, directly in the prediction of flood frequency characteristics. It has not been the intention to provide accurate simulations of real catchments at this stage. The model simulations have shown that reasonable predictions of water balance components can be achieved over long periods of time, even with arbitrarily chosen soil parameter values. However, the water balance is largely

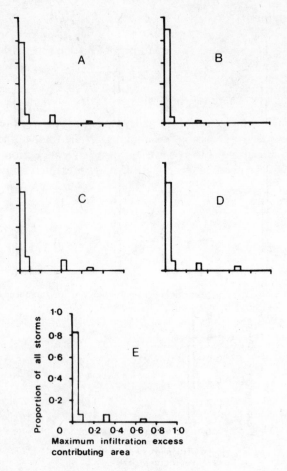

Figure 9.6 Histograms of values of proportion of surface flow at hydrograph peaks for all simulated storms: (A) upland Wales, (B) Massachusetts, (C) California, (D) Georgia, and (E) Kansas. $\bar{K}_0 = 0.003$ m/h for all cases.

constrained by the specified inputs and this is only a minor criterion of success. It has not been possible to check the accuracy of the time distribution of predicted discharges during storm periods, although for the upland Wales regime previous versions of the model have provided accurate simulations of both continuous discharges (Beven et al., 1984) and flood frequency characteristics (Beven, 1985a).

Thus, although further checks of the model are required by making applications to actual catchments (which may lead to further modifications), the results presented in this chapter suggest that a viable structure is now available for making predictions of flood frequency characteristics in a way that reflects the differences in hillslope hydrology and runoff production mechanisms in different climatic regimes.

Acknowledgment

This work has been funded by the Ministry of Agriculture, Fisheries and Food, United Kingdom.

Addendum

After this chapter had reached the galley stage, further work revealed a coding error in the interstorm period calculations. While this will have affected the results, and in particular may have led to an overprediction of contributing areas, it does not invalidate the methodology presented above.

References

Beven, K. J., Infiltration into a class of vertically non-uniform soils, *Hydrological Sciences Journal, 29*, 425–434, 1984.

Beven, K. J., Towards the use of catchment geomorphology in flood frequency predictions, *Earth Surface Processes and Landforms*, in press, 1985a.

Beven, K. J., Runoff production and flood frequency in catchments of order n: an alternative approach, *Journal of Hydrology*, in press, 1985b.

Beven, K. J., and P. F. Germann, Macropores and water flow in soils, *Water Resources Research, 18*, 1311–1325, 1982.

Beven, K. J., and M. J. Kirkby, A physically-based variable contributing area model of basin hydrology, *Hydrological Sciences Bulletin, 24*, 43–69, 1979.

Beven, K. J., M. J. Kirkby, N. Schofield, and A. Tagg, Testing a physically-based flood forecasting model (TOPMODEL) for three UK catchments, *Journal of Hydrology, 69*, 119–143, 1984.

Beven, K. J., and E. F. Wood, Catchment geomorphology and the dynamics of runoff contributing areas. *Journal of Hydrology, 65*, 139–158, 1983.

Cordova, J. R., and I. Rodriguez-Iturbe, Geomorphologic estimation of extreme flow probabilities, *Journal of Hydrology, 65*, 159–173, 1983.

Diaz-Granados, M. A., J. B. Valdes, and R. L. Bras, A physically-based flood frequency distribution, *Water Resources Research, 20*, 995–1002, 1984.

Dunne, T., Field studies of hillslope flow processes, in *Hillslope Hydrology*, edited by M. J. Kirkby, pp. 227–294, John Wiley, Chichester, 1978.

Eagleson, P. S., and T. E. Tellers, Ecological optimality in water-limited natural soil–vegetation systems. 2. Tests and applications, *Water Resources Research, 18*, 341–354, 1984.

Freeze, R. A., A stochastic conceptual analysis of rainfall–runoff processes on a hillslope. *Water Resources Research, 16*, 391–408, 1980.

Gilman, K., and M. D. Newson, Soil pipes and pipeflow: A hydrological study in upland Wales, *BGRG Research Monograph 1*, Geobooks, Norwich, 1980.

Hebson, C., and E. F. Wood, A derived flood frequency distribution, *Water Resources Research, 18*, 1509–1518, 1982.

Kirkby, M. J., Hydrograph modelling strategies, in *Processes in Physical and Human Geography*, edited by R. Peel, M. Chisholm, and P. Haggett, pp. 69–90, Academic Press, London, 1976.

Luxmoore, R. J., and M. L. Sharma, Runoff responses to soil heterogeneity: experimental and simulation comparison for 2 contrasting watersheds, *Water Resources Research, 16,* 675–684, 1980.

Morel-Seytoux, H. J., and J. Khanji, Derivation of an equation of infiltration, *Water Resources Research, 10,* 795–800, 1974.

Sklash, M. G., and R. N. Farvolden, The role of groundwater in storm runoff, *Journal of Hydrology, 43,* 45–65, 1979.

Smith, R. E., and R. H. B. Hebbert, A Monte Carlo analysis of the hydrologic effects of spatial variability of infiltration, *Water Resources Research, 15,* 419–429, 1979.

10

A two-dimensional simulation model for slope and stream evolution

M. J. Kirkby

Abstract

A computer simulation models landform evolution by wash and creep/splash over a two-dimensional 64×64 grid. The same process law is used for all points to allow the development of a valley network minimally constrained by model rules to represent erosion on a surface of constant composition. The initial surface consists of a uniform slope perturbed by subtracting a low-relief fractal surface from it. Fractal perturbation was chosen to minimize scale constraints; and some form of perturbation in either the process law or in the initial surface is necessary, as symmetry otherwise prevents the formation of channels at all.

Three model runs are presented, differing mainly in the critical unit area (i.e., area per unit contour width) at which wash transport becomes equal to creep/splash. Critical unit areas of 5000, 1280, and 320 m gave respectively no channels and average channel spacings of 2200 and 560 m. Contour crenulations indicating an eroded channel occurred at unit areas of about twice the critical values. Simulation results are compared with the theoretical criterion for unstable growth of surface hollows, and good agreement is found, supporting the theory and suggesting ways of making it operational.

The interaction between streams and slopes

The single most important parameter to describe a fluvially produced landscape is almost certainly some measure of its drainage texture. The most commonly used measure is drainage density, expressed as total channel length (however defined) per unit area; its reciprocal is then a measure of average stream spacing. Drainage density varies from less than $1.0 \ km/km^2$ for some humid areas to about $1000 \ km/km^2$ in some areas of badland dissection.

Although drainage density has been shown to be well correlated with vegetation cover (Melton, 1957), there have been few attempts to explain this or other dependencies in terms of the processes operating on hillslopes and/or channels.

The most fundamental approach to the problem has been by Smith and Bretherton (1972), who established the conditions under which small random depressions in the surface topography would grow into valleys in an initially unstable fashion rather than fill in. This condition is met for conditions of transport-limited removal (i.e., when sediment transport is at its capacity rate) when converging flows of sediment have a combined transporting capacity in excess of their previous summed capacities; that is, when transporting capacity increases more than linearly with drainage area for a given gradient. In mathematical notation, there is instability if and only if

$$\partial S/\partial a > S/a \tag{1}$$

where S is the sediment transporting capacity of the flow per unit width, and a is drainage area per unit width (Smith and Bretherton, 1972; Carson and Kirkby, 1972, p. 394).

Thus, for example, soil creep and rainsplash processes, which depend on gradient but not on distance or drainage area, do not produce instabilities but produce smooth landforms without growth of depressions. Wash processes, on the other hand, with a more than linear dependence of transport rate on overland flow discharge and hence on drainage area, tend to enlarge small depressions into macroscopic valleys. Equation (1) is also the condition for hillslopes to become convex in profile under conditions of downcutting at a constant rate for all points.

This argument was applied to the stable length for individual slope profiles by Kirkby (1980), who argued that it would be efficient for valleys to adjust towards a state of minimum drainage density in which slope lengths are close to their maximum stable value. Suppose that, for example, transport-limited erosion occurs with a combination of splash and wash, at rate

$$S = k(1 + a^2/u^2)g \tag{2}$$

where k is the constant of proportionality for splash or creep, u is the distance beyond which wash transport exceeds splash, and g is the tangent slope gradient along which sediment is assumed to flow. The form of this expression assumes a linear dependence on gradient for splash, creep, and wash. A variety of forms have been used, most commonly power laws with exponents ranging from about 0.7 to 2, but an adequate fit to many data is found for the exponent of unity used here (Carson and Kirkby, 1972, p. 209–213). The dependence on unit area a is also commonly expressed as a power law. Unit area is used as a surrogate for overland flow discharge per unit width, and the simple square law used here is adequate where the rate of production of overland flow may be assumed constant, so that overland flow discharge increases more or less linearly with area. This assumption is approximately met in semiarid areas, but is far less satisfactory for humid areas where overland flow is generated by saturation excess mechanisms. A fuller discussion of an appropriate and compatible transport law for humid areas may be found in Kirkby (1978). With the semiarid transport law (2), it may be seen that instability occurs for all drainage areas $a > u$ irrespective of gradient, whereas for humid slopes there is a marked increase in the critical a value as gradients are decreased. It follows that the semiarid slopes should have lengths not exceeding u so that drainage density is of the order of $1/(2u)$, and approaches this value if drainage density tends towards its minimum. There is then a simple approximate relationship between the process rates from which u is estimated and the drainage texture.

To pursue this relationship further, it is necessary to resort to representations of the landscape with two horizontal dimensions, as the growth or infilling of hollows cannot be represented with a single dimension, even though the stability criterion may be evaluated. The development equations, although easily stated, do not offer an analytical solution in any but the simplest and most symmetrical cases. It has therefore been necessary to proceed through a computer simulation.

There have been a number of two-dimensional slope simulations, but none to date has satisfactorily addressed the problem of drainage texture. The most relevant are those of Sprunt (1972), Ahnert (1976), Armstrong (1976), and Cordova et al. (1983). Sprunt (1972) worked within a 32×32 grid, allowing sediment transport in all downslope directions according to a flow law that was linear in area drained and less than linear in gradient (Horton, 1945, p. 321). An initially uniform gradient received randomly distributed net rainfall in each time interval, generating flow, sediment transport, and a network of valleys over time. The location of stream heads was arbitrarily assigned to points where accumulated runoff first exceeded $6 + 40/(\text{iteration number})$. Stream density was thus somewhat arbitrary, particularly as the sediment transport law provides neutral stability according to Smith and Bretherton's criterion (1). Sprunt's simulation nevertheless produced valuable results.

Ahnert's model is limited by a rather small grid size (10×10) and by introducing geological and other patterns into his model, which force the location and spacing of the resulting drainage pattern. Although such areal differences are of interest and relevance, they side-step the theoretical question of where drainage is located in the absence of evident control. Armstrong's work is for a larger grid (40×40), but begs the question of stream spacing by assuming a fixed network of channels and different transport laws for slopes and channels. Hugus and Mark (1984) also used different transport laws for streams and slopes but chose a critical drainage area for the transition from one to another, though with a discontinuous change from a stable slope sediment law to a stream law that is stable or unstable depending on the discharge exponent used.

Cordova et al. (1983) came much closer to solving the problem, in that they used a uniform area, or one with random perturbations only, and a uniform transport law. However, they employed a rather small grid size (16 × 16) and, more seriously, a transport law based solely on hydraulic transport. In effect, their flow law is equivalent to a power law of both area and gradient, with exponents greater than unity. The process modeled is therefore inherently unstable in the Smith and Bretherton sense, so that hollows tend to grow everywhere, without a divide area of smooth slopes. Their simulation, as might be expected, generated a valley network that obeys Horton's law of stream numbers. They also demonstrated that a uniform slope with no initial irregularities at all will develop no channels, and that a slope which is irregular in having a notch in the centre of its bottom edge will develop a single central channel. This result might also be argued from considerations of symmetry, although perhaps less convincingly! They recognized the vital importance of some pattern of irregularities, however small, to initiate perturbations that can grow into valleys, and their final simulation is based on a random assignment of the process rate parameter, which then remains constant at each point over time.

Model formulation

A rather simple model has been used here. It is based on transport-limited removal at a rate given by (2) constrained by a mass balance calculated for each point of a two-dimensional grid. Initial irregularities have been introduced through small perturbations of the initial topography, whereas the process has been held spatially uniform. The initial perturbations have been generated from a pseudo-fractal surface. These choices are discussed below.

There are good *a priori* reasons, though only slight empirical backing, for assuming that soil creep and rainsplash transport are proportional to slope gradient. For both processes, a proportionality constant of 10^{-3} m²/yr is of the correct order of magnitude. These processes are thought to be dominant near divides. It is therefore important to simulate these processes, if divides as well as channels are to be modeled.

For wash processes, a good fit to many data is obtained from empirical relationships of the form

$$C \propto q^m g^n \tag{3}$$

for constant exponents m and n, and a satisfactory fit may commonly be obtained with $m = 2$ and $n = 1$. These exponents are used here for annual over-land flow totals. A slightly higher exponent of discharge (2.5–3) is appropriate for individual storm discharges. Runoff in individual storms may be adequately modeled, in the absence of detailed antecedent moisture and intensity data, from daily rainfall data, which is commonly distributed, at least for each season, as

$$N(r) = N \exp(-r/r_0) \qquad (4)$$

where $N(r)$ is the number of days with rainfall greater than r, $N = N(0)$ is the total number of rain days, and r_0 is the mean rain per rain day = total annual rainfall divided by total number of rain days = R/N. The runoff on any day is estimated as a proportion λ of any rainfall exceeding a threshold r_c. Integrating over the frequency distribution, the total annual runoff production per unit area is then given by

$$F = \sum f = \lambda R \exp(-r_c/r_0) \qquad (5)$$

The effect of raising Q to the mth power may be compared with a storm-based sum

$$\sum (f^m) = m! \, (\lambda r_0)^m N \exp(-r_c/r_0) \qquad (6)$$

It may be seen that the λr_0 terms on the right-hand side of (5) are raised to the mth power, whereas the remainder are not. The effect is, in practice, similar to using a somewhat lower exponent for the annual total than for the sum of storm totals. Appropriate values for λ and r_c have been derived from Meginnis (1935) for some Mississippi loess soils and are shown in Table 10.1.

Conversion of runoff production to area has been done on the assumption of uniform production, so that total discharge per unit width is directly proportional to area per unit width, and unit discharge $q = aF$, where F is obtained from (5). Wash erosion is then proportional to $q^2 g$, with a constant empircally evaluated at about 0.017 yr/m^2. This gives a power law for the wash processes that is similar in its effect to the law used by Cordova et al., (1983) but contains an explicit summation over the frequency distribution of storm rainfalls.

The transport law used in this simulation takes the form of (2), which may be seen to be sum of a creep or splash term and a wash term, based on the

Table 10.1 Estimated daily rainfall before any overland flow runoff (r_c) and proportion of subsequent rainfall flowing overland (λ) for Mississippi loess soils.

Vegetation cover	λ	r_c (mm)
bare soil	0.60	4.7
grass	0.12	36.6
forest	0.025	37.6

Source: Meginnis (1935)

values quoted above. It may be seen that the distance u at which wash overtakes creep and splash processes is given by

$$u = 240 \, \exp(r_c/r_0)/(\lambda R) \tag{7}$$

where the annual rainfall R is expressed in millimeters and the critical distance u is in meters. This relationship was given by Kirkby (1976a) for the case of $\lambda = 1$. For an annual rainfall of 500 mm falling on 50 days per year, the values of λ and r_c tabulated in Table 10.1 give u values of 1.3, 155, and 825 m, respectively, representing a substantial part of the range of values implicit in the variety of drainage textures. During individual major storms, the tendency for hollow enlargement is expected to occur at unit areas much smaller than u, but periods of low intensity or no rainfall allow creep and splash to refill any rills thus formed. If the creep/splash and wash terms are disaggregated by storms, the short-term history or frequency distribution of cut and fill may be examined. The headward end of valleys is thus thought to consist of smooth slopes with a detailed cut-and-fill stratigraphy representing the relevant local historic sequence of large storms and intervening periods of slope infilling.

More complex models for hillslope runoff may be used, to allow for the development of partial saturated areas in humid soils. Some suggestions have been made (Kirkby, 1976a, 1978, 1980) in which the unit areas of (2) are replaced by terms of the form

$$\left(\int_0^x w\phi(x/g) \, dx \right) \Big/ w \tag{8}$$

for a suitable slope hydrology function ϕ, where w is the width of the profile strip at distance x from the divide. The case considered here corresponds to the simplest possible case of $\phi \equiv 1$. These more general cases may also be investigated within the framework adopted here.

Initial perturbations may be introduced either in the initial surface or in the process parameters. Each produces conditions that are inherited by the subsequent landforms. In the case of a perturbed topography, small perturbations generate an initial distribution of unit drainage areas that determine where erosion is initially most rapid. Erosion then smooths or enhances the topographic irregularities, and inheritance is therefore conditional on this evolutionary process. This topographic perturbation is preferred here because the effect of initial disturbances is not reimposed at subsequent times, so that the area modeled is treated as if it were fully uniform. If process parameters are varied over the model area, then systematically or randomly (as in Cordova et al., 1983) distributed differences are reimposed indefinitely. This is thought to be appropriate for describing geological differences, as in Ahnert's (1976) model, but appears to violate the assumption of a uniform area for landscape evolution.

A second problem with introducing initial perturbations lies in avoiding an initial pattern that deliberately or accidentally imposes a scale on the resulting drainage pattern. The solution adopted here is to use a deliberately scale-free perturbation, derived from a fractal surface, which is then superimposed at very low relief on a smooth initial surface.

Computational procedure

This model has been implemented, with some approximations, on a 32 kbyte microcomputer over a 64 × 64 grid. Figure 10.1 gives a skeleton flow diagram for the computational procedure. Elevation and other values are stored as 2-byte values, giving a resolution of about 4 mm, over a total elevation range of 250 m. The initial surface generated is in fact a pseudofractal obtained by successive mid-point interpolation with a suitably reducing variance about linearly interpolated values. Strictly this surface is only scale-free at the 2:1 scales used to generate it, but it is thought to be an adequate approximation for the present purpose. This surface is generated within a square, or distorted square (rhomboidal), frame at elevation zero; in terms of a central height, a standard deviation k, and an excess dimension h. The central point (32,32) is either specified or drawn from a normal distribution (generated as a sum of rectangularly distributed random numbers) with zero mean and standard deviation k, which is then reduced to

$$k' = k(1 - 2^{-2h})^{1/2} \tag{9}$$

The next set of four points is obtained by linear interpolation from the four corners of an upright square: for (16,16) from (0,0), (0,32), (32,0), and (32,32), and similarly for (48,16), (16,48), and (48,48). The interpolation is weighted inversely by distance to each corner if the square is distorted. To the interpolated value is added a normally distributed random variate of zero mean and standard deviation k'. The standard deviation is then reduced at this and successive interpolation scans in the ratio $2^{-(1-h)/2}$. The next four points are obtained by interpolating from the corners of a diagonal square: (32,16) from (16,16), (48,16), (32,0), and (32,32), and so on. Alternate scans thus interpolate alternately from the corners of the nearest upright and diagonal squares around them, with steadily reducing variance corresponding to the reducing interpolation span.

This initial surface is then linearly combined with a specified regular surface. In the examples shown, this surface is a uniform slope, falling to zero at the base. The range of elevations on the slope is scaled over the range 0 to 65 535 to give maximum resolution in the subsequent simulation. Each point in the grid is considered to be the mid-point of a square or rhombus. The following boundary conditions have been adopted. The upper edge of the grid ($y = 0$) is taken as a divide, with symmetrical removal on the opposing slope. The slope base ($y = 64$) is considered to be a base level of fixed position and zero elevation, at which all material is removed. The two sides of the slope ($x = 0$

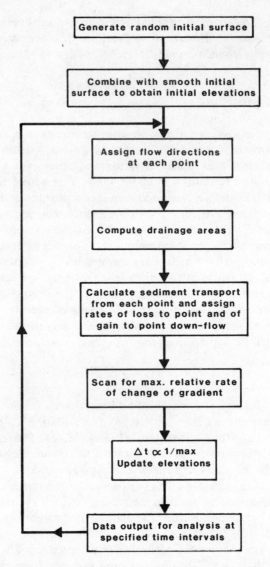

Figure 10.1 Outline flow diagram for two-dimensional simulation model.

and $x = 64$) are allowed to connect to one another in a circular fashion, so that drainage lines may go off at $x = 0$ and continue at $x = 64$ on the other edge. This condition is thought to provide less forcing of stream spacing than a condition of reflecting boundaries at the sides.

At each model time step, the elevation of each point is compared to that of all its immediate neighbors, and the direction of flow is assigned to the lowest of these points. All flow is then assumed to follow the path to the lowest neighbor or to sink underground in a closed depression. This approximate algorithm greatly speeds computation, but forces the drainage to adopt a dendritic network without distributaries, as flow is never divided between alternative routes. Using this procedure, surfaces on which flow is generally divergent tend to have a large number of internal sources, so that drainage areas on them are nowhere large, and wash processes tend to be insignificant using (2), in accordance with expectation.

In early runs of the model the simulation was carried out in a square grid, and flow was assigned from a point to any of its eight immediate neighbors or to a sink at the point itself. Although this provides a maximum number of possible flow directions, the simple algorithm used takes no account of the greater distance to the diagonal corners, which may therefore be assigned flow in cases where the gradient to some other point is greater. In later runs this difficulty was avoided by distorting the grid into a $60°$ rhombus. Flow was then assigned to one of the six equally near neighbors in the surrounding hexagon.

A second approximation, which is difficult to avoid in this type of model, is the assumption that flow occupies the entire width of an individual grid square. It may be seen, from the form of (2), that if flow becomes concentrated over a width much less than the dimension of a grid element, then the total sediment transported within that element may become arbitrarily large. It is therefore essential to maintain a grid element that is of the correct order of magnitude for observed channel widths. A grid element width of 10 m has been adopted here as a compromise between meeting this criterion and simulating over an adequate total area $640 \times 640 \text{ m}^2$.

Drainage area at each point in the grid is evaluated by starting at each element, and adding unity to the area of this element and to all elements to which it successively drains down the length of the slope. This procedure was continued from each starting point either to the base of the slope ($y = 64$) or to any enclosed depression (treated as a sinkhole for water). Sediment transport may now be evaluated from each grid element, using (2) or an alternative. Each element provides a rate of erosion for the element itself and deposition in the element to which it immediately drains. No sediment was considered to leave the lowest point of an enclosed depression, which therefore inevitably fills up. The computational procedure prevents the formation of enclosed depressions, and their treatment is only required to allow for the elimination of any which might be present in the initial surface.

This explicit solution procedure is stabilized by choosing a time step that allows gradient to change by a maximum of 50 percent (or a lower threshold) at the worst point. To allow slope reversals associated with divide migration and stream capture, this condition is not applied where points differ by less than 4 mm (i.e., gradient is calculated as zero for the computed resolution). This procedure for stabilizing the solution is a little less stringent than that suggested by the Neumann analysis, but gives a criterion of the same order of magnitude and meets the Neumann criterion at almost all points. The Neumann criterion is approximately given, for the two-dimensional case by the inequality

$$\Delta t/(\Delta x)^2 \leqslant 1/\{4k[1 + (a/u)^2]\}$$

This condition is met for the average iteration time for all values of a less than 6600, 6900, and 3940 m in runs 1, 2, and 3, respectively (i.e., well beyond the theoretical limit of unstable hollow growth). It is argued that the criterion used is in fact at least as safe as the Neumann criterion, as may be seen from examining the worst possible case of a divide or sink with equal slopes to or from all directions. The Neumann criterion is then exactly equivalent, in one or two dimensions, to the condition for no reversals of slope (i.e. a threshold of 100 percent). By selecting any lower threshold for rate of change of gradient, this essential component of stability is maintained, and the solution is in fact thought to be stable at all points. Comparison with exact solutions for one-dimensional slope profiles shows good agreement as gradient thresholds are altered and also shows that rounding errors associated with the 2-byte representation of values have been kept small.

Elevations are finally updated, and the next time step begun. At chosen intervals, the data matrices are stored for analysis of erosion rates, drainage areas and network patterns.

Example simulations

The results of three runs are presented here to illustrate the way in which the model behaves. In the first run, the grid consists of square cells, with flow possible to eight directions. In the other two runs the grid is rhomboidal, with flow possible to six directions in a hexagonal pattern. In each case the creep/splash rate constant (k in (2)) has been taken as 0.001 m^2/yr, the unit cell dimension is 10 m, and the initial surface consists of a perturbed uniform slope with a relief of 255 m, giving an initial gradient of 21.7° (40 percent) in run 1 and of 24.6° (46 percent) in runs 2 and 3. The main difference in the runs was in the critical threshold distance (u in (2)) beyond which transport due to wash is greater than that due to creep/splash, and beyond which hollows should grow in size. In the three runs, u has the values 5000, 1280, and 320 m, so that drainage from 500, 128, and 32 cells, respectively, is required to reach the threshold value.

Figure 10.2 shows the initial surface for run 1. Figure 10.2A shows the fractal surface with elevations from −20 to +35 m. In Figure 10.2B a uniform gradient with 255 m of total relief has been perturbed by subtracting 10 percent of the fractal values to give a surface with cross-relief of about 4 m. Drainage lines are also shown for cells with catchments larger than the critical value u. It may be seen that the topography has little obvious relationship to the position of the drainage lines, but that the random combination of lines

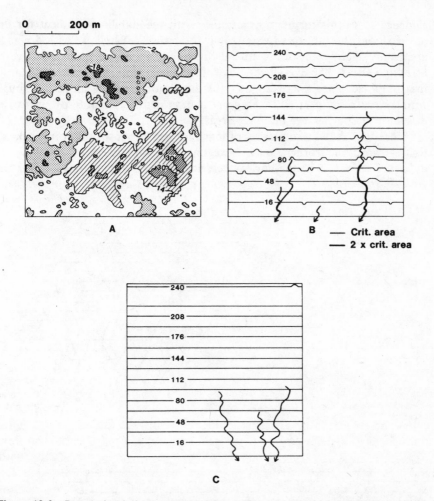

Figure 10.2 Run 1 simulation: (A) fractal surface used to perturb initial surface; (B) initial surface perturbed by subtracting 10 percent of the fractal in A; and (C) surface after 1.6 million years (176 iterations) and 0.86 m average denudation.

leads to areas of potential instability, even though this critical unit area (5000 m) is about eight times the average (640 m) at the slope base. In Figure 10.2C it may be seen that, after 1.6 million years the irregularities of the initial surface have been very much reduced, although the remaining small differences still define substantial drainage areas which frequently shift in position through processes similar to the micropiracy described by Horton (1945, p. 333–339) for rilled surfaces of low cross-relief. The general trend toward smoothing is also shown by the halving of the largest drainage areas present. A total of 0.86 m of denudation has taken place, largely through rounding of the upper convexity, with 15 m lowering of the upper divide. The evidence from this run is in accordance with the stability argument for the upper slopes, but appears to show that the growth of hollows with drainage areas only slightly above the critical threshold is too slow to be effective in forming valleys. The overall course of the slope evolution is closely similar to that obtained from the analogous one-dimensional profile model, although with a very slightly raised average near-base denudation rate owing to the existence of some larger drainage areas.

Figure 10.3 shows a similar set of maps for run 2, but with flow in six directions toward the corners of a hexagon, and with the grid consequently deformed into a 60° rhombus, as described above. Figure 10.3A shows the

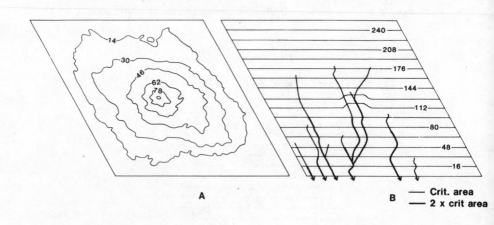

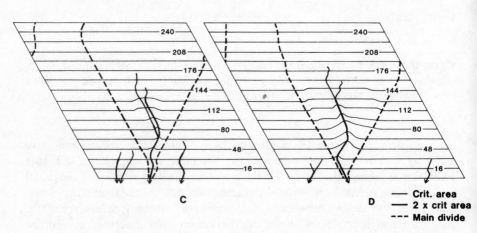

Figure 10.3   Run 2 simulation: (A) fractal surface with marked central peak; (B) initial surface perturbed by subtracting 5 percent of the fractal in A; (C) surface after 40 000 years (52 iterations) and 0.08 m average denudation; and (D) surface after 240 000 years (287 iterations) and 0.43 m denudation.

original fractal surface with a marked central peak of 100 m elevation. In
Figure 10.3B the initial uniform gradient has been perturbed by substracting
5 percent of the fractal surface, producing an initial surface with a 5 m central
depression. The most significant difference in conditions for run 2 is the
reduction of the critical unit area u to 1280 m, about a quarter of its value in
run 1. In consequence, the initial network defined by drainage areas above the
critical value is denser and clearly follows the central depression. In Figures
10.3B and 10.3C, after respectively 40 000 and 240 000 years and 0.08 m and
0.43 m denudation, the process of upper slope smoothing has proceeded as
before, but at the same time the central valley has gradually developed into a
subtantial erosional feature with strong channel incision. There has also been

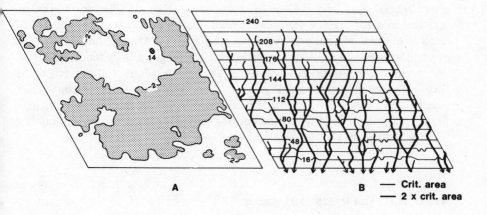

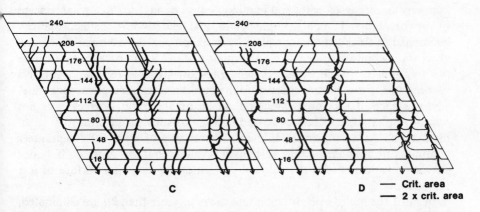

Figure 10.4 Run 3 simulation: (A) fractal surface of low relief; (B) initial surface
perturbed by subtracting 5 percent of the fractal in A; (C) surface after 20 000 years
(118 iterations) and 0.34 m denudation; and (D) surface after 39 000 years (238
iterations) and 0.63 m denudation.

some evolution of side-walls with valley widening and the establishment of convexities along the valley rim. Some smaller streams have been eliminated, and at the same time there has been a slight enlargement of the largest catchment area. Examining detailed cross-profiles, clear evidence of crenulation may be seen only for cells with drainage areas greater than about *twice* the critical unit area. This criterion is equally applicable for all times at which crenulations are clearly present (from about 20 000 years in this case).

In Figure 10.4 (run 3) the critical unit area has once more been quartered to 320 m. Flow directions are again hexagonal, as in run 2. In Figure 10.4A an irregular fractal surface with 30 m total relief is used to perturb the initial surface (Fig. 10.4B), giving it a maximum cross-relief of 2 m. The lower critical area produces a dense network of channels, as defined by initial drainage areas. After 20 000 years and 0.34 m of denudation (Fig. 10.4C), and after 39 000 years and 0.63 m of denudation (Fig. 10.4D), smoothing of interstream areas may be seen together with incision of the largest streams at or close to their original positions, and a reduction in the number and density of small streams as a result of capture of their drainage areas by the largest ones. As in run 2, the threshold for observable crenulation is about twice the theoretical critical unit area. Figure 10.4 also shows one clear example of a channel and valley bifurcation. Given the 40 percent initial gradient, it is perhaps not surprising that stream courses show a strong downslope trend that reduces tendencies toward bifurcation.

Analysis of simulated landscapes

The example runs above are far from exhaustive, but show some clear patterns that may be compared with, and in minor ways extend, the theoretical concept of valley stability. The only simple operational definition of channels that can be applied to these simulations must be based on contour crenulations. In all cases this definition appears to correspond to unit areas of about twice the theoretical value (u). For the sediment transport law used (equation (2)), this corresponds to a wash transport of four times the splash/creep transport rate. The existence of unit areas greater than $2u$ on the initial surface does not appear to guarantee effective incision because of subsequent piracy of catchment areas. As a rough rule of thumb a unit area of $4u$ usually guarantees continuity of drainage lines to develop into valleys, but this criterion is a poor substitute for examining the detailed pattern of capture on a surface of low cross-relief.

In run 1 the initial cells draining unit areas of more than $2u$ are eliminated, and no areas drain as much as $4u$: little or no coherent pattern may be seen in maps of erosion. In run 2 the main valley has a basal unit area of more than $10\,240\text{ m}^2$. A second catchment, flowing out near the center of the base, initially exceeds $5120\text{ m}^2(4u)$. After 40 000 years it still survives with a crenulation, but with a halved catchment area as a result of headwater capture by the main stream. After 240 000 years further capture has reduced its unit catchment area to 640 m^2, and the crenulation has filled. This example may be seen as a model for headwater stream capture: diversion of unchanneled divide areas leads to a loss of collecting area for the disadvantaged stream head, which falls below the threshold for hollow enlargement and so fills with slope sediment with a consequent loss of stream length. Other examples of this process may be seen in run 3, although less clearly because of the greater density of channels. A probable candidate for subsequent capture may be seen near the lower left corner of Figures 10.4B, C, and D.

The continuity equation following a flow-line in the downslope direction x may be written in the form

$$-\partial z/\partial t = \partial S/\partial x - S/\rho \qquad (10)$$

where z is elevation, t is elapsed time, S is sediment transport along the flow path, and ρ is the radius of contour curvature, taken to be positive in hollows. The left-hand side of the expression is thus the rate of lowering at any point. Substituting a slope transport law

$$s = f(a)g \qquad (11)$$

which is linear in gradient g for a suitable function f of unit area a, and substituting the geometrical identity $\partial a/\partial x = 1 + a/\rho$ leads to

$$-\partial z/\partial t = g\partial f/\partial a + f\partial g/\partial x + (g/\rho)(a\partial f/\partial a - f) \qquad (12)$$

On the right-hand side of this expression, the first term is always zero or positive and the second is positive on profile convexities and negative on concavities. The sign of the final term depends on both the direction of plan curvature (ρ) and the term $a\partial f/\partial a - f$: overall it is positive for unstable growth of hollows and negative for stable infilling. Because small perturbations in the surface influence curvature very much more strongly than area or gradient, the final term in (12) provides a derivation for the stability criterion. The magnitude of this term determines the rate at which hollows erode within the unstable area or, in other words, the sensitivity of the landscape to the degree of instability. For the transport law used here in (2), then

$$f(a) = k(1 + a^2/u^2) \qquad (13)$$

giving for the stability term

$$a\partial f/\partial a - f = k(a^2/u^2 - 1) \qquad (14)$$

which takes the value of $-k$ on divides ($a = 0$), zero for unit area u, and $+3k$

for unit area $2u$, which has been seen above to be a good working definition for a channel head demonstrating crenulation. For the initial near-uniform slopes used in the simulations, the rate of smoothing near divides appears low, but may be seen to be strongly dominant over other terms, particularly until marked convexities develop. Downslope the hollow enlargement term needs to be substantial to dominate the first term on the right-hand side of (12). As initially small hollows grow, enlargement of the resulting stable valley is limited by the increase in profile concavity, which acts through the second term on the right-hand side. Any subsequent loss of drainage area as a result of slope capture then leads to partial or total infilling of the valley.

Figure 10.5 shows some aspects of the course of evolution of the model landscapes. In all three runs there is a reduction in drainage density over time, as measured by the length of channels draining unit areas of at least $2u$. For run 1, the channel length finally drops to zero, after some fluctuations, by 600 000 years. For run 2, the length appears to have stabilized quickly, perhaps because of the relatively definite valley in the initial surface. The final channel length is estimated as 350 m. This length has been converted to a drainage density by allocating it the area (2280 out of the 4096 cells) of the two catchments containing the streams, as headwater divide areas typically underestimate density, giving a density of 1.77 km/km^2, or an average stream spacing of 560 m. In run 3 drainage density also declines over time, but more slowly, and tends toward an estimated total length of 2600 m. This represents a drainage density (over the full area) of 7.33 km/km^2, or an average stream spacing of 136 m. These stream spacings are in ratio of 4.1 : 1, which is close to the ratio of 4 : 1 for the critical unit areas (u). This result gives support to the concept that stream spacing should be directly proportional to unit critical area. The empirical constant of proportionality of 0.43 rather than unity is thought to reflect, first, the problems of using an operational definition of stream heads and, second, the catchment shape, which for these rather early stages of development may remain strongly influenced by the overall downslope gradient tending to elongate drainage basins. If the constant of proportionality is used to extrapolate for run 1, then the expected stream spacing of 2200 m is consistent with an absence of channels in the area simulated (640 m × 640 m).

Average denudation rates are also shown in Figure 10.5. Initial differences between the runs reflect the different cell shape for run 1, but also the difference in the average basal sediment transport rates forecast by (2) and summed over the distribution of unit areas. Comparing values for runs 2 and 3, the initial denudation rates are compatible with a coefficient of variation for slope—base unit area of about 2.5 around its mean of 640 m. In detail, the distribution of unit areas is similar to that predicted by the random network model (Kirkby, 1976b, Table 1) applied to all cells, with truncation of the distribution for large magnitudes to satisfy the constraint of finite mean magnitude. As the model landscape evolves, Figure 10.5 shows a very slight increase in denudation rate for run 1, and a marked decrease for runs 2

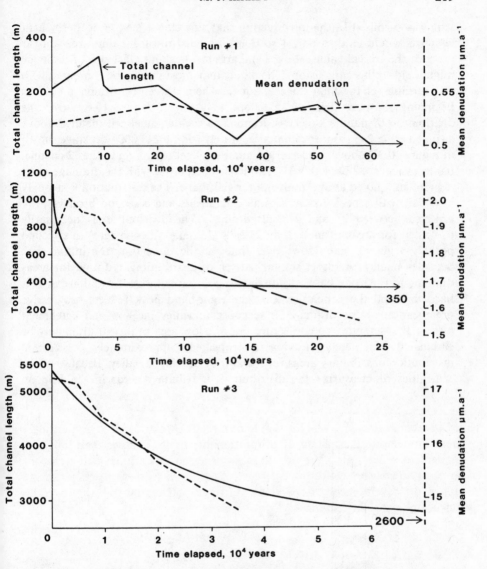

Figure 10.5 Changes in effective channel length (unit area $\geqslant 2u$) and mean denudation rate over time in (A) run 1, (B) run 2, and (C) run 3.

and 3. In run 1 it is thought that the overall smoothing of the landscape has steepened the gradients of random initial depressions, so that with the wash term everywhere negligible this infilling of depressions is the dominant influence. In runs 2 and 3 the development of marked valleys produces lower gradients along their axes, so that the initially high sediment outputs from the highest drainage areas, which dominate total output, progressively decrease over time.

It was surmised in the introduction that side slopes leading to permanent channels would tend to adjust so that all tended toward a unit area not far short of the critical value, and general arguments about efficiency of drainage might tend to this conclusion. The model may most readily be compared with this surmise for the final stage of run 2, where the central drainage basin is beginning to show signs of side-slope adjustment. Figure 10.6 shows the distribution of numbers of cells draining to each channel cell (defined above as unit area $\geqslant 2u$) from each bank separately for the 240 000 year stage shown in Figure 10.3D above. The average number of cells draining to the 72 channel banks (36 cells $\times$ 2 banks) is 31.7, which is compatible with the drainage spacing (560 m) quoted above, allowing for cell shape. The distribution is strongly bimodal, with a peak for very small catchments and a second broader peak centered between 2^5 and 2^6 cells drained. The distribution is necessarily truncated for areas of more than 2^9 cells. Initially (at time zero) this second peak is negligible and grows over time; at the same time the initial peak becomes smaller. A rather similar pattern of distributions is found for areas draining to the slope base, although with a broader peak of small areas. It is hard to avoid the conclusion that the right-hand peak reflects progressive organization of the drainage in response to valley incision and catchment piracy. It is tempting to conjecture that the low-area peak will ultimately be eliminated completely, leaving a truncated, approximately log-normal distribution of tributary areas for a maturely developed valley. Because of the difficulties of measuring the distribution of tributary areas in the field on

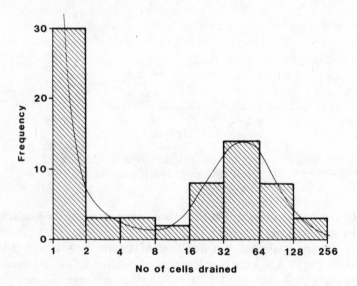

Figure 10.6 The distribution of catchment areas tributary to effective channels for the 240 000 year stage of run 2.

slopes which necessarily have only very slight cross-relief, a derivation through modeling may have considerable relevance to problems of rill location and overland flow hydrology (for the transport equation adopted here).

Conclusions

The simulations carried out support Smith and Bretherton's (1972) theory of slope and valley stability as an effective mechanism for the formation of valleys. It appears to lead to a stable drainage pattern with a texture that may be related to sediment transport laws through the theory. For a given unit critical area u, the unit area needed to support an eroded channel is about $2u$, at which point wash is four times more effective than splash/creep for the process law used. The mean distance of overland flow streams stabilizes at about $0.22u$. It is conjectured that this may be distributed approximately log-normally when the catchment is fully integrated.

There remain a number of uncertainties about the interpretation of the model used here, even though it appears to perform realistically overall. The most serious problems are concerned with the significance of the unit cell size, and with the assumption of a dendritic network throughout by carrying flow from each cell to only one other. The choice of unit cell size inevitably conceals some distribution of unit areas within each cell, which may, have substantial effects if flow threads are very much narrower than the individual cell dimension. Some confidence is given by the constant of proportionality of drainage spacing to unit critical area, but the problem must remain as a possible source of error. The use of small cells plainly increases accuracy, but at the expense of the total area that may realistically be modeled.

The assumption of dendritic drainage patterns throughout has proved advantageous in simplifying drainage area algorithms and is thought to be effective in practice, but may be a source of error in establishing the size and development of hillslope areas. In practice the dendritic assumption is usually seen to be correct for drainage areas large enough to carry stream flows and substantial sediment loads, so that its adoption will not introduce errors in macroscopic channels. For an alluvial fan context too, the dendritic assumption will suppress distributaries but will allow channel shifting, so that a fan may be built up in a reasonable way. The biggest possible error is for unchanneled slopes where the dendritic assumption may severely overestimate tributary areas and so distort the kind of distribution shown in Figure 10.6. This problem remains unresolved in the present context and is a topic for further research.

The transport law used here (equation (2)) is thought to be appropriate for semiarid rather than humid areas. It may be seen that under these assumptions the drainage density is strongly influenced by climate, together with hydrological properties probably dominated by vegetation influences, while gradient has little or no influence. If the same arguments are carried through to well vegetated humid areas where subsurface flow is important, drainage density should respond most sensitively to soil properties and decrease as slope gradient increases, other things being equal (Kirkby, 1980). The present paper is thought to make a small contribution to the problem of how slope and channel processes interact to scale the landscape. Much remains to be done, both at a theoretical level and in the field. The topic is a central one in process geomorphology, and merits a greater research effort.

References

Ahnert, F., Brief description of a comprehensive three-dimensional process–response model of landform development, *Zeitschrift für Geomorphologie Supplement Band, 25,* 29–49, 1976.

Armstrong, A. C., A three-dimensional simulation of slope forms, *Zeitschrift für Geomorphologie Supplement Band, 25,* 20–28, 1976.

Carson, M. A. and M. J. Kirkby, *Hillslope Form and Process,* 475 pp., Cambridge University Press, Cambridge, 1972.

Cordova, J. R., I. Rodriguez-Iturbe, and P. Vaca, On the development of drainage networks, Recent Developments in the Explanation and Prediction of Erosion and Sediment Yield, *International Association for Scientific Hydrology Publication 137,* 239–249, 1983.

Horton, R. E., Erosional development of streams and their drainage basins; hydrophysical approach to quantitative morphology, *Bulletin of the Geological Society of America, 56,* 273–370, 1945.

Hugus, M. K. and D. M. Mark, Spatial data processing for digital simulation of erosion, *Technical Papers of the Fall Convention American Society of Photogrammetry/American Congress on Surveying and Mapping,* 683–693, 1984.

Kirkby, M. J., Hydrological slope models: the influence of climate, in *Geomorphology and Climate,* edited by E. Derbyshire, pp. 247–267, John Wiley, Chichester 1976a.

Kirkby, M. J., Tests of the random network model, and its application to basin hydrology, *Earth Surface Processes, 1,* 197–212, 1976b.

Kirkby, M. J., Implications for sediment transport, in *Hillslope Hydrology,* edited by M. J. Kirkby, pp. 325–363, John Wiley, Chichester, 1978.

Kirkby, M. J., The stream head as a significant geomorphic threshold, in *Thresholds in Geomorphology,* edited by D. R. Coates and J. D. Vitek, pp. 53–73, George Allen and Unwin, London, 1980.

Meginnis, H. G., Effect of cover on surface run-off and erosion in the loessial uplands of Mississippi, *United States Department of Agriculture Circular 347,* 1935.

Melton, M. A., An analysis of the relations among elements of climate, surface properties, and geomorphology, *Technical Report 11, Office of Naval Research,* Project 389–042, Department of Geology, Columbia University, New York 1957.

Smith, T. R., and F. P. Bretherton, Stability and conservation of mass in drainage basin evolution, *Water Resources Research, 8,* 1506–1529, 1972.

Sprunt, B., Digital simulation of drainage basin development, in *Spatial Analysis in Geomorphology,* edited by R. J. Chorley, pp. 371–389, Methuen, London, 1972.

Part III:
Gravitational processes

11
Controls on the form and development of rock slopes in fold terrane

B. P. Moon

Abstract

Recent studies of the form and development of rock slopes have focused on the properties of materials as geomorphic controls. In the absence of structural controls, rock mass strength has been found to be an important determinant of slope form. In the fold mountains of the southern and southwestern Cape Province, South Africa, rock slopes on quartzite have developed to forms controlled by rock mass strength (strength equilibrium slopes), structure, and denudational processes. Contingency table analysis indicates that the rock properties of greatest significance as determinants of the form of strength equilibrium slopes are intact rock strength and the spacing of joints and other geological discontinuities. On undercut, unbuttressed rock masses and on buttressed (dip slope) forms, parting roughness is a key factor in slope development by virtue of the effect of parting roughness on shear strength. Roughness measurements, tilt tests, and field observation indicate that the effective friction angle of the quartzite is approximately 55°.

Introduction

Within the general theme of slope studies a subfield of inquiry has developed aimed at understanding the morphology and development of slopes formed on bedrock. The work of Lehmann (1933) and Bakker and Le Heux (1946, 1952) furthered the pioneering work of Fisher (1866) in producing models of rock slope morphology. Considerations of rock slope development are incorporated in the more general models of Penck (1924), Wood (1942), and King (1953); and Klimaszewski (1971) has considered rock face development in relation to various climatic conditions. In these examples the emphasis is on processes as controls of rock slope development. It is only very recently, however, that geomorphologists have followed the lead provided by Terzaghi

(1962) in turning their attention to rock mass properties as controls on rock slopes.

In general, a variety of rock properties has been demonstrated to be of geomorphic importance. For example, rock hardness has been found to be significant in the development of the landscape (Day, 1981) and in determining the susceptibility of limestones to weathering (Day, 1978). In addition compressive strength, elasticity, and the degree of rock weathering have been shown to exert control on such morphometric variables as drainage density, hypsometric integrals, and the gradient of valley-side slopes (Cooks, 1979, 1981, 1983). More pertinent to the study of rock slopes, however, is the recent development of a technique for the field determination of rock mass strength (Selby, 1980). Based on engineering rock mass strength classifications, the technique allows for the expression of rock mass strength (or the resistance of a rock mass to surface processes) on a scale from 25 to 100 (Table 11.1). The

Table 11.1 Geomorphological classification and ratings (r) of rock mass strength determinants.

Parameter	1 Very strong	2 Strong	3 Moderate	4 Weak	5 Very weak
Intact rock strength (N-type Schmidt hammer 'R')	100–60 $r:20$	60–50 $r:18$	50–40 $r:14$	40–35 $r:10$	35–10 $r:5$
weathering	unweathered $r:10$	slightly weathered $r:9$	moderately weathered $r:7$	highly weathered $r:5$	completely weathered $r:3$
spacing of joints	> 3 m $r:30$	3–1 m $r:28$	1–0.3 m $r:21$	300–50 mm $r:15$	< 50 mm $r:8$
joint orientation	very favourable; steep dips into slope; cross-joints interlock $r:20$	favourable; moderate dips into slope $r:18$	fair; horizontal dips or nearly vertical (hard rocks only) $r:14$	unfavourable; moderate dips out of slope $r:9$	very un-favourable; steep dips out of slope $r:5$
width of joints	< 0.1 mm $r:7$	0.1–1 mm $r:6$	1–5 mm $r:5$	5–20 mm $r:4$	> 20 mm $r:2$
continuity of joints	none continuous $r:7$	few continuous $r:6$	continuous; no infill $r:5$	continuous; thin infill $r:4$	continuous; thick infill $r:1$
outflow of groundwater	none $r:6$	trace $r:5$	slight (<25 l/min per 10 m^2) $r:4$	moderate (25–125 l/min per 10 m^2) $r:3$	great (>125 l/min per 10 m^2) $r:1$
total rating	100–91	90–71	70–51	50–26	< 26

Source: Selby (1980).

rock mass strength rating is gained from the measurement or numerical rating of the following rock properties: the strength of intact rock, the degree of weathering, the spacing, orientation, width, and continuity of joints, and the movement of groundwater out of the rock mass. For simplicity, geological partings of any description are referred to as joints, unless otherwise specified. Applications of the technique in studies of rock slope form in a variety of environments and on many different lithologies have demonstrated the widespread occurrence of strength equilibrium slopes—that is, rock slopes developed to gradients that are in equilibrium with the mass strength of the

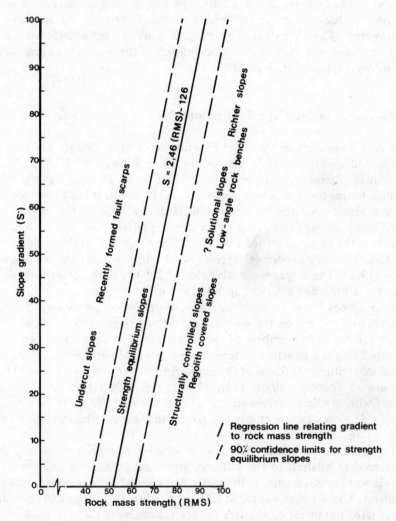

Figure 11.1 The relationship between gradient and rock mass strength. Strength equilibrium data plot between the broken lines, by definition.

slope-forming material (Selby, 1980, 1982a, 1982b; Moon and Selby, 1983). In addition, measurements of rock mass strength facilitate the interpretation of the dominant control of slope form, whether process, structure, or rock mass strength (Fig. 11.1).

The application of rock mechanics theory to the development of bedrock slopes has been taken still further. Selby (1982c) identified the controls of slope morphology in rock slopes of several different types: (1) those in strength equilibrium; (2) slopes on buttressed and unjointed rock masses that develop toward strength equilibrium as cross-joints open; and (3) rock slopes that are characterized by joints having a steeper dip than the slope gradient and are subject to failure along critical joints. The present chapter describes a similar analysis of rock slopes in the fold mountain terrane of the southern and southwestern Cape Province. This analysis leads to a better understanding of rock resistance and joint roughness as controls of the morphology and stability of different types of rock slopes.

Rock slopes in the Cape fold mountains

The fold mountains of the Cape Province have developed on the Paleozoic shales and quartzitic sandstones of the Cape Supergroup (Fig. 11.2). The Hercynian (Permo-Triassic) orogeny resulted in comparatively gentle folding along a north–south axis in the west; but in the south, folding has been more intense along east–west axes. Erosion since the Triassic has etched out a mountain landscape consisting of anticlinal quartzite ridges and homoclinal ridges on the larger breached anticlines. The fidelity with which the relief of the Cape mountains reflects the geology and fold structures has been noted by King (1963). The ridges reach altitudes of 2000 m in the Swartberg Range, which is a breached anticline up to 15 km wide.

Rock slopes in the Cape mountains are developed predominantly on the quartzite ridges, but on the semiarid northern margin, rock slopes have also developed on shale members of the succession. The bedrock slopes may be classified into five structure-dependent categories (Moon, 1984): the scarp and dip slope (buttressed) forms of the homoclinal ridges (Figs. 11.3A and 11.3C), low-angled erosional slopes (Fig. 11.3B), eroded dip slope (unbuttressed) forms with gradients exceeding the dip of the strata (Fig. 11.3D), and the steep side slopes of the transverse gorges that dissect the mountain ranges (Fig. 11.3E).

Twenty-three mountain slopes were selected as representative of slopes developed in relation to the different structural conditions observed. Each profile was surveyed with a clinometer and divided into its constituent slope elements. This division was based primarily on breaks of slope and secondarily (in isolated instances) on changes in rock characteristics across an element of uniform gradient. On each slope element the rock mass strength rating was determined (Fig. 11.4).

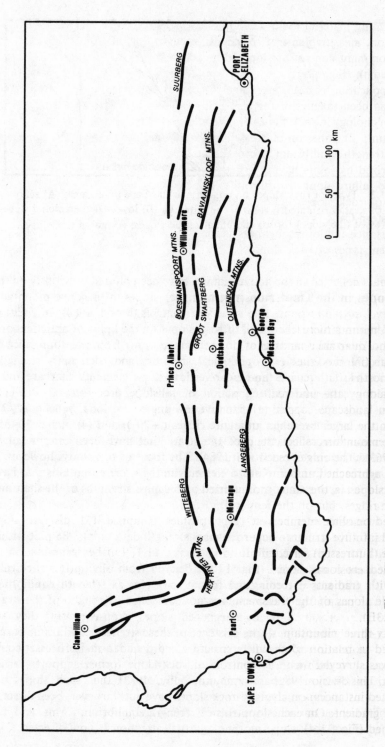

Figure 11.2 The major mountain ranges of the southern Cape Province, South Africa.

SUURBERG

PORT
ELIZABETH

BAVIAANSKLOOF MTNS.

BOESMANSPOORT MTNS.

Willowmore

GROOT SWARTBERG

OUTENIQUA MTNS.

Prince Albert

Oudtshoorn

George

Mossel Bay

LANGEBERG

WITTEBERG

Montagu

HEX RIVER MTNS.

Clanwilliam

Paarl

CAPE TOWN

km

100

50

0

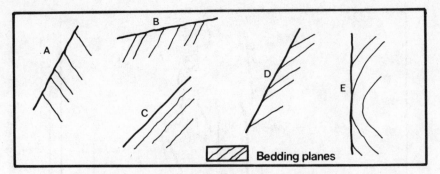

Figure 11.3   Types of rock slopes on quartzite in the Cape mountains: (A) scarp slopes with strata orientated favorably for stability; (B) low-angled erosional slopes; (C) buttressed (dip-slope) forms; (D) unbuttressed slopes; (E) gorge side slopes.

The data collected in the above manner have been plotted previously on the graph in Figure 11.1 to demonstrate that rock slopes in the Cape mountains have developed in response to a variety of controls (Moon, 1984). In addition to the obvious structural control of rock slopes on the limbs of anticlines and synclines, there are a number of slope forms resulting from the dominance of denudational processes over geological structure and rock mass strength. These include (1) vertical and near-vertical slope elements that are over-steepened by the undercutting action of hillslope processes; (2) fluvially undercut gorge side slopes; (3) extensive low-angled erosional benches (2–7°) and footslopes of the major quartzite ridges (~20°); and (4) planar 30–35° slopes eroded across the trend of the strata that have been interpreted as Richter denudation slopes (Moon, 1984). The focus of this study, however, is on the significant number of slope elements in the Cape mountains that have developed in response to control exerted by the mass strengths of the shale and quartzite.

The twenty-three surveyed slope profiles comprise 151 discrete slope elements. Of the 151 gradient–rock mass strength data points, 86 plot within the limits for strength equilibrium in Figure 11.1. Further examination of these 86 data points reveals that 17 are derived from obviously structurally controlled (dip slope) forms, and fortuitously plot as strength equilibrium slopes. The remaining 69 elements are not controlled by the dip of the strata nor are they erosional or oversteepened slopes. Consequently, they are strength equilibrium slopes. The existence of these strength equilibrium forms in a fold mountain region, where structure and denudational processes could be expected to dominate as controls of rock slope forms, supports earlier contentions (Selby, 1982c; Moon and Selby, 1983) that such slopes are widespread. Previous analyses of rock slopes have not, however, been directed at establishing the nature of controls on strength equilibrium forms, and it is to this aspect of rock slope morphology that attention is now directed.

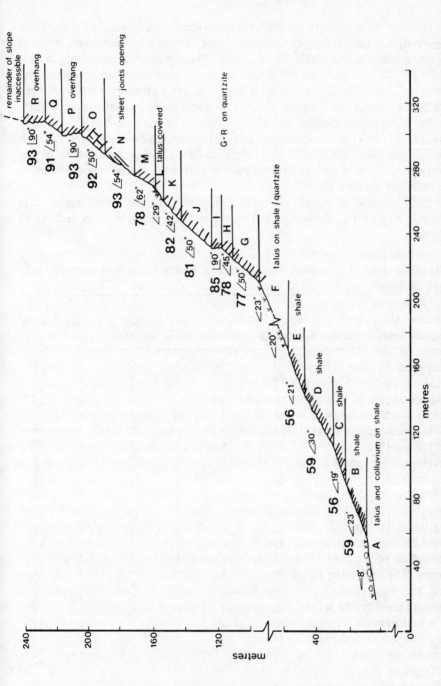

Figure 11.4 A mountain slope profile in the Swartberg Range. The rock mass strength rating for each element is shown in bold figures.

Rock resistance as a control of form

Slope elements in strength equilibrium are evident on almost all of the profiles surveyed, but they are predominant on rock masses with joints and bedding planes orientated favorably for stability. These elements range in gradient from 19° to 90° and are developed in rocks with mass strengths varying between 56 and 93. There is a high degree of correlation ($r_s = 0.81$) between gradient and rock mass strength, which is comparable to that found in other studies of strength equilibrium slopes. Changes in gradient are all related to changes in rock mass strength. The data derived from the 69 strength equilibrium elements (Table 11.2) were analyzed to ascertain which of the factors contributing to mass strength are important as morphological controls in the Cape mountains.

Given the ordinal nature of rock mass strength data, contingency table analysis was used to determine the degree of association between individual

Table 11.2 Rock mass strength data from strength equilibrium slope elements in the Cape mountains.

Gradient	Intact strength	Weathering	Joint spacing	Joint orientation	Joint width	Joint continuity	Ground water	Total strength rating
88	18	9	24	14	5	6	5	91
74	14	9	21	14	5	6	5	74
90	15	9	29	14	5	6	5	83
54	13	9	19	14	4	5	5	69
60	14	9	21	14	4	5	5	72
47	11	9	21	14	4	5	5	69
68	14	9	21	18	4	6	5	77
69	14	9	26	14	4	6	5	78
74	16	10	24	14	4	6	5	79
85	16	10	30	14	2	6	5	83
47	13	10	18	14	4	6	5	70
85	14	10	30	14	2	6	5	81
47	13	10	18	14	4	6	5	70
86	14	10	30	14	2	6	5	81
68	16	10	19	18	6	6	5	80
78	18	10	18	18	6	6	5	81
72	17	10	19	14	6	6	5	77
70	18	10	18	14	6	6	5	77
85	17	10	30	18	6	6	5	92
78	19	10	20	14	6	6	5	80
85	18	10	24	14	6	6	5	83
23	6	9	8	18	7	6	5	59
19	5	7	8	18	7	6	5	56
30	6	9	8	18	7	5	6	59
21	5	7	8	18	7	5	6	56

Table 11.2 Continued.

Gradient	Intact strength	Weathering	Joint spacing	Joint orientation	Joint width	Joint continuity	Ground water	Total strength rating
50	11	9	22	18	5	6	6	77
45	16	9	22	14	5	6	6	78
90	19	9	22	18	5	6	6	85
50	18	9	20	18	5	5	6	81
62	15	9	21	18	5	5	6	78
90	19	10	30	18	5	5	6	93
90	19	10	30	18	5	5	6	93
30	12	9	18	14	5	5	6	69
70	19	9	29	18	4	6	6	91
48	12	9	23	18	5	6	6	79
85	19	9	26	18	5	6	6	89
24	11	9	16	14	6	6	6	68
80	17	9	29	18	6	6	6	91
63	19	9	24	18	6	6	6	88
90	18	9	26	18	6	6	6	89
80	17	9	30	14	5	6	6	87
80	19	9	30	14	7	6	6	91
80	18	9	26	14	6	6	6	85
26	7	9	21	14	6	6	6	69
23	5	9	14	20	6	6	6	66
90	9	9	30	20	6	6	6	86
38	6	9	26	14	6	6	6	73
33	7	9	13	14	6	6	6	61
32	7	9	13	14	6	6	6	61
53	13	7	29	9	5	6	6	75
49	15	7	30	5	5	5	6	73
43	17	9	14	18	6	6	6	76
60	17	9	11	20	6	5	6	74
65	17	9	11	20	6	6	6	75
70	17	9	22	20	6	6	6	86
53	18	9	29	9	6	6	6	83
48	19	9	21	9	5	7	6	76
51	16	9	26	9	6	6	6	78
73	19	9	21	9	6	6	6	76
64	19	9	30	9	6	6	6	85
90	19	9	29	18	6	6	6	93
65	18	9	21	5	6	5	6	70
90	19	9	28	5	6	6	6	79
62	18	9	28	18	6	6	5	90
68	17	9	16	18	6	6	6	78
87	17	9	28	9	7	6	6	82
41	17	9	18	5	6	6	6	67
51	19	9	19	5	6	6	6	70
67	19	9	29	5	6	7	6	81

rock mass properties and slope gradient. The analysis was complemented by calculation of Spearman's rank correlation coefficient r_s between individual rock properties and gradient. For contingency table analysis, gradients were divided into gentle, intermediate, and steep classes, and strength data were classified into weak, intermediate, and high strength categories. The analysis was conducted using the Statistical Analysis Systems program, which measures the degree of association between gradient and strength variables using the chi-square statistic, the ϕ coefficient, and a contingency coefficient.

The chi-square test was found to be inappropriate as most of the contingency tables have more than 20 percent of their cells with expected counts of less than 5. Consequently values of the ϕ coefficient were used in preference, and these are ranked in Table 11.3.

The analysis indicates a low degree of association between gradient and all of the strength variables except intact rock strength and joint spacing. These results reflect both the general uniformity of weathering and groundwater conditions on slopes of varying gradients and the low degree of association between gradient and joint orientation. That strength equilibrium slopes develop preferentially where joints are orientated favorably for stability reduces the importance of both joint width and joint continuity because under these conditions even continuous, open joints will not lead to slope failure. On slopes where all joints are orientated favorably for stability, variation in that orientation should not be expected to be an important control of form. With these points in mind it is reasonable to conclude that in the Cape mountain ranges, intact rock strength and the spacing of partings are the dominant influences on the form of strength equilibrium slopes.

That joint orientation does not emerge as an important control of the form of the slopes studied has implications for the method of determining rock mass strength. It is suggested that further work be directed at determining whether

Table 11.3 Results of contingency table and correlation analyses of strength–gradient data.

	Relationship to gradient	
Strength variable	ϕ coefficient	r_s
intact strength	0.826	0.597
joint spacing	0.675	0.655
weathering	0.458	0.403
joint width	0.440	n.s.[a]
joint continuity	0.351	n.s.
joint orientation	0.304	n.s.
groundwater	0.263	n.s.
total rock mass strength rating	0.911	0.813

[a] n.s. denotes that r_s is not significantly different from zero at the 0.05 level.

strength ratings for joint orientation should be based on a single value for partings orientated favorably for stability. The remaining values on the scale could then reflect progressively weaker conditions as partings become orientated at steeper gradients out of the slope. Another attribute of joints that is treated inadequately in rock mass strength determinations in geomorphology is joint roughness. In Selby's (1980) original presentation of the technique, roughness is considered only cursorily in conjunction with joint orientation (Selby, 1980, Table 4). The rock slopes on buttressed and unbuttressed rock masses in the Cape mountains provide an opportunity to assess the role of joint roughness in the morphology of rock slopes.

Joint roughness as a control of slope morphology

Shear strength along geological discontinuities is widely recognised as a determinant of rock slope stability. Joint shear strength is enhanced by roughness of the joint walls in the absence of infilling material (Barton and Choubey, 1977; Patton, 1966). The consideration of joint roughness is essential to understanding the Cape mountain rock slopes. Moreover, the proposition that the stability of an unbuttressed rock mass is determined by the shear strength along the weakest continuous joint (Selby, 1982c) can be tested. Many of the significant partings on rock slopes in the southern Cape are bedding planes. It is possible to demonstrate the effects of failure along these planes on the form of both buttressed and unbuttressed slopes.

In order for sliding to occur along a single plane a number of conditions must be satisfied (Hoek and Bray, 1977): (1) the plane on which sliding occurs must strike parallel to, or nearly parallel to, the slope face; (2) the dip of the failure plane must be less than the gradient of the slope face; (3) there must be weaknesses that determine the lateral boundaries of the slide; and (4) the dip of the failure plane must be greater than the angle of friction along the plane. On unbuttressed slopes in the Cape mountains the first three of these conditions are met, although weaknesses (cross-joints) determining lateral boundaries are absent from some elements. Whether failure of such slopes occurs therefore depends upon the opening of joints across the strike and, more particularly, upon the gradient of bedding planes relative to the angle of friction.

The angle of friction (ϕ) for sawn samples of quartzite has been determined to be between $30°$ and $40°$. However, naturally occurring partings in rock are seldom planar, and Patton (1966) has shown that frictional forces along joints are increased due to the roughness of the joint walls. Joint roughness is considered at two scales: major undulations in the joint plane are referred to as first-order roughness, and the small irregularities on the plane are roughness of the second order. Both orders of roughness are measured by their angle (i) relative to the general dip of the plane. Second-order roughness has higher i values than those of the first order. Data presented by Barton (1973)

Table 11.4 Values of $(\phi + i)$ under low normal stress.

Type of surface	Normal stress (kg/cm^2)	$(\phi + i)$ (deg)	Tested by
limestone: slightly rough	1.57	77	Goodman (1970)
bedding surfaces	2.09	73	
	6.00	71	
limestone: rough bedding	3.05	66	Goodman (1970)
surfaces	6.80	72	
shale: closely jointed	0.21	71	Goodman (1970)
\ seam in limestone	0.21	70	
quartzite: gneiss and	–	80	Paulding (1970)
amphibolite discontinuities			
beneath natural slopes			
beneath excavated slopes	–	75	
granite: rough undulating	1.5	72	Rengers (1971)
artificial tension fractures	3.5	69	

Source: Barton (1973).

demonstrate that under low normal stress the effective angles of friction are greatly increased due to roughness (Table 11.4). It may be assumed that no shearing of the second-order roughness occurs at low normal stress levels (Hoek and Bray, 1977). As normal stress increases the second-order projections are sheared and the first-order roughness (with lower i values) becomes the control of the effective friction angle. As normal stress further increases first-order projections are sheared and the effective roughness angle (i) is reduced to zero. The angle of friction along the joint plane $(\phi + i)$ is thus reduced to ϕ.

An angle of sliding frictional resistance of approximately $38°$ has been estimated previously for the Cape supergroup quartzites, and it has been stated that stable slopes on these rocks will not be steeper than that angle where bedding planes outcrop (Selby, 1982c). More detailed examination of un-buttressed rock masses reveals, however, that apparently stable slopes exist where the dip reaches $50°$ and more. Consequently the nature of the bedding planes has been examined with a view to establishing a more accurate estimate of the angle of friction of the quartizite.

The first-order roughness of bedding planes has been measured using a contour gauge, and this has yielded first-order i values in the range $14–37°$ with mean of $22°$ (Fig. 11.5). These figures indicate a minimum effective friction angle for the quartzite of $50–70°$, given a basic friction angle of $35°$ (Barton, 1973; Hoek and Bray, 1977). It has been suggested (Patton, 1966) that only first-order irregularities contribute to the shear strength along joints beneath natural slopes, as creep and weathering would probably cause failure of small-scale asperities. However, all scales of roughness are likely to be

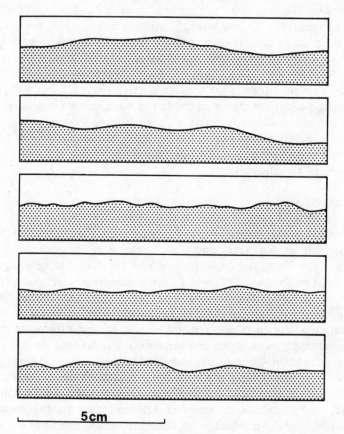

Figure 11.5 Typical roughness profiles of bedding planes in Cape Supergroup quartzite.

important below the weathered zone or where joints are tightly closed and unweathered.

In order to test the effectiveness of second-order roughness as a contributor to joint shear strength, tilt tests were performed on a sample of ten joint blocks ranging in length between 200 and 500 mm. These tests, under conditions of extremely low normal stress (slab thicknesses $\leqslant 110$ mm) revealed effective friction angles of $80-85°$ along joints with first-order i values of $10-15°$. It is therefore second-order roughness that accounts for the high angles of friction.

The effective friction angles determined experimentally do not necessarily apply on natural slopes in the Cape mountains, and there are a number of reasons why lower values should apply in a natural situation. First, there are significant scale effects (Bandis et al., 1981; Barton, 1981). Large joint blocks have a lower effective friction angle than a number of smaller blocks covering a comparable area, and this scale effect is most marked on rough undulating joints. The reduction in shear strength with increasing block size is explained by the reduction in contact area across a joint. Small blocks have a greater capacity to rotate slightly and maintain contact of small-scale roughness features. Second, slopes with bedding planes dipping steeply out of the rock mass will be stable while the partings remain tightly closed, but when shearing of the second-order roughness and dilation have taken place as a result of slight movement or when roughness has been affected by weathering, the effective friction angle is reduced significantly. Reference to examples of buttressed and unbuttressed slopes in the Cape mountains illustrates the geomorphic significance of the above.

Unbuttressed slopes

On an unbuttressed slope in the Swartberg Range (Fig. 11.6) the elements G, H, and J are bedding-plane controlled in two senses. First, they owe their form to the attitude of the strata and the resistance of the rock forming their surfaces; and, second, they have formed as a result of failure along bedding planes. Slope elements A to E and I have gradients steeper than the dip but are strength equilibrium slopes. The high friction angles along the joint planes contribute to the stability of these elements, as evidenced by the lack of scree accumulation at the slope base. On elements G, H, and J the angle of friction has been reduced by weathering and shearing to less than the dip angle to allow failure to take place. Consequently, the bedding planes on which those slope elements are formed were critical to the strength of the entire rock mass. Elsewhere in the Cape mountains small structurally controlled elements similar to G, H, and J exhibit slickensided surfaces; that is, the second-order roughness has been sheared allowing failure to occur. The form and development of such slopes are therefore dependent upon partings along which low strength conditions exist. Consequently, the proposition that the stability of a slope on an unbuttressed rock mass is determined by the shear strength along the weakest continuous joint is valid.

There is some difficulty in predicting stable angles for slopes developed on unbuttressed rock masses. It has not been possible to ascertain effective friction angles for large masses of rock *in situ*. Equally difficult is the assessment of the degree to which a joint plane is weathered and the extent to which second-order roughness is sheared or otherwise reduced. Although effective friction angles of 70° and higher have been suggested for the quartzite, examination of dip-slope (buttressed) forms provides an indication of a realistic value.

Buttressed slopes

Buttressed slopes are developed on all of the major fold features in the Cape mountains where strata dip into a confining mass. The largest are the dip slopes of hogbacks formed on major anticlines, but buttressed slopes also

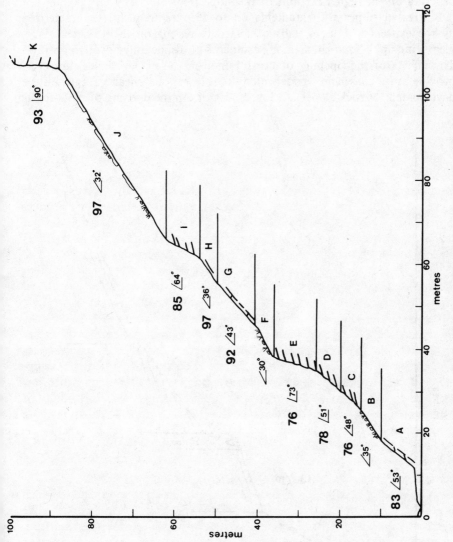

Figure 11.6 Profile of an unbuttressed slope in the Swartberg Mountains.

form on minor fold features. Gradients of the bedrock slopes range from 30°
to vertical. Buttressed slopes may have either a simple planar form where they
are developed on a single bedding unit or a stepped profile where dip-slope
elements are separated by an element of gentler gradient resulting from erosion
of strata on the higher sections of the slope (Fig. 11.7).

Buttressed slopes with gradients up to 55° are frequently characterized
by the existence of large slabs of quartzite resting on them. These slabs,
measuring up to 30 m² in area, are remnants of disintegrated overlying strata.
The effects of the opening of cross joints on the planar slopes, however,
provide more convincing evidence of the role of joint shear strength on the
development of rock slopes. At low angles of dip the opening of cross-joints

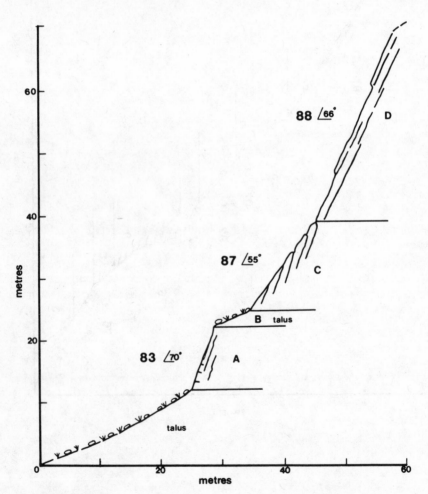

Figure 11.7 Profile of a buttressed (structurally controlled) slope in the
Boesmanspoort Mountains.

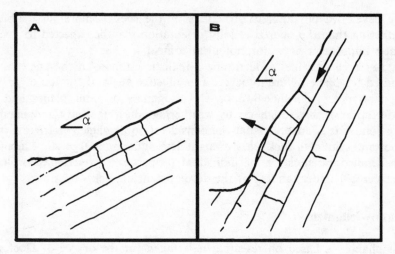

Figure 11.8   (A) Stable dip slope where $\alpha < \phi + i$; (B) plane failure where $\alpha > \phi + i$.

has little, if any, effect on slope form, despite the increase in shear strength as joint blocks are shortened and the presumed enhancement of weathering effects along bedding planes (Fig. 11.8A). On more steeply dipping rock masses the opening of cross-joints has a noticeable effect on slope stability and form. The cross-joints weaken the rock mass, and if the angle of friction ($\phi + i$) is less than the gradient, buckling of the strata and slope failure will occur (Fig. 11.8B).

Observations in the Cape mountains indicate that weakening of the rock mass due to cross-joint opening is significant on buttressed slopes with gradients greater than approximately $55°$. At gradients less than this value the strata maintain their buttressing function and do not buckle even after cross-joints have formed. The inference drawn is that despite weathering and possible shearing along bedding planes, the friction angle for the quartzite remains at about $55°$.

Conclusion

It is clear from the analysis of rock slopes in the Cape mountains that geological factors other than attitude of the strata exert control on the form and development of rock slopes in fold terrane. The existence of strength equilibrium forms in a situation where structural controls could be expected to dominate is evidence of the importance of rock mass strength as a control of form. The analysis of the Cape mountain data has shown that on strength equilibrium slopes, joint spacing and intact rock strength are the prime determinants of slope form. In other situations, where strength equilibrium slopes have developed on different structures or the rock is more susceptible to weathering than the quartzite, joint orientation may be expected to assume greater importance as a morphological control.

It has been possible to determine a realistic estimate of the angle of dip required for plane failure to occur. The effective angle of friction of 55° for the quartzites is dependent upon the roughness of joint planes and the modification of that roughness by weathering, shearing, and the opening of cross-joints. It is apparent that the consideration of angles of friction is an important adjunct to rock mass strength determination in rock slope analysis. The detailed examination of individual properties of rock masses clearly enhances our understanding of the development of rock slopes.

Acknowledgments

This chapter is based on research undertaken for the degree of Doctor of Philosophy at the University of the Witwatersrand. I thank Philip Stickler and Pat Moon for assistance with illustrations, the South African Geographical Society for permission to reproduce Figures 11.1, 11.3, 11.7, and 11.8, and Peter Fridjohn for advice on the statistical analysis. Financial support from the Council for Scientific and Industrial Research and the University of the Witwatersrand is gratefully acknowledged.

References

Bakker, J. P., and J. W. N. Le Heux, Projective-geometric treatment of O. Lehmann's theory of transformation of steep mountain slopes, *Koninklijke Nederlandsche Akademie van Wetenschappen, B49*, 533–547, 1946.

Bakker, J. P., and J. W. N. Le Heux, A remarkable new geomorphological law, *Koninklijke Nederlandsche Akademie van Wetenschappen, B55*, 399–410 and 554–571, 1952.

Bandis, S., A. C. Lumsden, and N. R. Barton, Experimental studies of scale effects on the shear behaviour of rock joints, *International Journal of Rock Mechanics and Mining Sciences and Geomechanics Abstracts, 18*, 1–21, 1981.

Barton, N., Review of a new shear strength criterion for rock joints, *Engineering Geology, 7*, 287–332, 1973.

Barton, N., Some size dependent properties of joints and faults, *Geophysical Research Letters, 8*, 667–670, 1981.

Barton, N., and V. Choubey, The shear strength of rock joints in theory and practice, *Rock Mechanics, 10*, 1–54, 1977.

Cooks, J., Die verband tussen litologie en landvorme, *South African Geographer, 7*, 127–135, 1979.

Cooks, J., Rock quality measured by seismic wave velocity as a factor in landform development, *South African Journal of Science, 77*, 517–521, 1981.

Cooks, J., Geomorphic response to rock strength and elasticity, *Zeitschrift für Geomorphologie, 27*, 483–493, 1983.

Day, M. J., The morphology of tropical humid karst with particular reference to the Caribbean and Central America, Ph.D. thesis, 611 pp., Oxford University, 1978.

Day, M. J., Rock hardness and landform development in the Gunong Mulu National Park, Sarawak, Eastern Malaysia, *Earth Surface Processes and Landforms, 6*, 165–172, 1981.

Fisher, O., On the disintegration of a chalk cliff, *Geological Magazine, 3*, 71–82, 1866.

Goodman, R. E., The deformability of joints, in *Determination of the in situ modulus of deformation of rock*, American Society for Testing and Materials, Special Publication 477, 174–196, 1970.

Hoek, E., and J. W. Bray, *Rock Slope Engineering*, Institute of Mining and Metallurgy, London, 1977.

King, L. C., Canons of landscape evolution, *Bulletin of the Geological Society of America, 64*, 721–751, 1953.

King, L. C., *South African Scenery*, Oliver and Boyd, Edinburgh, 1963.

Klimaszewski, M., A contribution to the theory of rock face development, *Studia Geomorphologica Carpatho-Balcanica, 5*, 139–151, 1971.

Lehmann, O., Morphologische Theorie der Verwitterung von Steinschlagwänden, *Vierteljahrsschrift der Naturforschende Gesellschaft in Zürich, 78*, 83–126, 1933.

Mallows, C. L., Some comments on CP, *Technometrics, 15*, 661–675, 1973.

Moon, B. P., The forms of rock slopes in the Cape fold mountains, *South African Geographical Journal, 66*, 16–31, 1984.

Moon, B. P., and M. J. Selby, Rock mass strength and scarp forms in southern Africa, *Geografiska Annaler, 65A*, 135–145, 1983.

Patton, F. D., Multiple modes of shear failure in rock and related material, Ph.D. thesis, 293 pp., University of Illinois, Urbana, 1966.

Paulding, B. W., Coefficient of friction of natural rock surfaces, *Proceedings of the American Society of Civil Engineers, Journal of the Soil Mechanics and Foundations Division, 96(SM2)*, 385–394, 1970.

Penck, W., *Die morphologische Analyse. Ein Kapitel der physikalischen Geologie*, Engelhorns, Stuttgart, 1924. English translation: Czech, H., and C. K. Boswell, *Morphological Analysis of Landforms*, Macmillan, London, 1953.

Rengers, N., Unebenheit und Reinbungswiderstand von Gesteinstrennflächen, Thesis, Technische Hochschule Fridericiana, Karlsruhe, Institut für Bodenmechanik und Felsmechanik Veröffentlichung, 47, 1–129, 1971.

Selby, M. J., A rock mass strength classification for geomorphological purposes: with tests from Antarctica and New Zealand, *Zeitschrift für Geomorphologie, 24*, 31–51, 1980.

Selby, M. J., Rock mass strength and the form of some inselbergs in the central Namib Desert, *Earth Surface Processes and Landforms, 7*, 489–497, 1982a.

Selby, M. J., Form and origin of some bornhardts of the Namib Desert, *Zeitschrift für Geomorphologie, 26*, 1–15, 1982b.

Selby, M. J., Controls on the stability and inclinations of hillslopes formed on hard rock, *Earth Surface Processes and Landforms, 7*, 449–467, 1982c.

Terzaghi, K., Stability of steep slopes in hard unweathered rock, *Geotechnique, 12*, 251–270, 1962.

Wood, A., The development of hillside slopes, *Proceedings of the Geological Association, 53*, 128–140, 1942.

12
Influence of scree accumulation and weathering on the development of steep mountain slopes

I. Statham and S. C. Francis

Abstract

This chapter reviews the development of scree slopes and considers three aspects: debris input from the headwall, scree slope accumulation and form, and subsequent long-term behavior through weathering. Supply of debris from the headwall is largely controlled by weathering along discontinuities, though climatic processes act as immediate trigger mechanisms. Consequently, any study of headwall slope development must take account of the rock structure if meaningful relationships between form and process are to be made. The subscree rock slope profile is theoretically convex in form, resulting from recession of the headwall and protection of the buried face. In practice, previous slope form often masks this theoretical trend. Models of scree slope accumulation must explain frequently observed slope characteristics, principally a straight profile with basal concavity, and downslope fine to coarse particle sorting. A rockfall model is consistent with these characteristics, the basal concavity resulting from an exponential distribution of downslope travel distances. Sorting is due to greater frictional resistance of smaller particles. The frictional resistance of a moving particle is dependent on particle shape and surface characteristics. Weathering of scree deposits may cause considerable changes in particle properties, such as particle size, grading, and shape. Relatively small changes in material strength may result, however, and the most important effect of weathering will be to reduce hydraulic conductivity. Ultimately this may lead to instability, which, for a number of reasons, is most likely to occur at the top of the scree slope, in contrast to regolith-covered hillsides.

Introduction

Scree slopes form by the accumulation of rock debris falling from outcrops in single-particle rockfalls or larger-scale rock instabilities. The presence and importance of screes in mountain landscapes are immediately apparent, as virtually every headwall in that environment not subject to basal erosion has at least some accumulation of scree beneath it. Consequently, scree processes are major controls in the development of steep, competent rock slopes. It is less obvious, perhaps, that scree processes can be significant in the development of slopes in relatively subdued upland areas; the hills of western England and central Wales are good examples in the United Kingdom. There, relatively rapid stream incision has produced steep rock valleys, with slopes too steep for detached particles to accumulate as regolith. In addition, many steep slopes show the characteristic scree slope form but are completely mantled by coarse regolith. On such slopes downslope transport of detached blocks has ceased or is sufficiently intermittent to allow scree degradation.

The interface between scree processes and regolith development on lower-angled, but still relatively steep, rock slopes is particularly interesting. There is undoubtedly a slope angle limit, below which scree processes cannot operate and regolith development is dominant. The division between the two is complex and is almost certainly interactive.

This paper first reviews scree accumulation processes and form and then considers their interaction with weathering trends on steep, competent rock slopes. This chapter is specifically confined to slopes where the movement of debris on the scree is mainly controlled by rockfall dynamics. Other situations, such as semiarid boulder control slopes and debris flow cones, are not considered.

Headwall process and rock slope form

Scree development relies on the headwall supply zone being weathering-limited (Carson and Kirkby, 1972, p. 104) with the potential for transport of loose blocks exceeding their rate of release by weathering. The processes of release and transport from the headwall are not necessarily continuous, as some short-term sediment storage may take place on the face, provided that detached blocks ultimately become unstable. Although most screes appear to be composed of large, fresh rock fragments, the majority possess a considerable percentage of interstitial fine material that washes or falls into the macropores. Consequently, weathering of the headwall may occur on two scales dominated by joints and cracks on the macro- and microscales, respectively. The presence of fine material is extremely significant and influences slope processes and the interaction of regolith-forming processes on lower-angled rock slopes (see below).

Input processes: importance of discontinuities

The most important rock strength characteristic that influences input to scree slopes is the discontinuity system, a point that has long been recognized by engineers (e.g., Terzaghi, 1962; Hoek and Bray, 1977). The importance of discontinuities is illustrated by Table 12.1 and Figure 12.1. Table 12.1 lists conjectural critical heights for vertical faces in some common rock types and highlights the effect of jointing. Figure 12.1 indicates that even quite low-angled and low-relief slopes may be unstable.

Instability processes may be classified according to scale (e.g., Statham 1977 a, p. 87–90), although in practice it is relatively unusual for falls to be well enough observed for an accurate classification to be made. Magnitude of input process is important in scree slope development, however, and this is discussed in the following section on rockfall accumulation.

The smallest scale may be taken as rockfall itself, involving one or several discontinuity-defined blocks separated from the face by weathering and becoming unstable. It is important to recognize that this process is part of the general weathering system and does not indicate overall slope stability. At an intermediate scale there is a set of processes that are indicative of overall slope instability and are directly controlled by discontinuity aspect with respect to the face. Faces containing such discontinuities were called strength equilibrium slopes by Selby (1982) and are normally classified into three types, i.e., slab, wedge, and toppling failures, depending on the orientation of discontinuities with respect to the face. These processes were reviewed in Hoek and Bray (1977), who also discussed in detail the mechanics and methods of analysis.

The importance of discontinuity orientation and dip with respect to the headwall is readily apparent and often explains why some rock slopes in a region have very large screes beneath them and others very little accumulation. The largest scree slopes frequently occur as cones below pronounced gullies in the headwall. These almost always develop where discontinuities are closely spaced, for example near faults, and headwall recession is rapid.

Large-scale processes, loosely termed rock avalanches, require consideration

Table 12.1 Influence of discontinuities on stability of rock slopes[a].

Rock type	Intact rock		Jointed rock		
	Compressive strength (kg/cm^2)	Critical height (m)	Cohesion (kg/cm^2)	ϕ (deg)	Critical height (m)
granite	1000–2500	4000–10000	1–3	30–50	12–65
sandstone	200–1700	1000–9000	0.5–1.5	30–45	9–40
shale	100–1000	400–4000	0.2–1.0	27–45	4–20
limestone	300–2500	900–8000	0.25–1.0	30–50	5–25
quartzite	1500–3000	6000–11000	1–3	30–50	12–65

[a] Stability analyses for a vertical cliff using the Culmann method, assuming isotropic rock mass and discontinuity distribution.

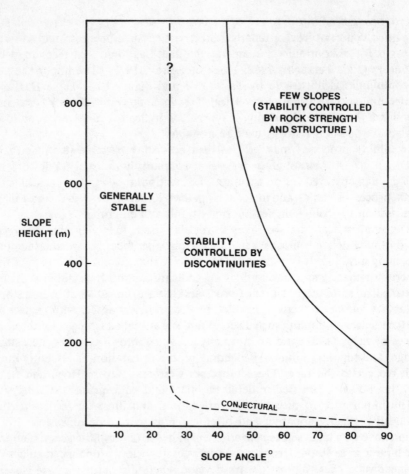

Figure 12.1 Rock slope stability controls. (Adapted from Hoek and Bray (1977).)

of a great number of factors other than discontinuities, for instance intact rock strength, groundwater conditions, and tectonism. Some can be analyzed approximately as circular arc failures (Hoek and Bray, 1977), but it appears that the mechanics of many large-scale rock failures is only poorly understood.

Selby (1982) modified the Norwegian Geotechnical Institute's rock mass strength designation for tunneling (e.g., Barton et al., 1974) to classify rock slopes retreating by mass failure. The system is based upon eight parameters, which are subjectively assessed and given non-dimensional numerical ratings. These ratings are then summed to produce a mass strength rating for the rock. The parameters used include intact rock strength, weathering, joint orientation, and joint spacing. Selby correlated rock mass strength rating with cliff slope angle for cliffs composed of jointed rocks that are unbuttressed and lack major discontinuities of overriding importance to their stability.

However, the limitations of this approach should be clearly realized: it is both subjective and conveniently dumps the operation of failure processes into a numerical index without any regard to the way in which they operate. For example, a moderately strong, moderately weathered, and closely jointed sandstone could possess the same numerical classification as a weak unweathered widely fissured mudstone, though the processes of recession on cliffs composed of these different rock types may be quite dissimilar. In short, there is no substitute for a careful and detailed analysis of the mechanics of the input processes in the headwall, irrespective of the field difficulties in carrying this out. The rock mass strength index methodology does not help to understand the input from the headwall.

Climatic and microclimatic controls on rockfall

It is fairly well established that rockfall and probably medium-scale instabilities are often triggered by climatic events.

Rapp (1960) and Luckman (1976) distinguished between primary falls, blocks freshly detached from the face, and secondary falls that result from the dislodgment of previously fallen debris which had accumulated on ledges. In practice it is rare for a rockfall to be so clearly observed that its primary or secondary character can be determined; hence this distinction cannot be effectively considered in the context of control mechanisms.

Observations on rockfall are either indirect, based upon long-term accumulation or on damage records (e.g., Rapp, 1960; Bjerrum and Jorstad, 1968), or they may be a direct inventory of falls over a specific time period (Gardner, 1967, 1969, 1971; Stock, 1968; Luckman, 1976; Francon, 1982). Most observations have been made in high alpine or subarctic environments where freeze–thaw is apparently a major control. Rapp (1960), Stock (1968), Bjerrum and Jorstad (1968), and Francon (1982), for example, have noted a spring maximum in activity, when most freeze–thaw oscillations occur. An autumn peak is sometimes noted (Rapp, 1960), but it is less widely reported. In addition to the seasonal variations, a diurnal pattern may also occur related to headwall microclimate. Gardner (1969, 1971), working in the Canadian Rockies, found that about three-quarters of all falls occurred on the colder, northwest-facing slopes, with maximum frequencies in the early morning and midday. The maxima appeared to correspond quite well with the diurnal temperature curve on the rock faces.

Luckman (1976) cautioned against an oversimplistic interpretation of the interaction between climate and rockfall. He also worked in the Canadian Rockies and found that, although freeze–thaw certainly is a major release mechanism, snowstorms, heavy rains, and even short showers may be responsible for triggering falls.

Luckman also drew attention to marked differences in response to climatic events on faces of different morphology and aspect (hence microclimate) and geology. Although broad generalizations may be made about the relationship of climate to rockfall, they can undoubtedly be masked by characteristics peculiar to the individual rock face.

The accumulation of snow and ice on the free face, for example, may have a marked effect on temperature fluctuations and hence rockfall activity. As Luckman (1976) concluded, it is doubtful if general statements can be improved upon without a more systematic approach, where specific hypotheses relating release triggers to rockfall events may be put to the test.

Subscree rock slope form

The concept of scree protection in the development of hard rock slopes was introduced by Fisher (1866), who introduced a simple mathematical model to describe the development of a rock profile by means of progressive disintegration of a vertical cliff (Fig. 12.2). His model assumes parallel headwall retreat, producing a volume of scree equal to the volume of headwall lost. This accumulates at the base of the cliff as a straight repose slope and the portion of headwall buried is protected from subsequent erosion. Progressive headwall retreat results in a convex buried rock slope, since a progressively smaller headwall contributes to an increasingly long scree.

Fisher's basic model has been extensively refined in a number of papers, incorporating factors such as bulking, basal removal, and headwall angles (Lehmann, 1933), and hillslope form and cliff retreat mechanisms (Bakker and Le Heux, 1946, 1947, 1950, 1952). With one exception the subscree profile is convex, with increasing convexity downslope, as found with hillslope simulation modeling (Kirkby and Statham, 1975; Kirkby, 1983). The exceptional case occurs with complete removal of scree from the system and results in a rectilinear bedrock slope, whose angle is controlled by the avalanche process.

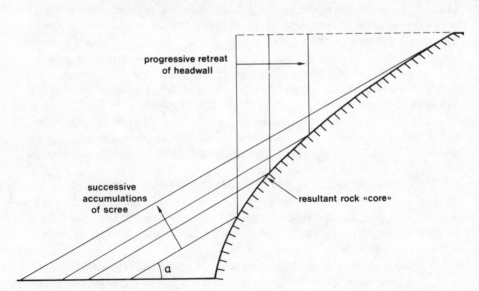

Figure 12.2 The development of a rock core beneath an accumulating scree.

It is certainly the case that original rock slope form strongly influences subsequent development, and initial forms are often markedly concave, particularly in glaciated valleys. Further, where a rock slope is being buried by material not simply derived from the headwall immediately above it, the subscree profile will not bear any model relationship to the overlying scree. This occurs in areas of scree cones, where the model would only apply along the axis of the supply gulley and not elsewhere.

Field evidence of subscree rock slope form is scarce. Some rock profiles, exhumed by later gullying, were examined by Statham (1973) in south Wales and the Isle of Skye, and very little evidence for convexity was found; indeed, most profiles were rectilinear or concave. In contrast, the simulation studies mentioned earlier supported the Fisher model and predicted convex buried rock slopes. In simple systems a convex profile probably does occur. However, it is likely that so-called simple systems are exceptional and that other geometric factors nearly always exist to modify the subscree form.

Rockfall accumulation: dynamics and scree slope form

A simple rockfall model
It has often been pointed out that scree slopes accumulating by rockfall possess characteristics that are not generally present on debris slopes accumulating by other processes. These include a straight, steep upper slope generally less than the angle of repose, a basal concavity, and well-defined downslope particle sorting from fine to coarse. Table 12.2 summarizes the characteristics

Table 12.2 Major characteristics of some debris slopes.

	Debris flow cone	Avalanche boulder tongue	Rockfall scree slope	Boulder-controlled slope
Profile	concave	concave	straight/concave	straight with long concavity
Angle	low	low	high (generally lower than repose)	high (generally lower than repose)
Sorting	fining downslope (slight)	fining downslope (slight)	fining downslope	fining downslope
Boulder tails	—	downslope of large obstructions	upslope of large obstructions	—
Surface character	irregular with levees and debris flow trails	irregular with avalanche trails and channels	fairly smooth	fairly smooth
At-a-point size dispersion	high	high	moderate	high

of the major types of debris slope and highlights the differences due to dominant surface processes (also see White, 1981). Any model of the rockfall process must seek to explain the observable characteristics of rockfall scree, and these are discussed below.

Stones falling from a cliff onto a scree will bounce, slide, and roll over the surface before coming to rest. This simple process is difficult to model but as a first approximation it can be treated as though the stone falls freely through space and then slides on a rough, inclined surface after impact (Statham, 1973; Kirkby and Statham, 1975; Statham, 1976, 1979). The distance x that stones travel down a scree after falling a height h is given by

$$x = \frac{h \sin^2 \alpha}{\cos \alpha \tan \phi_{\mu d} - \sin \alpha} \tag{1}$$

where α is the slope angle and $\phi_{\mu d}$ is the dynamic angle of sliding friction for the stone. This expression requires many assumptions to be made, which are discussed by Kirkby and Statham (1975) and Statham (1976, 1979).

Equation (1) is similar to an expression proposed by Gerber and Scheidegger (1974):

$$x = \frac{\frac{1}{2}u_s^2}{g(\cos \alpha \tan \phi_{\mu d} - \sin \alpha)} \tag{2}$$

where

$$u_s = 2gh \sin \alpha \tag{3}$$

is the initial stone velocity on the slope, and g is the acceleration due to gravity. However, they applied the expression to the motion of a dry frictional slide over a scree surface, on the assumption that the slide behaves as a single, rigid block. It is considered that the model is better applied to the movement of single particles, as a mass-slide is unlikely to behave as a rigid body.

Carson (1977) raised objections to Kirkby and Statham's rockfall model based on observations made on quarry stockpiles. He proposed that quarry stockpiles were good models for scree accumulation and that as their characteristics and accumulation processes were different from the scree slope simulations made by Statham (1973, 1976), the rockfall model must be in error. In fact, Carson failed to mention that the stockpiles also did not look much like real screes.

It is apparent from Carson's paper that several aspects of stockpiles, particularly the nature of their input zone, rate of input, grain size characteristics, and ratio of fall height to scree height, were quite atypical of real scree systems. Carson also asserted that Statham (1973, 1976) had not taken edge effects into account in laboratory experiments involving measurements of angle of repose or accumulation, an assertion that is quite untrue. Francis

(1984*a*) presented comprehensive evidence of the influence of edge effects in angle of repose experiments that supports this statement. Statham (1977*b*) dealt comprehensively with the shortcomings of Carson's arguments.

The rockfall model expressed in (1) relies on many assumptions that must be considered before proceeding further. First, as the model stands, it can only predict mean travel distances (or velocities) on the scree. This point is discussed more fully in the following section. Second, it has not been practicable to allow for air resistance, which would tend to reduce impact velocities. Nonvertical headwalls have also not been considered. In theory the frictional effects of steeply inclined rock surfaces could easily be incorporated into the model, if required. In practice, however, the limitations of obtaining measurements of the frictional interaction between headwalls and falling particles are considerable. Third, large particles tend to crater into the scree and proportionally much more of their energy is directly absorbed. They may also shatter on impact. Finally, the scale of the input process has not been considered, and it is assumed that the rockfalls involve one or several particles at a time, which do not interact significantly with one another. We shall return to the influence of scale below.

Observed movement of particles on screes

The simple model assumes the scree slope acts as a smooth, frictional surface and that sliding particles are retarded under the influence of a steadily applied frictional force. Actual particle movement is more complex. Observations of a great many laboratory and small-scale field tests have shown that movement generally occurs as a series of bouncing and rolling impacts with the scree, which may retard the particle, accelerate the particle, or stop it instantly.

Motion pictures made of a medium-scale rockfall (300–500 m^3) at Lecco, Italy, show some similarities with, but also some marked differences from, the small-scale tests mentioned above. A mass of rock was blasted from an almost vertical face, whence it fell about 190 m, before impacting on a scree inclined at 34–36°. Considerable breakage occurred on impact and also some cratering by large particles. It was estimated that only 14–25 percent of the total kinetic energy prior to impact remained unabsorbed and therefore contributed to downslope movement. Beyond the impact zone, particles 'flew,' then bounced or rolled (according to shape) before stopping. As noted in the small-scale experiments, stopping was frequently an abrupt rather than a gradual process.

The observations made at Lecco show that scale of input process influences the behavior of particles on the slope and, by inference, slope development. For the time being, however, we will return to small-scale rockfalls involving a few particles that are not large enough for cratering to be a serious effect.

In Figure 12.3 the cumulative distribution of downslope travel distance is plotted for the motion of stones dropped from many different heights of fall onto laboratory-simulated and field scree slopes. The results represent many observations, each symbol representing 20–50 drops using a variety of stone

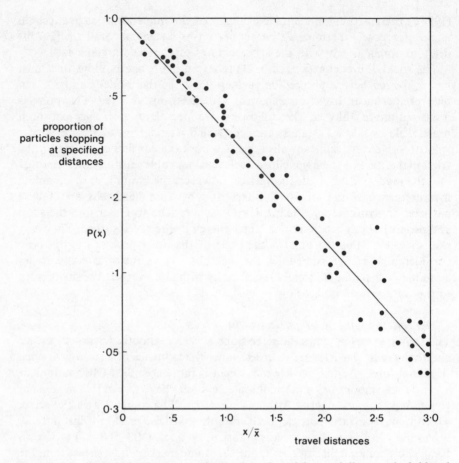

Figure 12.3 Cumulative frequency distribution of particle travel distances in field and laboratory experiments.

sizes and scree slope angles. The experiments always show a similar pattern of an exponential distribution of travel distances:

$$P(x) = e^{-x/\bar{x}} \tag{4}$$

where $\bar{x}$ is the mean travel distance for any experiment and $P(x)$ is the proportion of particles traveling at least a distance x. The negative exponential distribution arises because, at any distance x, a proportion of the particles has already stopped and only those still moving are subject to a constant probability of stopping. This mechanism produces a wedge of material that runs out beyond the straight upper slope of the scree and forms the basal concavity.

The constant probability of stopping is from (1):

$$\frac{1}{\bar{x}} = \frac{\cos \alpha \tan \phi_{\mu d} - \sin \alpha}{h \sin^2 \alpha} \tag{5}$$

The modified model, in which particles are generally brought to rest downslope but at any time may stop instantaneously, provides a much closer fit to the style of movement observed in experiments (Kirkby and Statham, 1975). It does, however, imply a change in the meaning of $\phi_{\mu d}$, as discussed in the following section.

Experimental investigation of $\phi_{\mu d}$

From the preceding section it is clear that $\phi_{\mu d}$ should be redefined as the angle at which the stone continues to move with constant velocity. In this way, all processes of retardation are included without the need to distinguish between them.

Results of many laboratory and field experiments (Statham, 1973; Kirkby and Statham, 1975; Statham, 1979) indicate that the relationship between height of fall h and mean travel distance $\bar{x}$ is linear, as predicted by (1). Furthermore, calculated values of $\phi_{\mu d}$ from the experimental results have also shown a reasonable fit with the model. One of the most important controls on $\phi_{\mu d}$ for a fallen particle seems to be the ratio of its diameter d_* to the size d of particles on the slope. Figure 12.4, in which the results of many field and laboratory tests are plotted, indicates that

$$\tan \phi_{\mu d} = \tan \phi_0 + Kd/d_* \tag{6}$$

where ϕ_0 is the intercept of the line with the vertical axis and corresponds to $\phi_{\mu d}$ for particles that are very large in comparison to those on the scree, and K is a constant related to particle characteristics. But as particles become proportionally smaller they are more easily retarded in the depressions on the slope and hence $\phi_{\mu d}$ increases.

This effect is largely responsible for the typical fine to coarse sorting down scree slopes mentioned in Table 12.2 and reported by many workers (e.g., Andrews, 1961; Tinkler, 1966; Gardner, 1968; Bones, 1973; Statham, 1973, 1976). Sorting is, of course, a self-propagating mechanism, as particles tend to come to rest most quickly in material of similar size.

The magnitudes of ϕ_0 and K are certainly dependent on properties of the scree particles and the falling stones. It would seem reasonable that ϕ_0 may be related to surface frictional properties of particles while K is influenced by particle shape, Kd/d_* being effectively a dilatation angle for the contact. This is to some extent illustrated by the field and laboratory tests shown in Figure 12.4. The field scree consisted of angular boulders of 30 cm or more in size. Shearbox tests on the fine fraction gave a loosest state angle of internal shearing resistance (ϕ'_{cv}) for this material of $40°$. The laboratory material, on the

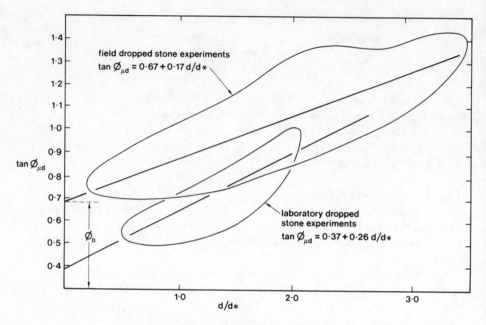

Figure 12.4 Influence of particle size on $\phi_{\mu d}$.

other hand, was a limestone gravel with a ϕ'_{cv} of 35°. One may expect ϕ_0 to be related to ϕ'_{cv} (though certainly not equal to it), as both may be linked to particle surface frictional properties. However, K is probably controlled by shape and packing. A lower value is expected for the field scree, as the particles are markedly flatter (Zingg sphericities of 0.8 and 0.65 were calculated for the laboratory and field materials, respectively) and tend to be imbricated, resulting in less particle interlocking. Imbrication is a characteristic of some screes and is best developed where platy particles are present (Taylor, 1971; Statham, 1976; McSavenney, 1972). Camponouvo (1977) presented results from a model scree which show that particle travel velocities increase with increasing sphericity and d/d_*, also implying that $\phi_{\mu d}$ is controlled by particle size and shape.

Development of slope form on rockfall screes
The results of laboratory scree accumulation models and computer simulations of single-particle rockfall have been reported by Kirkby and Statham (1975) and Statham (1979). These results indicate that initial accumulation is quite low-angled and relatively concave, but the concavity becomes progressively less significant and confined to the base as particle fall height becomes small in comparison with the length of scree. The concavity is replaced with a straight slope, initially slightly less than ϕ'_{cv} for the material but quickly approaching it as particle fall height diminishes. Figure 12.5 shows

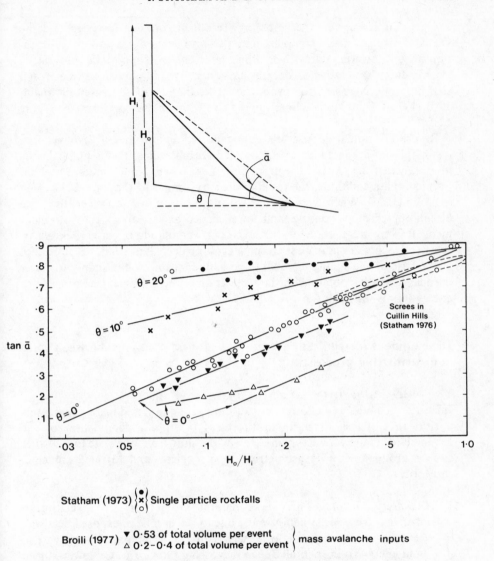

Figure 12.5 Mean slope angle and height of fall in simulated screes.

the decline of concavity, expressed in terms of mean slope angle ($\tan \bar{\alpha}$) with increasing H_0/H_i, where H_i is the initial headwall height and H_0 the height of the accumulated scree against the headwall at any time. The results are from laboratory simulations and clearly show increasing mean slope angle, approaching ϕ'_{cv} for the material as headwall diminishes. The angle θ of the initial accumulation slope also influences the trend, with steeper mean scree angles on inclined accumulation zones for any stage of development.

Field data to support the laboratory results are difficult to collect because site geometry is usually complex, and also problems of space–time substitution arise. However, data for Cuillin screes tend to support the laboratory experiments. They suggest decreasing concavity and increasing mean and straight slope angles as the proportional height of fall decreases (Statham, 1973, 1976). The Cuillin screes appear to fall in the range shown in Figure 12.5.

Results of similar laboratory tests reported by Broili (1977) have been reworked and are included in Figure 12.5. Broili's experiments highlight the importance of the scale of the input processes. Where model screes were built by small mass additions, the results were very similar to those of Kirkby and Statham (1975). Where very large events were substituted, however, the scree became much more concave and lower-angled at similar states of development. It is apparent, therefore, that screes accumulating by large rockfalls or rock avalanches are much less steep, suggesting decreased frictional resistance at the base of the sliding mass. It is possible that this is due simply to fewer particles being in contact with the scree at any time and therefore the rate of energy loss is much slower.

Development of regolith-covered rock slopes: a balance between scree and weathering processes

The onset of inactivity on scree slopes

Inactivity in a scree is a relative, not absolute, condition and merely represents a shift in emphasis from input processes to *in situ* degradation of the accumulated material. Factors influencing the shift from one state to the other can be grouped into (1) geometric, (2) geological, and (3) environmental categories.

1. *Geometric factors.* There is a natural progression from activity to inactivity in a scree system, simply due to the decline in exposed headwall area as the scree accumulates. There is also an accompanying decrease in input energy from rockfall due to decreasing fall height. On lower-angled headwalls the balance is further shifted toward inactivity because more work is required to dislodge particles and set them moving onto the scree slope. At the lower limit of activity there is a progression between scree systems and *in situ* regolith development and some degree of interaction is to be expected about this transition.

2. *Geological factors.* Rock type will undoubtedly change the rate at which the progression between activity and inactivity occurs, as softer rocks degrade more quickly and cease to behave as scree. There may also be changes in the discontinuity pattern within the headwall as it erodes, resulting in a speeding up or slowing down of input events due to changes in overall cliff stability.

3. *Environmental factors.* Probably the most important environmental change that could affect the activity of a scree is climate. As discussed above, climatic triggers are certainly important in releasing rockfalls from the headwall, and any change in climatic conditions may have a profound effect on debris supply. Basal erosion may also be considered in the context of environmental factors. Removal of material from the base of the scree will lead to a steepening of the slope, increased downslope transport and ultimately re-exposure of the headwall with renewed erosion.

As discussed above, the upper-bound stable angle for scree material is ϕ'_{cv}. Typically, this is about 38–40° for coarse, hard-rock debris, though to some extent this depends on grading. Francis (1984*b*), for example, found that ϕ'_{cv} can be in the range 33–39° for experimental sandstone regolith particles. The gradings resulting in the lower values, however, are unlikely to occur on active screes subject to downslope sorting and are probably confined to *in situ* regoliths.

Input of particles to the scree tends to increase its angle toward ϕ'_{cv}. This is partly counteracted by downslope particle movement as discussed above and also by rock waste slides (Drewry, 1973) brought about by instantaneous dilatation of surface layers following particle impacts.

Even in the active state, therefore, screes tend to be stable, with slope angles in the range 33–38°. As they move toward inactivity, other processes (such as animal disturbance, snow and slush avalanching, and surface water flow) tend to reduce the slope angle further and, therefore, increase the factor of safety against landsliding, assuming that porewater pressures do not develop at this stage. This stability is often accompanied by colonization of the scree surface by vegetation.

Inactive screes, therefore, are usually very stable landforms and, consequently, remain as dominant slope facets for a considerable length of time. For any subsequent slope development to occur, there must be an *in situ* change in the scree material itself brought about by weathering.

Weathering trends in screes

Weathering trends in hard-rock scree debris. A model for mechanical weathering of strong rock regoliths has been put forward by Francis (1984*b*) and is to some extent applicable to scree. The model is based on trends in rock particle properties as a rock breaks down from *primary units*, defined by significant discontinuities (bedding, joints), to *secondary units*, which consist of individual sediment grains in the case of sedimentary rocks and mineral grains in the case of igneous rocks. The existence of subprimary units *en route* between the end members is recognized, and these may be thought of as smaller pieces of rock, defined perhaps by less prominent discontinuities or microfractures opened up during weathering. The extent to which subprimary clasts are produced during weathering will determine the development of the grain size curve.

Where most disintegration is by granular breakdown the grain size curve will become significantly bimodal during weathering. On the other hand, in fissile rocks where splitting of clasts into subprimary units occurs, the resulting sediment will become uniformly graded. The terms discontinuous and continuous weathering have been applied to these two trends, respectively.

Figure 12.6 is an attempt to show how the weathering trends may occur in the scree system. In the active state there is a certain amount of breakdown from primary to subprimary units, but very little significant weathering. There is, however, a degree of segregation of sorting of particles along the scree that is significant in subsequent weathering and slope development. In the inactive state weathering may follow the discontinuous or continuous trends, depending on the rock type. It will probably be most significant at the surface, influenced by the weather and the establishment of vegetation, and at depth changes may be initially restricted to washing in of fines from above.

Influence of weathering on ϕ_{cv}'. Francis (1984b) identified the following factors as important in influencing ϕ_{cv}' of a weathering sediment: (1) particle size, (2) particle shape, (3) grading curve, (4) surface frictional properties of particles, (5) void ratio, (6) particle strength, and (7) normal stress during shear.

On geometric grounds, particle size should have little effect on ϕ_{cv}', other factors being equal. However, it should be remembered that the smaller particles are often different in composition and may therefore have different surface properties. Perhaps the most extreme example of this effect is "screes" forming by rockfall from rapidly degrading hard clay cliffs, where the large clasts very quickly break down to very soft clay on exposure to form mudslides (e.g., Hutchinson and Bhandari, 1971). In this case the process of scree formation and weathering take place simultaneously to produce a low-angled clay slope without an intermediate, steep scree slope phase. In general, however, the changes in ϕ_{cv}' due to decreasing particle size in scree weathering can be regarded as small.

A considerable amount of experimental evidence from hard sandstone regoliths was reported by Francis (1984b) showing significant trends in particle shape and grading during regolith weathering. Despite these trends, however, his results showed that only very small changes in ϕ_{cv}' are to be expected as a hard-rock regolith develops by weathering. Other workers have noted an initial measurable increase in ϕ_{cv}' (or related strength parameters) as percentage of fines increased (e.g., Holtz and Gibbs, 1956; Leps, 1970; Kawakami and Abe, 1970; Statham, 1974), followed by a decrease with further increase of fines (Lupini et al., 1984). However, these studies mainly considered rather crude gradings, which consisted of two part mixtures of coarse and fine particles and may not be realistic models of weathering trends. Using artificial soils, Statham (1974) showed substantial variations of void ratio with grading,

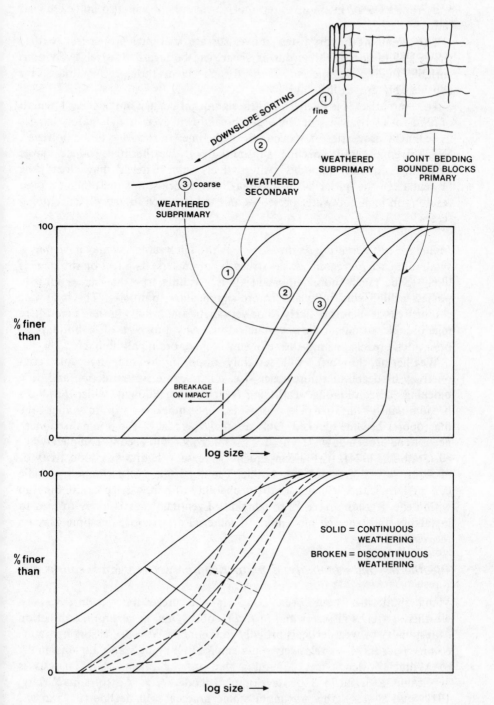

Figure 12.6 Weathering and grading in scree systems.

and Francis (1984*b*) demonstrated that particle shape will also influence void ratio.

The remaining factors listed above, surface frictional properties, particle strength, and normal stress during shear, can be regarded as relatively non-variable due to weathering of hard-rock scree, though they may vary considerably due to lithology.

To summarize, it is likely that little significant change in the shear strength of hard-rock scree will occur by *in situ* breakdown, even though major changes in sediment properties may occur through time. In the long term, however, shear strength could decline substantially if weathering produces large percentages of clay minerals. Alternatively, fine material may already be present from the initial headwall breakdown process or from other sources, resulting in high porewater pressures and failures even in actively developing screes.

Influences of weathering on drainage.   Although weathering may have only a small influence on regolith shear strength parameters, its effect on drainage is likely to be much more important. It is certainly true that screes initially possess a high void ratio and, more significantly, pore size. The hydraulic conductivity is therefore likely to be very high, and usually in the range where macroscopic turbulent flow would occur. Any prospect of a buildup of porewater pressure within the sediment is therefore highly unlikely.

Weathering, however, will certainly result in declining hydraulic conductivity by decreasing pore-space size, due to grain size reduction and pore blocking. Discontinuous weathering (granular breakdown), which produces predominantly fine particles, provides the quickest way in which the macropores become blocked. Only about 12 percent of the original sediment needs to be broken down to fines for the macropores to become completely filled (Statham, 1974), with a consequent decrease in hydraulic conductivity of probably several orders of magnitude. The importance of declining hydraulic conductivity is that it increases the probability of porewater pressure buildup within the degrading scree through time. Eventually this is likely to lead to instability and renewed movement of material downslope, this time as mass movements.

Weathering and instability: a comparison between in situ *regoliths and screes*

Many similarities have been noted in the preceding sections between weathering trends in screes and *in situ* regoliths, and in general the initiation of instability by weathering is broadly comparable. With decreasing hydraulic conductivity there is a relatively rapid change in limiting stable angle from ϕ'_{cv} to a much lower value. Assuming that infinite planar slide analysis is applicable, as usual in slope development models (e.g., Carson and Kirkby, 1972, p. 178–184), the maximum stable gradient will decline to $\frac{1}{2} \tan \phi'_{cv}$. Typically this would be about 16–21°.

There are, however, some important contrasts between the way in which scree and *in situ* regolith will respond to instability. First, *in situ* regoliths are likely to be stability-limited in areas of rapid downcutting. Scree slopes rarely are, however, and may be expected to undergo changes in hydraulic conductivity much less quickly than *in situ* regoliths.

There are also likely to be major differences in the location of instabilities on screes, in comparison to *in situ* regoliths. As there are no consistent trends in particle size downslope or necessarily in thickness in a regolith, the buildup of porewater pressure through time is likely to be most marked toward the base of the slope. As a general rule, therefore, instability is likely to be initiated at the slope base and to progress upslope through time.

Many factors may change this pattern, such as local changes in soil properties and slope hydrology, which will result in a scattering of failures across the slope.

A quite different pattern of instability is likely on scree slopes. Mention has already been made of the commonly observed fine to coarse sorting downslope, and at the top of the scree, blocking of the macropores by fines from the headwall is more likely. In addition, screes are generally very much thinner at the top, which is just where the input of surface water from the headwall is most concentrated. In short, impeded drainage and development of porewater pressures is most likely at the top of a scree slope, in contrast

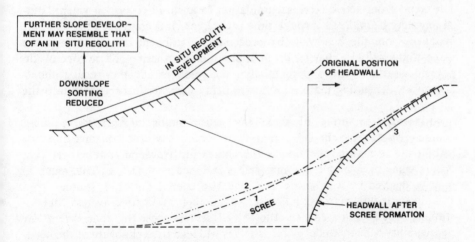

Figure 12.7 Degradation and slope development in an inactive scree. Profile 1: Scree at time of headwall burial; steep rectilinear upper facet and basal concavity. Profile 2: Slope profile after upslope instability. The profile tends toward a rectilinear slope at an angle of approximately $\phi'/2$ as a result of landslipping. Gully erosion may cause further angle reduction, with the infilling of the original basal concavity by weathered scree transported from upslope. The lower slope is now mantled with relatively fine debris. Profile 3: The upper headwall is re-exposed. If it is stable, *in situ* regolith will develop.

to regoliths produced by *in situ* weathering. Instability on scree slopes, therefore, is most likely to be initiated near the top. This pattern of failure is often observed, the instability taking the form of shallow landslides or, quite frequently, highly mobile debris flows, which travel for considerable distances over the steep slope (Statham, 1973, 1976; Gardner, 1968, 1983; Rapp, 1960; Prior et al., 1970; Whitehouse and McSavenney, 1983).

These differences in the location of instability hold some implications for trends in slope development on screes, as opposed to *in situ* regoliths. In a regolith the changes in slope angle will probably be gradual, confined into relatively small events in space and time, perhaps with restricted re-exposure of bedrock. Intermediate slope gradients, between $\tan \theta'$ and $\frac{1}{2}\tan \theta'_{cv}$, are therefore likely within the landscape at common states of development of the slopes. As envisaged by Kirkby (1983), these intermediate angles are not thresholds.

In contrast, degradation of screes is likely to cause considerable re-exposure of the upper headwall and eventually replacement of the scree with an over-riding, lower-angled debris accumulation, as illustrated in Figure 12.7. In this case, intermediate slope angles are not probable and successive stages will be represented by complex, two-phase slopes.

Conclusion

The trend from active scree accumulation to an inactive system where land-sliding may be initiated through time is complex. It is controlled by materials supplied from the headwall, by processes of sediment distribution on the slope, and finally by changes brought about by weathering. These controls do not necessarily operate sequentially. To some extent they occur simul-taneously, depending mainly on how quickly fines are incorporated into the system.

Although weathering of scree can cause major changes in sediment characteristics, these changes generally have relatively little influence on slope stability. The major exception is the change in hydraulic conductivity that occurs when appreciable fine material is included in the scree. This material may be derived by weathering, but may also come from other sources.

Unlike regolith-mantled hillslopes, a number of factors suggest that in-stability will be initiated near the top of scree slopes, the zone where low permeability is likely to develop first. This will lead to re-exposure of the head-wall and the replacement of the scree with an overriding, lower-angled debris slope.

References

Andrews, J. T., The development of scree slopes in the English Lake District and Central Quebec Labrador, *Cahiers de Geographie, Quebec, 10*, 219–230, 1961.

Bakker, J. P., and J. W. N. Le Heux, Projective geometry treatment of G. Lehmann's theory of the transformation of mountain slopes, *Koninklijke Nederlandsche Akademie van Wetenschapen, Amsterdam, 49*, 532–547, 1946.

Bakker, J. P., and J. W. N. Le Heux, Theory on central rectilinear recession of slopes, *Koninklijke Nederlandsche Akademie van Wetenschappen, Amsterdam, 50*, 959–966, 1154–1162, 1947.

Bakker, J. P., and J. W. N. Le Heux. Theory on central rectilinear recession of slopes, *Koninklijke Nederlandsche Akademie van Wetenschappen, Amsterdam, 53*, 1072–1084, 1364–1374, 1950.

Bakker, J. P., and J. W. N. Le Heux, A remarkable new geomorphological law, *Koninklijke Nederlandsche Akademie van Wetenschappen, Amsterdam, 55*, 399–410, 554–571, 1952.

Barton, N., R. Lein, and J. Lunde, Analysis of rock mass quality and support practice in tunnelling, and a guide to estimating support requirements, *Norwegian Geotechnical Institute Report 54206*, 1974.

Bjerrum, L., and F. Jorstad, The stability of rock slopes in Norway, *Norwegian Geotechnical Institute Publication 79*, 1–11, 1968.

Bones, J., Process and sediment size arrangement on high arctic talus, Devon Island, NWT, Canada, *Arctic and Alpine Research, 5*, 29–40, 1973.

Broili, L., Relations between scree slope morphometry and dynamics of accumulation processes, *Istituto Sperimentale Modelli e Strutture Report 90*, 11–24, 1977.

Camponuovo, G. F., ISMES' experience of the model of San Martino, *Istituto Sperimentale Modelli e Strutture Report 90*, 25–38, 1977.

Carson, M. A., Angles of repose, angles of shearing resistance and angles of talus slopes, *Earth Surface Processes, 2*, 363–380, 1977.

Carson, M. A., and M. J. Kirkby, *Hillslope Form and Process*, 475 pp., Cambridge University Press, Cambridge, 1972.

Drewry, D. J., Rock waste sliding—a surface transportation mechanism on screes, paper presented to *Institute of British Geographers Annual Conference*, Birmingham, 1973.

Fisher, O., On the disintegration of a chalk cliff, *Geological Magazine, 3*, 354–356, 1866.

Francis, S. C., The limitations and interpretation of the "angle of repose" in terms of soil mechanics: a useful parameter?, in *Assessing BS5930, 20th Regional Conference of the Engineering Group, Geological Society of London*, edited by A. B. Hawkins and C. R. I. Clayton, pp. 213–229, University of Surrey, Guildford, 1984*a*.

Francis, S. C., The geotechnical properties of weathering sandstone regoliths, Ph.D. thesis, 517pp., University of London, 1984*b*.

Francon, B., Chute des pierres et éboulisation dans les parois de l'étage périglaciale: observations faites d'octobre 1979 à juin 1981 dans la combe de Laurichard (Hautes Alpes), *Revue Geographie Alpine, 70*, 279–300, 1981.

Gardner, J. S., Notes on the avalanches, ice falls and rockfalls in the Lake Louise District, July and August, 1966, *Canadian Alpine Journal, 50*, 90–95, 1967.

Gardner, J. S., Debris slope form and processes in the Lake Louise District, a high mountain area, Ph.D. Thesis, 263 pp., McGill University, Montreal, 1968.

Gardner, J. S., Rockfall; a geomorphic process in high mountain terrain, *Albertan Geographer, 6*, 15–20, 1969.

Gardner, J. S., A note on rockfalls and north faces in the Lake Louise Area, *American Alpine Journal, 17*(2), 317–318, 1971.

Gardner, J. S., Rockfall frequency and distribution in the Highwood Pass Area, Canadian Rocky Mountains, *Zeitschrift für Geomorphologie, 27*, 311–324, 1983.

Gerber, E., and A. E. Scheidegger, On the dynamics of scree slopes, *Rock Mechanics, 6*, 25–38, 1974.

Hoek, E., and J. Bray, *Rock Slope Engineering*, 402 pp., Institute of Mining and Metallurgy, London, 1977.

Holtz, W. G., and H. J. Gibbs, Triaxial shear tests on pervious gravelly soils, *Journal of the Soil Mechanics and Foundations Division, Proceedings of the American Society of Civil Engineers, 82*, Paper 867, 1–22, 1956.

Hutchinson, J. N., and R. K. Bhandari, Undrained loading, a fundamental mechanism of mudflows and other mass movements, *Geotechnique, 21*, 353–358, 1971.

Kawakami, H., and H. Abe, Shear characteristics of saturated gravelly, clays, *Transactions of the Japanese Society of Civil Engineers, 2*, 259–298, 1970.

Kirkby, M. J., Modelling cliff development in South Wales: Savigear reviewed, *Leeds University School of Geography Working Paper 351*, 17 pp., 1983.

Kirkby, M. J., and I. Statham, Surface stone movement and scree formation, *Journal of Geology, 83*, 349–362, 1975.

Lehmann, O., Morphologische theorie der verwitterung von steinschlagwande, *Viertel jahrschrift der Natürforschenden Gesellschaft im Zurich, 78*, 83–126, 1933.

Leps, T., Review of shearing strength of rockfill, *Journal of Soil Mechanics and Foundation Engineering Division, Proceedings of the American Society of Civil Engineers, 94*, 1159–1170, 1970.

Luckman, B. H., Rockfalls and rockfall inventory data; some observations from Surprise Valley, Jasper National Park, Canada, *Earth Surface Processes, 1*, 287–298, 1976.

Lupini, J. F., A. E. Skinner, and D. R. Vaughan, The drained residual strength of cohesive soils, *Geotechnique, 31*, 181–213, 1981.

McSavenney, E. R., The surficial fabric of rockfall talus, in *Quantitative Geomorphology, Some Aspects and Applications*, edited by M. Morisawa, pp. 189–197, Publications in Geomorphology, Binghampton, NY, 1972.

Prior, D. B., H. Stephens, and G. R. Douglas, Some examples of modern debris flows in North East Ireland, *Zeitschrift für Geomorphologie, 14*, 275–288, 1970.

Rapp, A., Recent developments of mountain slopes in Karkevagge and surroundings, Northern Scandinavia, *Geographiska Annaler, 42*, 65–200, 1960.

Selby, M. J., Controls on the stability and inclinations of hillslopes formed on hard rock, *Earth Surface Processes, 7*, 449–467, 1982.

Statham, I., Process form relationships in a scree system developing under rockfall, Ph.D. thesis, 272 pp., University of Bristol, 1973.

Statham, I., The relationship of porosity and angle of repose to mixture proportions in assemblages of different sized materials, *Sedimentology, 21*, 149–162, 1974.

Statham, I., Debris flows on vegetated screes in the Black Mountains, Carmarthenshire, *Earth Surface Processes, 1*, 173–180, 1976.

Statham, I., *Earth Surface Sediment Transport*, 184 pp., Oxford University Press, Oxford, 1977*a*.

Statham I., Angles of repose, angles of shearing resistance and angles of talus slopes—a reply, *Earth Surface Processes, 2*, 437–440, 1977*b*.

Statham, I., A simple dynamic model of rockfall; some theoretical principles and model and field experiments, *Proceedings of the International Society of Rock Mechanics Colloquium on Physical Geomechanical Models*, 237–258, 1979.

Stock, R., Morphology and development of talus slopes at Ekalugad Fjord, Baffin Islands, NWT, B.A. thesis, 84 pp., University of Western Ontario, London, Ontario, 1968.

Taylor, A., Scree development in the Isle of Skye, B.Sc. dissertation, University of Bristol, 1971.

Terzaghi, K., The stability of steep rock slopes on hard unweathered rock, *Geotechnique, 12*, 251–270, 1962.

Tinkler, K. J., Slope profiles and scree in the Eglwyseg Valley, North Wales, *Geographical Journal, 132*, 379–386, 1966.

White, S. E., Alpine mass movement forms (non-catastrophic): classification, description and significance, *Arctic and Alpine Research, 13*, 127–137, 1981.

Whitehouse, I. E., and M. J. McSavenney, Diachronous talus surfaces in the Southern Alps, New Zealand, and their implications to talus accumulation, *Arctic and Alpine Research, 15*, 53–64, 1983.

13
Flow behavior of channelized debris flows, Mount St. Helens, Washington

Thomas C. Pierson

Abstract

Between 1981 and 1983 field measurements of dynamic and physical properties of channelized debris flows were obtained at two sites at Mount St. Helens. Motion picture photography, time lapse photography, acoustic rangefinding, timing drift, and hand sampling of flows were used to obtain velocity, depth, and compositional data for 10 debris flows.

All flows observed moved as surges downchannel, having a steep, bouldery flow fronts that impeded the flow of the more fluid slurry behind the fronts. Flows ranged in magnitude from about 1 to about 50 m^3/s. Front velocities were as much as 4.4 m/s and thalweg surface velocities as much as 5.9 m/s for flow depths up to 2.3 m on channel slopes between 7° and 22°. Velocity of the boulder fronts appeared to be controlled largely by channel gradient and depth. Fluid velocity appeared to be influenced by slurry sediment concentration, as well as by depth and slope. Less concentrated flows tended to move at higher velocities than more concentrated flows. The unsteady flow of the debris slurries was at times supercritical and at times turbulent. The development of rigid plugs, a function of slurry yield strength, appeared to be controlled by the relative concentration of boulder-sized particles.

Sediment concentration of debris flow samples (slurry matrix), which was highest near the heads of flows, ranged from 76 to 86 percent by weight (55 to 70 percent by volume). When concentrations dropped below 74 percent by weight, slurry coherence was lost and flow transformed to hyperconcentrated streamflow.

Introduction

Debris flows are extremely mobile, highly concentrated dispersions of very poorly sorted sediment (up to boulder-sized particles) in water. Sediment concentrations of debris flow matrix slurries (cobbles and boulders excluded) typically range from about 75 to 90 percent by weight (wt percent), which for most sediments is about 53 to 77 percent by volume (vol percent). The consistency of such mixtures is closely similar to wet concrete. Macroscopically, a debris flow slurry tends to behave as a coherent single-phase system: the pore fluid (water and fines) is, to a large degree, trapped in a framework of coarser grains and moves *en masse* with the liquefied flowing sediment. Slurries poor in coarse particles (< 50 percent gravel) are often termed mudflows (Varnes, 1978).

Debris flows are non-Newtonian fluids that possess a plastic yield strength, a high bulk density (generally about twice that of water), and a viscosity that is much greater than that of water (Johnson, 1970). These properties allow debris slurries to flow with much less turbulence than clear water, to flow in a channel at velocities considerably higher than is possible with clear water, and to transport large boulders (which can exert extremely high impact forces) in suspension. The shear strength exhibited by debris flows is due to a combination of (1) cohesive forces provided by the electrochemical interaction of very fine particles, and (2) frictional forces (sliding and interlocking) between coarser particles.

When debris flow slurries are diluted, shear strength is decreased (Kang and Zhang, 1980). At the point where shear strength becomes negligible the sediment−water mixture takes on markedly different fluid properties and is termed hyperconcentrated streamflow (Beverage and Culbertson, 1964). In this type of flow the coherency of the debris flow slurry has been disrupted, and two macroscopically separate phases have evolved: a fluid phase composed of water, clay, and silt, and a solid phase of dispersed coarser particles. No longer held in suspension by matrix strength, the coarser particles easily settle out of suspension, and the pore fluid is no longer trapped within the grain framework. However, hyperconcentrated flows experience turbulence much more readily than debris flows, and this turbulence may keep sand-sized (even fine gravel-sized) particles in suspension (Pierson and Scott, 1985).

Debris flows may occur on a variety of scales in channels or on open slopes. They can be as small as several centimeters wide and deep, flowing only several centimeters per second. During a heavy rain, steep, unvegetated slopes are often good sites to observe such miniature flows. Deposits left by such flows characteristically include lateral levees on the margins of the flow path and steep-fronted lobes of debris in the deposition zone. At the other extreme, they can be hundreds of meters wide, many tens of meters deep, flow at several tens of meters per second, and travel more than a hundred kilometers. Such catastrophic debris flows can be triggered by volcanic eruptions (Neall, 1976; Janda et al., 1981), by massive landslides (Plafker and Ericksen, 1978), or by rapid draining of lakes (Neall, 1976; Pierson and Scott, 1985). Debris flows at this scale also typically exhibit marginal levees and steep-fronted lobate deposits. Intermediate-sized flows are commonly triggered by landslides of water-saturated debris, usually during heavy rain or rapid snowmelt, and by small jökulhlaups (glacier outburst floods) mobilizing loose debris in channels. Because the same types of features may be seen in debris flow deposits on all scales, it is assumed that the same types of processes are active on all scales (Johnson, 1970).

The rheology of debris flows is poorly understood owing to (1) the infrequent occurrence of such flows, at least at locations where they can be studied, and (2) the difficulty in working in the laboratory with coarse sediment mixtures, whose physical properties do not remain constant unless energy is added to the system to keep it fully dispersed. Consequently, there is little quantitative information for coarse non-Newtonian slurry flows, particularly at the field scale, that can lead to prediction of such crucial aspects of flow behavior as velocity, depth, impact forces, and runout distances.

The objectives of this study were (1) to obtain a better qualitative understanding of the physical mechanisms involved in and controlling the flow of debris slurries, (2) to quantify the relationships between flow velocity and potential controlling factors such as depth, channel gradient, sediment concentration, particle size distribution, and slurry temperature, and (3) to record horizontal velocity distributions within flows. This research is still in progress; the results and conclusions presented here are based on the 1981, 1982, and 1983 field seasons only.

Location

Field observations and measurements of debris flow dynamics have been carried out at two locations on Mount St. Helens. The instrumented study site is a 500 m reach of stream channel that is incised about 30 m and has an average slope of about $8°$. The reach is immediately downstream from the terminus of the Shoestring Glacier on the southeast flank of Mount St. Helens (Figs. 13.1, 13.2, and 13.3) and is the head of the Muddy River. This ravine has conveyed debris flows and hyperconcentrated flood surges with frequencies as high as 1 to 2 per week during the snow-free season (August through October) since before the May 18, 1980, eruption. Small flows, with peak discharges less than 10 m^3/s, are triggered during warm weather by outburst floods from the glacier and by landslides of saturated debris off the glacier ice. Larger flows (up to 50 m^3/s) are triggered by rainstorms and originate from the glacier debris mantle or moraine sides. A meltwater stream with a discharge usually less than 1 m^3/s occupies a narrow shifting channel on the floor of the ravine. Channel gradients in the study reach range from 0.12 ($7°$) to 0.40 ($22°$). One major waterfall approximately 10 m high is contained within the reach (Fig. 13.2). In a few locations, the stream is confined to a smooth bedrock channel cut into andesite; elsewhere, it flows over unconsolidated pyroclastic deposits and coarse alluvium.

An auxiliary data set was fortuitously collected on April 30, 1981, when flows were observed and sampled at the base of a steep talus cone on the west

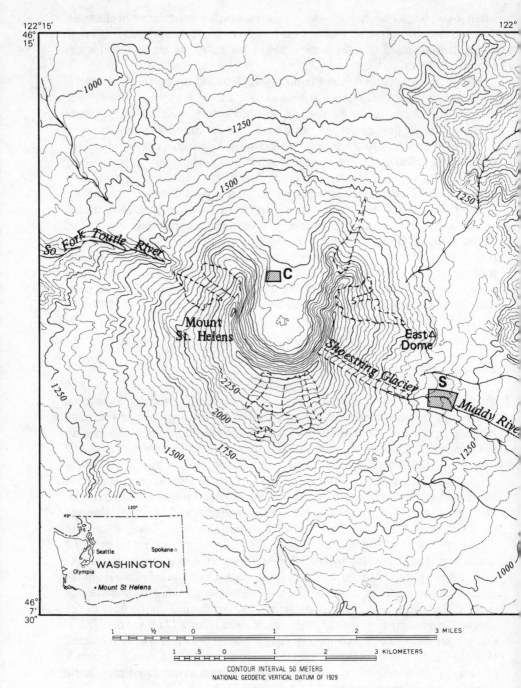

Figure 13.1 Locations of Shoestring (S) study site and the crater (C) data collection locality at Mount St. Helens, Washington.

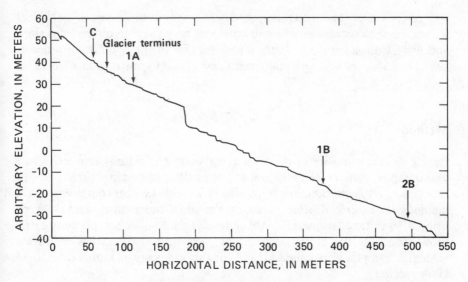

Figure 13.2 Longitudinal profile of monitored channel reach at the Shoestring study site. Measurement and sampling locations are indicated by arrows.

Figure 13.3 View upstream of monitored channel reach at Shoestring study site. Mounted movie camera (without weatherproof case) in foreground.

wall of the crater (Fig. 13.1). Apparently in response to fumerolic melting of ice and snow, a succession of small flows was released high on the talus slope and flowed down onto the crater floor. Because of the addition of volcanic heat, these slurries were unusually warm, estimated to range from 30 to 36°C.

Methods

Debris flow monitoring at the Shoestring study site utilizes motion picture photography, time lapse photography, acoustic rangefinding, timing of surface clasts, and hand sampling of passing flows with 3.6-liter containers. These methods are described below. Data on the small crater flows were collected using a 30-m tape, stopwatch, Abney level, and 3.6-liter sample containers. Flow fronts in the crater were timed over measured reaches to obtain velocities, samples were collected, and channel gradients measured over 10- to 20-m reaches.

A 16-mm movie camera was used to record the overall appearance of the flow, size and composition of the flow front, degree of turbulence in the flow, and velocity of the flow front and flow surface. Normally, it was mounted inside a weatherproof housing, which was fixed on a post by a ball-and-socket head (Fig. 13.3). When a debris flow was sighted, the camera was manually turned on and left to run unattended until the film ran out. At 16 frames per second, one 30-m roll of film lasts for slightly more than four minutes, which is sufficient time to record most of the flow features. Painted rocks and targets within the field of view were spaced at even intervals along the flow path, to permit computation of flow velocities. One difficulty encountered in field operation has been poor resolution owing to dense fog produced by low storm clouds and to raindrops on the lens window.

The time lapse system consists of a 35-mm camera with a wide-angle lens, an hour–minute–second digital data back, a motor drive, and an infrared pulse transmitter and receiver for wireless remote control. The camera and infrared receiver were housed in a weatherproof case and suspended from a lightweight cableway (Fig. 13.4) about 7 m directly above a bedrock reach of channel. The camera case was mounted on a cable car that was operated from the bank by a pulley system. When a flow passed beneath the camera, the operator used the infrared remote control to take periodic sets of exposures (5–8 frames) at a rate of 3.1 frames per second. The camera was suspended just downstream of the sample site, and photo sequences were taken immediately after samples were collected to ensure that recorded flow behavior and flow composition corresponded. The distance traveled by individual cobbles and boulders suspended in the debris flow was scaled off the sequential photographs, allowing determination of average surface velocities and horizontal velocity profiles. Total error in the velocity computation was assumed to be the root mean square of the independent partial errors

Figure 13.4 Cable car suspended over channel at Shoestring study site. The 35-mm camera (in box) and acoustic rangefinder (suspended below) attached to cable car.

(Benjamin and Cornell, 1970, p. 186), which result from (1) distance measurement errors in the field, (2) measurement errors in scaling distances from photographs, including lens distortion, and (3) time increment errors due to temperature effects on the camera motor drive. These errors were estimated, on the basis of the procedures used, to be 7, 4, and 6 percent, respectively. Total error is therefore estimated to be 10 percent. The system worked well. The only drawback was that a bulk film magazine could not be used together with the data back, which is essential for synchronizing photos with sampling.

The acoustic rangefinders were used to record stage hydrographs of the flows at the Shoestring study site. An ultrasonic transceiver was suspended over the channel on a cableway (Fig. 13.4). It emits high-frequency sound pulses and times the echos. Microprocessor circuitry converts the timing of the

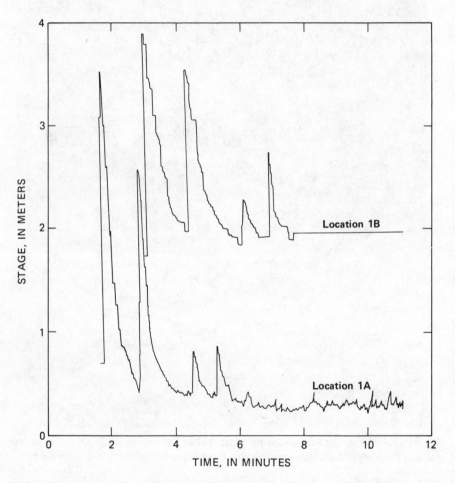

Figure 13.5 Stage hydrographs of multiple surges of Oct. 1, 1981b debris flow, recorded by ultrasonic acoustic rangefinders. Stations are shown in Figure 13.2.

echo to a distance and filters out background noise. The digital signal is converted to an analog voltage and is carried by wire to a stripchart recorder located in an instrument hut nearby. The rangefinders used in this study have an effective range of 0.1–6 m, emit signal pulses 16 times per second, and have a resolution of 3 cm on the recorders used. Two such devices are currently in use on cableways 113 m apart at the study site. In addition to recording a stage hydrograph, the rangefinders provide a way to measure the amount of channel downcutting or aggradation, shape of the flow fronts, peak attenuation or amplification, and average flow-front velocity between the two cableways (Fig. 13.5). The system produced good results, but major problems were encountered with the reliability of the electronics under rigorous field conditions, the loss of return signal due to wind, and unwanted reflections from channel sides, particularly at the lower cableway where the bedrock channel is narrow.

Material properties of the debris flows (sediment concentration and particle size distribution) were obtained from dip samples of slurry collected by hand at the edge of the flows at both the Shoestring site and the crater locality. The containers were 3.6-liter plastic jars with openings 95 mm in diameter. This sampling procedure provided only sieve-by-weight frequency data for the finer matrix of the slurry. In order to obtain the total size distribution, these data could be supplemented by grid-by-number samples of large clasts from photographs taken by the overhead camera (Kellerhals and Bray, 1971). Filled jars were labeled and sealed in the field with screw lids and tape before returning to the laboratory for analysis of sediment concentration and size distribution.

Results

Hydraulic properties were measured on ten channelized debris flows during the 1981, 1982, and 1983 field seasons (Table 13.1). Flow magnitude ranged from about 1 to about 50 m^3/s in peak discharge. Slurry dip samples were collected from nine of these flows; multiple samples were collected from eight. Additional debris flows were observed and recorded on some instruments, but insufficient data were obtained to allow comparisons to be made.

Flow events at the Shoestring study site and in the crater included both debris flow and hyperconcentrated streamflow. Determination of flow type was based on photographic evidence. Fully developed debris flow was recognized by the complete suspension of gravel-sized clasts and the nearly complete damping of surface turbulence. The hyperconcentrated flow exhibited extremely vigorous surface turbulence and was not sufficiently competent to suspend gravel-sized clasts.

Flow properties

The debris flows studied were surges of dense slurry moving downchannel and would be classified as rapidly to gradually varied flow. In most cases they were

Table 13.1 Hydraulic characteristics of flows studied[a].

Flow type[b]	Flow date and number	Measurement location[c]	Velocity (m/s)			Depth (m)		Top width (m)	Channel gradient	Slurry temperature[f] (°C)	Froude number
			Front	Surface	Mean[d]	At channel center	Mean[e]				
D	Aug. 21, 1981	S-1B	–	~2[g]	~1.8	~0.8	~0.6	~3	–	11.5	0.8
H	Sept. 9, 1981	S-1B	–	~1	–	0.2	0.2	3.4	0.123	12.0	–
D	Oct. 1, 1981a	S-1C	4.4	–	–	~1.5	~1.1	~5	0.404	7.0	–
D	Oct. 1, 1981a	S-1B	2.0	2.3	2.1	1.4	0.8	5.6	0.123	7.0	0.75
D	Oct. 1, 1981b/1	S-1A to 1B	3.5	–	–	2.8	2.1	–	0.184	–	–
D	Oct. 1, 1981b/2	S-1A to 1B	3.1	–	–	2.1	1.6	–	0.184	–	–
D	Oct. 1, 1981b/3	S-1A to 1B	2.8	–	–	0.6	0.5	–	0.184	–	–
D	Oct. 1, 1981b/4	S-1A to 1B	2.7	–	–	0.8	0.6	–	0.184	–	–
D	Sept. 3, 1982	S-2B	2.3	3.8	3.4	0.7	0.4	2.4	0.134	6.5	1.72
D	Sept. 9, 1982 presurge	S-2B	–	4.2	3.8	1.3	0.8	3.7	0.134	3.0	1.36
D	Sept. 9, 1982 surge	S-2B	–	5.0	4.4	1.5	0.9	4.3	0.134	3.0	1.48
D	Oct. 30, 1983	S-2B	–	5.9	5.3	2.3	1.3	9.0	0.134	4.0	1.49
H	April 30, 1981a	C	0.8	–	–	0.05	0.05	0.3	0.123	~33	–
D	April 30, 1981b	C	–	–	–	0.3	0.2	1.2	0.141	~33	–
D	April 30, 1981c	C	1.5	–	–	0.5	0.4	1.8	0.167	~36	–
D	April 30, 1981d	C	1.3	–	–	0.6	0.5	1.2	0.123	~30	–
D	April 30, 1981e	C	1.3	–	–	0.5	0.4	2.4	0.141	~33	–

[a] Measurements correspond to sampling done as close to peak flow as possible. When no samples taken, values represent peak flow.

[b] Debris flow = D, hyperconcentrated flow = H.

[c] Shoestring = S, crater = C; Shoestring locations shown in Figure 13.3.

[d] Mean velocity estimated by multiplying surface velocity by 0.75, as is commonly done for water flow.

[e] Obtained by multiplying channel center value by 0.9 (assumes flattened U-shaped channel cross section), except for Oct. 1, 1981a, Sept. 3, 1982 and Sept. 9, 1982 debris flows and the hyperconcentrated flows.

[f] Temperature values with ~ symbol estimated by feel and later comparison to known temperatures.

[g] Hydraulic values with ~ symbol were estimated from movie film.

over in less than 15 min. Although multiple surges are a common characteristic of debris flows, solitary surges were recorded in all cases except the Oct. 1, 1981b flow, which had four surges within 4 min. The flow appeared to be laminar in the cases where computed Froude numbers were less than 1 (Table 13.1). Vigorous turbulence was observed and photographed at margins of the rigid plug in two other flows that had Froude numbers greater than 1.

Geometry of flow bodies

The debris flows sampled had a number of physical features in common, and these have also been observed in debris flows elsewhere. Each flow characteristically had a very steep flow front, a "head" containing the densest slurry and achieving the highest stage of flow, and a progressively more dilute "tail" that accounts for the recessional limb of the slurry "flood wave" (Fig. 13.6). Compared to normal flood surges, the debris flows were extremely steep-limbed owing to their short duration (Fig. 13.5).

The steep, lobate flow front was composed predominantly of the coarsest particles available for transport. Such fronts were typically an openwork pile or ridge of boulders being bulldozed along by the flow. When matrix slurry did not fill all the interstices, the front was not liquefied and internal friction was much higher than in the rest of the debris flow. Motion picture photography of boulder movement on the surface of the Oct. 1, 1981a flow showed that boulders on the flow surface were moving 1.8 times faster than the front itself (Fig. 13.7). Surface particles moved to the flow front and tumbled down the steep leading edge. Cobbles and small boulders were over-ridden and reincorporated into the flow; the larger boulders were simply pushed ahead. This "conveyor-belt" behavior has also been observed in laboratory experiments (Hirano and Iwamoto, 1981). Consequently, the

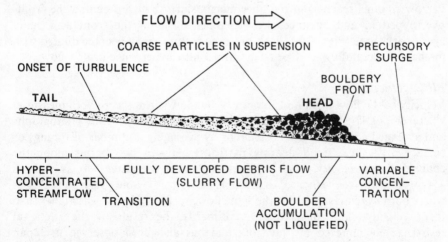

Figure 13.6 Schematic representation (vertically exaggerated) of a typical debris flow of the type studied (single surge), showing the principal component parts.

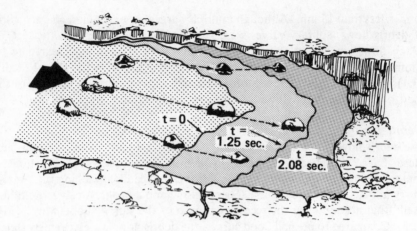

Figure 13.7 Stop-action movement traced from a movie film of surface boulders and flow front of the Oct. 1, 1981a debris flow at Shoestring study site.

coarsest particles accumulate at the front, and the boulder front tends to get bigger with distance downstream. However, this tendency is countered somewhat by the front boulders being continually shouldered aside by the slurry pushing from behind. This process contributes to the formation of lateral levees.

Boulder fronts act as moving dams and provide resistance to flow owing to their high internal friction and sliding friction against the channel bed. They can effectively slow down a debris flow, particularly when a narrow channel reach is encountered (Fig. 13.8). This damming action can cause peak discharge and peak depth to increase dramatically as the surge moves downstream. In one case (Oct. 1, 1981a) the coarse front became momentarily jammed in a narrow channel reach, and the finer matrix slurry built up behind the front, overtopped it, and advanced downchannel ahead of the front as a small precursory surge (Fig. 13.6). Such precursory surges may become diluted with in-channel streamflow and be transformed into hyperconcentrated flows.

Flow velocity

Velocity of the flow front and velocity of the flow surface were both measured wherever possible. Surface velocity was measured either from photographs and averaged across the cross section or by averaging a number of timings of surface drift (i.e., small floating particles) between two points along the channel. Front velocity values at Shoestring and in the crater were as much as 4.4 m/s and surface velocities as much as to 5.9 m/s (Table 13.1). In the cases where both were measured for the same flow, surface velocity was higher than front velocity. This differential is primarily the result of the frictional resistance encountered by the front and has also been observed in Japan (Watanabe and Ikeya, 1981). Flow at the head tends to back up behind the front, and surface velocity commonly increases slightly toward the tail of the

Figure 13.8 Advance of well developed, steep boulder front of a debris flow at the crater study site. Front is about 1 m high.

flow as cross section area decreases. This phenomenon was recorded for the Oct. 1, 1981a flow (Fig. 13.9) and observed in other flows as well.

Some debris flows exhibited rigid plugs (Johnson, 1970), whereas others did not. The very coarse flow at Shoestring on Sept. 9, 1982, had a rigid plug that occupied approximately 90 percent of the total width (Fig. 13.10). The flow was unsteady, with a 19 percent increase in surface velocity (corresponding to a 0.1 m increase in depth) observed over an interval of a few seconds. Shear was concentrated along a very thin zone (10–50 cm) at the flow margin, and boulders at the surface were interlocking in places (Fig. 13.11). The Oct. 1, 1981a flow at Shoestring, on the other hand, had shear distributed throughout the cross section (Fig. 13.12), although this was a slower shallower, debris flow.

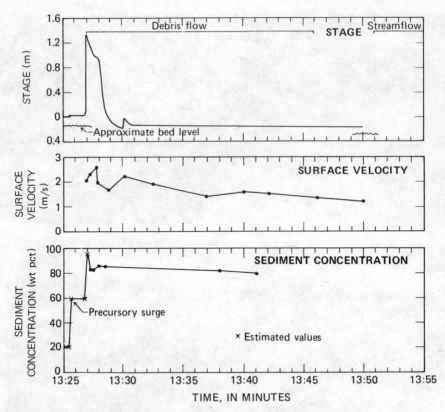

Figure 13.9 Variation in stage, surface velocity, and sediment concentration during passage of Oct. 1, 1981a debris flow at location 1B at Shoestring study site.

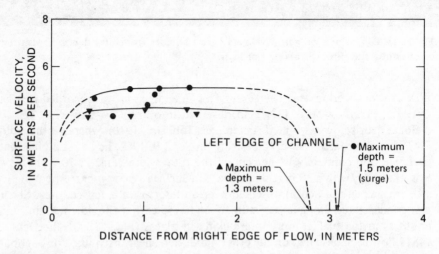

Figure 13.10 Half-width horizontal velocity profiles during peak flow of the Sept. 9, 1982 debris flow at location 2B (Fig. 13.2) at Shoestring study site.

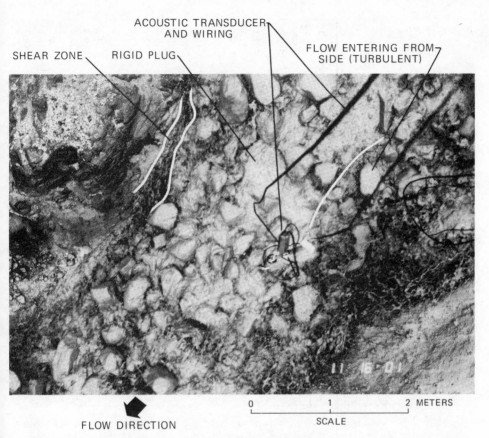

ACOUSTIC TRANSDUCER
AND WIRING

SHEAR ZONE   RIGID PLUG   FLOW ENTERING FROM
SIDE (TURBULENT)

0   1   2 METERS
SCALE

FLOW DIRECTION

Figure 13.11   Vertical overhead 35-mm photograph of peak flow of Sept. 9, 1982 debris flow at location 2B at Shoestring study site. Flow width averages 3.3 m.

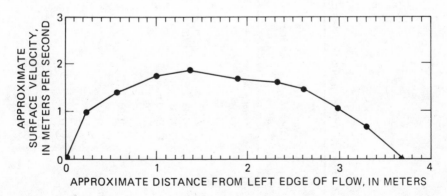

Figure 13.12   Approximate horizontal velocity profile of Oct. 1, 1981a debris flow, extracted from movie photography just above location 1B at Shoestring study site. Scale visually estimated.

Table 13.2 Physical properties of flow samples[a].

Flow	Time collected[b] (h:min:s)	Slurry temperature (°C)	Sediment conc. by weight[c] (percent)	Sediment conc. by volume (percent)	Bulk density (g/cm³)	Mean grain size (φ units)	Sorting coefficient (φ units)	Skewness	Silt and clay fraction (percent)
Aug. 21, 1981	13:00 p	11.5	82.9	64.5	2.08	0.87	3.23	−0.08	16.4
	13:05		79.9	60.0	1.99	1.23	3.09	−0.05	17.0
	13:10		79.8	59.8	1.99	1.47	3.02	−0.03	18.7
	13:15		78.8	58.4	1.96	1.17	3.06	−0.05	17.0
Oct. 1, 1981a	13:27:15	7.0	81.4	62.4	2.03	−0.47	3.61	−0.13	10.4
	13:27:35		82.2	63.6	2.05	−0.60	3.70	−0.07	10.6
	13:28:00 p		86.0	69.8	2.15	−0.77	3.47	+0.07	9.3
	13:28:30		85.7	66.2	2.09	−0.30	3.31	−0.03	10.1
	13:38:00		81.7	62.7	2.03	−0.20	3.58	−0.07	12.2
	13:41:00		79.4	59.3	1.98	0.23	3.54	−0.24	13.1
Sept. 3, 1982	11:25	6.5	73.5(H)	51.1	1.84	1.80	2.45	+0.04	17.4
	11:26 p		77.6	56.7	1.94	2.30	2.60	−0.01	23.3
	11:27		80.9	61.5	2.02	1.27	3.22	−0.07	19.4
	11:31		73.7(H)	51.4	1.85	2.47	2.29	+0.11	23.4
Sept. 9, 1982	11:10	3.0	64.5(H)	40.7	1.67	3.00	2.39	+0.03	28.7
	11:16 p		76.3	54.9	1.91	1.50	2.69	−0.01	16.6
	11:18		77.6	56.7	1.94	0.97	3.50	−0.17	16.3

Oct. 30, 1983	13:15	4.0	75.1	53.2	1.88	1.63	2.69	−0.07	15.7
	13:16 p		77.6	56.7	1.93	0.77	3.10	−0.10	13.3
	13:18		75.3(H?)	51.4	1.85	1.70	2.64	−0.05	15.6
	13:22		72.4(H?)	49.7	1.82	1.30	2.50	−0.09	12.9
	13:23		70.5(H?)	47.4	1.78	1.87	2.24	+0.03	14.4
	13:28		65.9(H?)	42.2	1.70	2.07	2.36	+0.04	16.6
April 30, 1981a	10 s	33	61.3	37.4	1.62	2.93	2.14	+0.01	29.9
April 30, 1981b	5 s	33	77.5	56.5	1.93	0.27	3.94	−0.24	17.1
	15 s		81.2	62.0	2.02	−0.13	3.45	−0.02	12.8
	30 s		83.2	65.1	2.08	−0.10	3.23	+0.09	11.7
April 30, 1981c	10 s	36	83.0	64.8	2.07	−1.20	3.65	+0.27	11.0
	20 s		83.5	65.6	2.08	−0.20	3.18	+0.12	9.9
	40 s		80.0	60.2	1.99	0.80	2.89	+0.09	15.3
April 30, 1981d	3 s	30	83.0	64.8	2.07	0.83	3.76	−0.29	19.1
	18 s		85.5	69.0	2.14	−1.50	3.37	+0.45	8.5
	33 s		77.6	56.7	1.94	1.13	3.02	−0.06	16.6
	48 s		79.1	58.8	1.97	0.57	2.94	+0.08	13.8

[a] Sampling sites are indicated in Table 13.1. Samples represent the matrix slurry; particles larger than about 9 cm in diameter could not be sampled.
[b] Sampling times denoted by p were close to time of peak flow. Times for crater samples (approximate) indicated as seconds after front had passed.
[c] Hyperconcentrated flow indicated by (H).

Physical composition of the slurries

The debris flows studied were composed primarily of normal-density dacite and andesite, with minor dacite pumice. The cobble- and boulder-sized clasts were subrounded to subangular in shape. Pebble-sized clasts were commonly very angular. Tephra deposits from the 1980 and more recent eruptions probably composed a large proportion of the finer fractions. Clay-sized particles, although not analyzed, are believed to be lithic fragments rather than clay minerals, due to the young age and limited weathering of the source material.

Sediment concentrations of observed Shoestring debris flows (matrix) ranged from 76.3 to 86.0 wt percent solids (Table 13.2). This is equivalent to a range in volume concentration of between 54.9 and 69.8 percent. Hyperconcentrated flows associated with the debris flows ranged from 61.3 to 73.5 wt percent. Concentrations of the crater site debris flows ranged from 77.5 to 85.5 wt percent. Temperatures of the sediment−water mixtures ranged from 3 to about 36°C.

A bouldery flow front undoubtedly has the highest sediment concentration of an entire flow (close to 100 wt percent), but it poses major sampling problems. All the flows observed were sampled immediately behind the bouldery front. In some cases this first sample had the highest concentration of the sample sequence, but in other cases, the second or third sample did (Table 13.2).

After the head of a flow is past, sediment concentration generally decreases toward the tail of the flow (Figs. 13.6 and 13.9). At some point in the tail of the flow, depending on particle size distribution, the macroscopically single-phase debris flow slurry will transform to a hyperconcentrated two-phase mixture. For the flows studied this sediment concentration threshold lies between 73.7 and 76.3 wt percent solids with a narrow transition zone. Even after discharge in the stream channel is nearly back to its original level, sediment concentration commonly remains in the hyperconcentrated range.

Particle size distributions for dipped samples from the Shoestring and crater sites are summarized as cumulative curve envelopes (Fig. 13.13). Grain size statistical parameters are summarized in Table 13.2. All the samples are very poorly sorted and most size distributions have symmetrical tails (cf. Folk, 1965). Sorting is only slightly better in the hyperconcentrated flow samples than in the debris flow samples. The distributions tend to be bimodal, with peaks in gravel and sand classes. This is particularly obvious when the data are plotted as frequency of individual size catagories, such as for two samples from the Oct. 1, 1981a flow (Fig. 13.14). This diagram shows that the positions of the modal peaks do not change; a decrease in the gravel fraction is simply compensated for by an increase in the sand peak.

All the samples are poor in fines. Clay-sized particle contents range from 0.5 to 2.5 wt percent. Total fines content (silt- and clay-sized particles) ranges from 8 to 24 wt percent for debris flows, and up to 30 wt percent for hyperconcentrated flows. Silt and clay content generally increases within individual flows as total sediment concentration decreases, a relative increase due to the

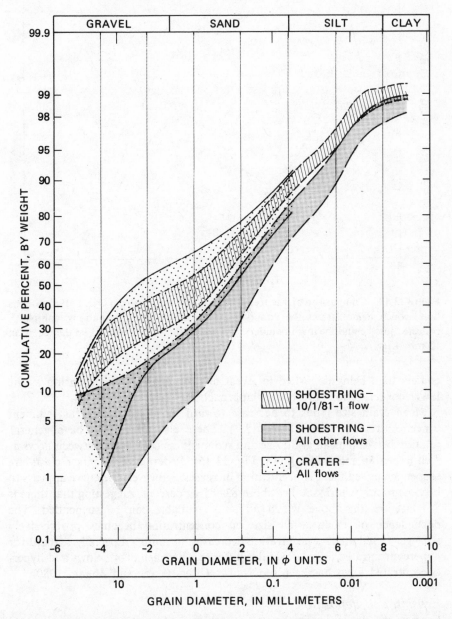

Figure 13.13 Envelopes of cumulative particle size distribution curves for Shoestring and crater slurry samples (matrix slurry, particles finer than 95 mm).

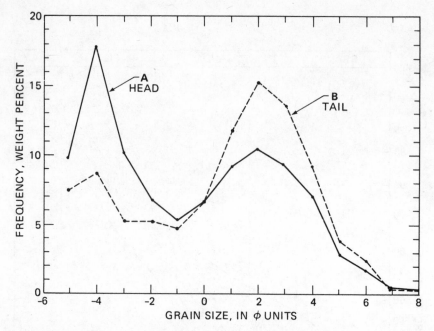

Figure 13.14 Comparison of size frequency for two samples from Oct. 1, 1981a debris flow, which demonstrates the bimodal nature of the size distributions: sample A collected just behind the main bouldery flow front; sample B collected on the recession 13 min after A.

coarser load being deposited in levees or dropping out of suspension as the lower concentration threshold is approached.

Mean grain size tends to decrease toward the tail of a flow as sediment concentration decreases (Table 13.2). There also appears to be a general relation between mean grain size and sediment concentration between flows as well as within individual flows (Fig. 13.15). Progressively coarser clasts are suspended as sediment concentration increases. However, the curve appears to become roughly horizontal at about 83–85 wt percent, suggesting that there is a concentration above which all sizes available can be suspended. The dependence of mean grain size on concentration becomes progressively stronger as sediment concentration decreases, becoming especially marked for hyperconcentrated flows. A similar difference between debris flows and hyperconcentrated flows has been reported in China (Kang and Zhang, 1980).

Erosion and deposition

Erosion and deposition in the study reach at the Shoestring site was observed in detail for only one event, the Oct. 1, 1981a debris flow. Prior to the arrival of the flow the active channel (where unconfined by bedrock) was mostly rectangular in shape, 0.5–1.0 m deep, and 4–6 m wide. The channel bed was lined with cobbles and boulders and was relatively rough.

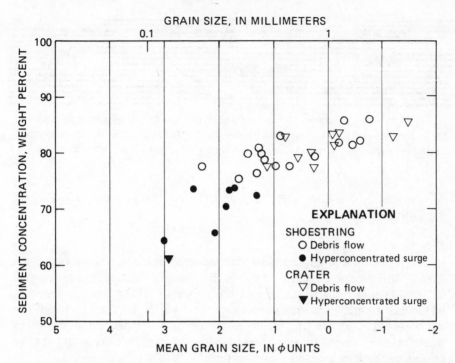

Figure 13.15 Relation between mean grain size and sediment concentration by weight for all flows.

The passage of the Oct. 1, 1981a debris flow left a channel that was much more U-shaped in cross section, creating this by a combination of deposition in the corners of the former rectangle and downcutting in the center (cf. Johnson, 1970). A thin (0.5–1.0 cm) compact, muddy sand coating was left by the flow on the surface of the bed. This accreted layer may be analogous to the "sole layer" described by K. M. Scott (personal communication, 1984) for much larger flows. Openwork lateral boulder levees formed intermittently where channel margins were flat enough to support them. At the location where samples were collected (Fig. 13.2, location 1B) the levees were composed of boulders 0.5–0.7 m in diameter left slightly below the high mudline, though some of the boulders protruded up to half their diameter above this line.

Slurry broke through the levees in a number of locations to be deposited as steep-fronted lobes of debris. Lobes composed of only sand and finer particles were less than 1 cm thick. Those containing clasts up to fine gravel size were about 4 cm thick. Boulders present in the debris mixture increased the thickness up to about 1 m.

After the head of the flow had passed it could be seen that 15–25 cm of downcutting had occurred in mobile-bed reaches (Fig. 13.9). Shortly after the transition from debris flow to hyperconcentrated flow occurred, bank under-cutting and widening of the channel was observed and the rectangular channel

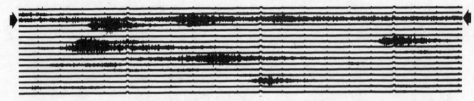

Figure 13.16   Seismic signal (between arrows) from East Dome seismograph at Mount St. Helens (1 km distant) recording a portion of Oct.1, 1981a debris flow in Shoestring channel. Interval between ticks is 10 s. Other signals are rockfalls in crater.

form began to be re-established. In places this new channel was incised within the U-shaped channel.

Seismic effects
In debris flows the grinding action of boulders against other boulders and the channel bed produces ground vibration that leaves a distinctive seismic signal (cf. Okuda et al., 1980). The signal from the Oct. 1, 1981a flow was recorded by the East Dome seismograph approximately 1 km distant (Fig. 13.16). When not obscured by wind noise, the other debris flows recorded at the Shoestring site left similar seismic traces. Areas of increased amplitude are assumed to be caused by the flow cascading over waterfalls.

Discussion

Factors affecting velocity
Velocity of debris flows should be some function of the independent variables flow depth, channel gradient, roughness imposed by the channel bed, and internal resistance to deformation which is controlled largely by the sediment concentration and particle size distribution.

In steady, uniform open channel flow of clear water, the functional relation between these factors can be represented by the Manning equation

$$\bar{v} = n^{-1}R^{2/3}S^{1/2} \tag{1}$$

where $\bar{v}$ is the mean velocity of the fluid (in metric units), n is the roughness coefficient, R is the hydraulic radius (approximately equal to mean depth for wide channels), and S is the energy slope (equal to the channel slope for uniform flow). Although the flow of the studied debris flows was neither uniform nor steady, the fluid was not clear water, and only surface velocity could be measured, this or some similar functional relationship should be operative for debris flows.

To evaluate the applicability of the Manning equation, estimated $\bar{v}$ was plotted against $R^{2/3}S^{1/2}$ for the five debris flows for which data were collected

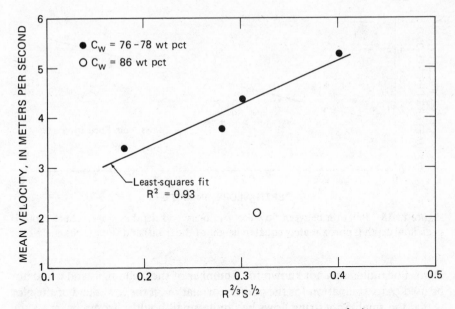

Figure 13.17 Relation between average fluid velocity and $R^{2/3}S^{1/2}$, where R is hydraulic radius and S is channel slope, for five debris flows (peak flow) from Shoestring study site. Sediment concentration C_w is indicated.

(Fig. 13.17). Four of the data points were for flows with matrix slurry concentrations between 76 and 78 wt percent, and these four suggest a good linear relationship between velocity and $R^{2/3}S^{1/2}$. For the least-squares fitted line $R^2 = 0.93$ and the slope of the line is $1/n = 8.81$. Thus $n = 0.114$, a higher value than the range of 0.065–0.085 that would be expected for streamflow (G. L. Gallino, personal communication, 1984). The fifth data point for a slurry with an 86 wt percent concentration plots well below this line, with a velocity of only about half what it might have been with a lower sediment concentration.

Although five points comprise a meager data set, the results suggest the Manning uniform flow equation may be appropriate as a predictive equation for debris flows with similar sediment concentrations. This has also been suggested by Antonius Laenen and R. P. Hansen's (personal communication, 1984) successful use of a modified Manning equation for computer routing large debris flows down stream channels at Mount St. Helens, where model-computed velocity and peak stage values were in good agreement with observed and indirectly computed values. It would also appear that slurry sediment concentration is an important factor in controlling debris flow velocity.

Front velocity of debris flows with bouldery fronts correlated better ($R^2 = 0.78$) with DS (hydraulic depth × slope) (Fig. 13.18) than it did ($R^2 = 0.74$) with $D^{2/3}S^{1/2}$. In this case D is the height of the boulder front.

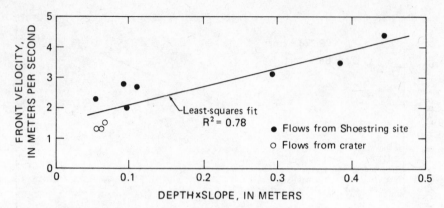

Figure 13.18 Relation between flow-front velocity and depth × slope, where depth is peak flow depth (approximately equal to height of the front) and slope is channel slope.

Hydraulic radius was not known for a number of these channels and could not be used. An explanation for the poorer correlation at the lower end of the plot is that the small Shoestring flows had quite small boulder fronts, whereas the crater flows of the same magnitude had unusually large fronts (Fig. 13.8). This relation may also form the basis for a useful predictive equation when more data points are included.

Estimating slurry shear strength for one flow

Johnson (1970, 1984) has developed a method for calculating total shear strength of a debris flow slurry based on the relative width of the rigid plug when such a plug occurs in debris flows, assuming a semi-elliptical channel and steady, unidirectional flow:

$$k = \frac{(W_p/2)\gamma_d \sin \beta}{(W/2d)^2 + 1} \qquad (2)$$

where k = shear strength (dyn/cm^2), W_p = width of plug (cm), W = width of flow (cm), d = depth of flow (cm), β = channel slope (deg), and γ_d = unit weight of slurry (sampled slurry matrix) (dyn/cm^3).

Averaging the irregular widths of the plug and flow for the Sept. 9, 1982, debris flow, which occurred in a trapezoidal channel, results in a range of shear strength values from 7.8×10^3 to 8.4×10^3 dyn/cm^2. This estimate of shear strength is 30–40 percent higher than that measured by Johnson (1970) for debris flows of similar size at Wrightwood, California, and it is a minimum value because unit weight would have been greater if the large clasts were included. It can be seen from photographs of the flow (Fig. 13.11) that the biggest particles in the Sept. 9, 1982, flow were quite large relative to flow width and closely packed together. Boulders up to 0.7 m in diameter were photographed. This evidence suggests that interlocking of these coarse particles is providing a large proportion of the shear strength responsible for plug formation.

Estimating viscosity for one flow

Bingham viscosity (μ_b) can be estimated for a semi-elliptical channel from the equation

$$\mu_b = \frac{kW_p[(W/W_p) - 1]^2}{4u_m} \tag{3}$$

where u_m is plug velocity (Johnson, 1970, 1984). Substituting values of strength and channel dimensions from the previous section into the equation, the estimated Bingham viscosity for the Sept. 9, 1982, debris flow is 80–120 P. This range is more than an order of magnitude lower than viscosities computed for the Wrightwood debris flows (Johnson, 1984).

This computation suggests that, although the Sept. 9, 1982, debris flow recorded at the Shoestring site had a relatively high shear strength, it had a relatively low viscosity for a debris flow. This conclusion is supported by photographs that show interlocking boulders composing the wide rigid plug and a very fluid matrix that was extremely turbulent where it was being sheared (Fig. 13.11). Samples of the flow matrix also had a relatively low sediment concentration for a debris flow slurry (Table 13.2). The properties of the Sept. 9, 1982 flow contrast with those of the Oct. 1, 1981a flow, which did not have a rigid plug, flowed at a slower velocity, had a higher sediment concentration, and had the general appearance of being more viscous.

Sediment concentration threshold

Observations and photographs of flowing sediment–water mixtures studied at Mount St. Helens have shown that slurry matrix samples with concentrations of 76.3 wt percent or greater had sufficient shear strength to suspend gravel-sized particles. Mixtures having concentrations of 73.7 wt percent or less could not suspend gravel-sized particles. Furthermore, the more dilute mixtures exhibited vigorous turbulence over their entire flow surfaces. These visual differences were the basis for differentiating between debris flow and hyperconcentrated flow in this study.

The threshold separating debris flow from hyperconcentrated streamflow may not, however, be the same as the boundary between Newtonian and non-Newtonian fluid behaviour. Very dilute slurries may have insufficient strength to suspend gravel but enough to suspend sand-sized particles, thereby remaining non-Newtonian. Such transitional slurries have been observed to occur over concentration ranges of 5 wt percent or more when a high percentage of fines is present in the mixture (Pierson, 1985). Furthermore, hyperconcentrated streamflow samples that appear to have no shear strength could still be non-Newtonian fluids because of nonlinear stress-strain rate relationships. Other mixtures poor in fines may remain Newtonian up to high concentrations (Fei, 1983). It is not possible to define from field observations the precise thresholds of rheological behavior, but it can be said that there was an abrupt change in fluid properties with only a slight change in sediment concentration (less than 3 wt percent) in the flows studied here.

Conclusion

The flow characteristics of 10 debris flows have been recorded at a field study site and at one other locality at Mount St. Helens over three field seasons. The study is ongoing, but on the basis of the results so far some preliminary conclusions about the flow behavior of channelized debris flows are possible.

1. The debris flows have a steep-fronted head characterized by very high sediment concentration. The tail, which is faster-moving and more dilute, feeds the head with slurry, causing flow to pile up there.

2. Progress of the debris flow head is impeded by frictional resistance provided by a concentration of coarse (commonly boulder-sized) particles at the flow front due to the conveyor-belt action of boulders rafting forward at the flow surface faster than the front is progressing. The bouldery front is continually being shouldered aside by the flow pushing from behind, and this process contributes to the formation of lateral levees.

3. Front velocity appears to be controlled largely by channel gradient and height of the boulder front. Fluid velocity, on the other hand, appears to be strongly influenced by sediment concentration, depth of flow (hydraulic radius), and slope. Less concentrated slurries tend to flow at higher velocities than more concentrated slurries.

4. At least for the flows studied, the Manning uniform flow equation appears to be valid as long as the flows have similar sediment concentrations. They exhibit higher roughness coefficients than would be expected for streamflows of similar magnitude.

5. Observed flow behavior suggests that a range of slurry shear strengths and viscosities may occur in different debris flows originating from the same general source area. The size and relative amount of large particles in the slurry appear to have an important effect on slurry shear strength.

6. Particle size distributions of slurry samples were bimodal with modes in the gravel and fine sand range, and very poorly sorted with tails of the distributions roughly symmetrical. Up to a point, sediment concentration increased as progressively coarser sediment was suspended.

7. A wide range of slurry temperatures was recorded, but warm flows did not appear to differ significantly in their flow behavior from cool ones.

8. The transition from hyperconcentrated flow to debris flow occurred between sediment concentrations of about 74 and 76 wt percent.

9. On 8° slopes, the debris flows were erosive and tended to convert rectangular channels to U-shaped channels. Hyperconcentrated flow toward the end of the flow tail caused bank undercutting and channel widening and sometimes further incision.
10. Significant seismic energy is generated by debris flows.

References

Benjamin, J. R., and C. A. Cornell, *Probability, Statistics, and Decision for Civil Engineers,* 684 pp., McGraw-Hill, New York, 1970.

Beverage, J. P., and J. K. Culbertson, Hyperconcentrations of suspended sediment, *Journal of the Hydraulics Division, Proceeding of the American Society of Civil Engineers, 90,* 117–128, 1964.

Fei, Xiang Jun, Grain composition and flow properties of heavily concentrated suspensions (English abstract), in *Proceedings of the Second International Symposium on River Sedimentation,* pp. 307–308, Water Resources and Electrical Power Press, Nanjing, China, 1983.

Folk, R. L., *Petrology of Sedimentary Rocks,* 159 pp., Hemphill's, Austin, Texas, 1965.

Hirano, M., and M. Iwamoto, Measurement of debris flow and sediment-laden flow using a conveyor-belt flume in a laboratory, in Erosion and Sediment Transport Measurement, pp. 225–230, *International Association of Hydrological Sciences Publication 133,* 1981.

Janda, R. J., K. M. Scott, K. M. Nolan, and H. A. Martinson, Lahar movement, effects, and deposits, in The 1980 Eruptions of Mount St. Helens, Washington, edited by P. W. Lipman and D. R. Mullineaux, pp. 461–478, *U.S. Geological Survey Professional Paper 1250,* 1981.

Johnson, A. M., *Physical Processes in Geology,* 577 pp., Freeman, Cooper, San Francisco, 1970.

Johnson, A. M., Debris flow, in *Slope Instability,* edited by D. Brunsden and D. B. Prior, pp. 257–361, Wiley, Chichester, 1984.

Kang, Z., and S. Zhang, A preliminary analysis of the characteristics of debris flow (English abstract), in *Proceedings of the International Symposium on River Sedimentation,* pp. 213–225, Chinese Society of Hydraulic Engineering, Beijing, China, 1980.

Kellerhals, R., and D. I. Bray, Sampling procedures for coarse fluvial sediment, *Journal of the Hydraulics Division, Proceeding of the American Society of Civil Engineers, 97,* 1165–1179, 1971.

Neall, V. E., Lahars as major geologic hazards, *Bulletin of the Association of Engineering Geologists, 14,* 233–240, 1976.

Okuda, S., H. Suwa, K. Okunishi, K. Yokoyama, and M. Nakano, Observations on the motion of a debris flow and its geomorphological effects, *Zeitschrift für Geomorphologie Supplement Band, 35,* 142–163, 1980.

Pierson, T. C., Effects of slurry composition on debris flow dynamics, Rudd Canyon, Utah, in *Proceedings of the Specialty Conference on Delineation of Landslide, Flash Flood, and Debris Flow Hazards in Utah,* in press, 1984.

Pierson, T. C., and K. M. Scott, Downstream dilution of a lahar: transition from debris flow to hyperconcentrated streamflow, *Water Resources Research,* in press, 1985.

Plafker, G., and G. E. Ericksen, Nevados Huascaran Avalanches, Peru, in *Rockslides and Avalanches, 1*, edited by B. Voight, pp. 277–314, Elsevier, Amsterdam, 1978.

Varnes, D. J., Slope movement types and processes, in *Landslides—Analysis and Control,* edited by R. L. Schuster and R. J. Krizek, pp. 11–33, National Academy of Sciences, Washington, 1978.

Watanabe, M., and H. Ikeya, Investigation and analysis of volcanic mudflows on Mount Sakurajima, Japan, in Erosion and Sediment Transport Measurement. pp. 245–256, *International Association of Hydrological Sciences Publication 133,* 1981.

14
Dynamics of slow landslides: a theory for time-dependent behavior

Richard M. Iverson

Abstract

Motion of large, complex, persistently active, earthflow-like landslides commonly varies in time and space. Unusually good documentation of such unsteady, nonuniform landslide motion is provided by a 12-year record of the behavior of Minor Creek landslide in northwestern California. Based on inferences drawn from Minor Creek landslide data and on deductions drawn from general physical principles, a mathematical theory of unsteady, nonuniform landslide motion is developed. The theory employs a generalized constitutive model that can represent landslide deformation styles ranging from "dilatant" viscoplastic flow to rigid–plastic frictional slip. A perturbation analysis, which embodies the generalized constitutive model and departs from equations that reflect steady landslide shear deformation, is used to investigate unsteady, nonuniform landslide motion. The analysis shows that transient, localized perturbations in landslide motion propagate slowly downslope as kinematic waves and spread rapidly outward by diffusion. The importance of perturbation propagation relative to perturbation diffusion is dictated principally by the value of a single dimensionless parameter, called the landslide Peclet number. This number can be expressed as an algebraic function of landslide physical properties, and its value appears to hold major implications for landslide behavior. The transient response of Minor Creek landslide to an episode of toe erosion can be explained by applying the perturbation theory and evaluating the landslide Peclet number.

Introduction

Large, complex areas of surficial mass movement are a common feature of steep, regolith-mantled terrain throughout the world. Particularly in areas subject to geologically recent tectonism and high seasonal precipitation, lithologic, hydrologic, and topographic factors combine to form environments conducive to large-scale hillslope instability. Such instability or gravitational disequilibrium may be manifest as simple, discrete landslides that move abruptly downslope to establish new equilibrium hillslope configurations. Commonly, however, the instability leads to development of terranes characterized by nearly continuous mass movement that varies in space and time. Such mass-movement terranes offer challenging geotechnical problems, and their geomorphological role as conveyors of sediment can be locally enormous (Swanson and Swanston, 1977).

Important features of many mass-movement terranes are large, complex landslides that are typically deep-seated (up to 50 m thick), slow-moving (usually less than 5 m/yr), and persistently active (for hundreds of years or more). Spatially varied, time-dependent movement patterns are characteristic of these complex landslides and are the object of the research described here.

Terminology

In the Coast Range of northwestern California, large, complex landslides are common and have traditionally been called earthflows (Kelsey, 1978). However, most classical earthflows are less than 2 m thick, occupy midslope portions of hillslopes, and appear to move primarily by boundary slip on discrete, slickened planes (Keefer and Johnson, 1983). The massive, complex landslides of northwestern California are, in contrast, typically more than 5 m thick, extend from ridges to stream channels, and appear to move primarily by boundary shear distributed through zones of varied thickness. In this chapter, therefore, large, complex, earthflow-like landslides similar to those of northwestern California will be referred to by the generic terms *landslide complex* or, simply, *landslide*. This terminology is intended to avoid semantic confusion. It also serves to emphasize the broad applicability of much of the analysis presented here to mass movements other than earthflows. Movement patterns of earthflow-like landslide complexes are, however, the object of particular emphasis in this chapter.

Movement patterns of landslide complexes

Superposed on the overall downslope motion of many landslide complexes are movements associated with surficial, seasonal soil creep, episodic surficial slumping and sliding, and effects of external agencies other than the steady action of gravity (Swanson and Swanston, 1977). Soil creep and episodic surficial movements contribute to continual mass redistribution within landslide complexes (Janda et al., 1980). They are also suggestive of the complex rheology and variety of deformation styles within the regolith mantle. Effects of external agencies other than gravity include erosion of the toes of landslide complexes by adjacent streams, which may initiate transient accelerated landslide movement (Kelsey, 1978; Iverson, 1984). Variable hydrologic conditions, regolith strength, rates of headscarp slumping, or human influences may also trigger locally or periodically accelerated motion (e.g., Blackwelder, 1912; Wasson and Hall, 1982). Thus movement patterns of landslide complexes are not simple; they respond continually to a complicated combination of environmental controls.

Downslope movements and deformation in landslide complexes are therefore neither uniform in space nor steady in time. Understanding this unsteady, nonuniform landslide motion is important because it holds large implications for projecting sediment yields and geologic hazards. Furthermore, unsteady, nonuniform behavior of large landslides reveals fundamental aspects of landslide dynamics that are not apparent in simple static or steady-state situations. To understand unsteady, nonuniform landslide behavior better it is necessary to measure the behavior accurately and to develop a physically viable theory for the behavior. This chapter summarizes field and theoretical research directed toward that end.

Behavior of Minor Creek landslide

An unusually complete data set that reflects long-term, unsteady, nonuniform landslide motion has been assembled for Minor Creek landslide in the Redwood Creek drainage basin, Humboldt County, northwestern California (Fig. 14.1). As part of a U.S. Geological Survey forest geomorphology research program, transverse stake lines and rain gauges were installed on Minor Creek landslide and several other landslides in the Redwood Creek basin in the early 1970s. Stake lines were surveyed repetitively to record landslide movements, and rain-gauge data were used to correlate temporal movement and precipitation patterns. Minor Creek landslide was subsequently selected for more detailed study, and two recording extensometers, two flumes to record gully discharges, and six inclinometer tubes were installed on or near the landslide between 1975 and 1978. Data obtained until 1982 from this instrumentation were used to characterize the overall motion and sediment discharge of Minor Creek landslide (Harden et al., 1978; Janda et al., 1980; Noland and Janda, 1985). In 1982 to facilitate assessment of stress states and deformation rates, additional instrumentation was installed on Minor Creek landslide. This instrumentation included over 50 piezometers, nine inclinometer tubes, and a longitudinal stake line that incorporated several strain rhombs.

Some important generalizations about the characteristics and behavior of Minor Creek landslide have emerged from analysis of data collected through the summer of 1984 (Iverson, 1984).

1. The landslide material consists primarily of poorly sorted, dense, low-plasticity, gravelly clayey sand. The clay fraction appears to dominate the material's behavior, however, and the material's hydraulic conductivity is about 10^{-7} m/s. Its residual friction angle is about $20°$.

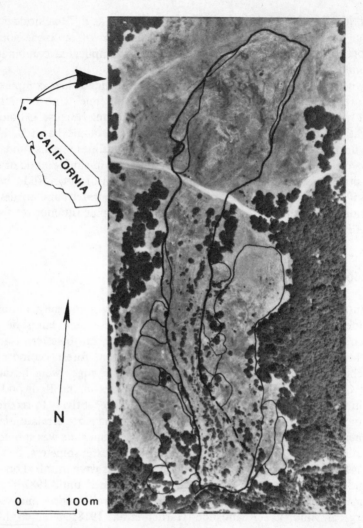

Figure 14.1 Vertical aerial photograph of Minor Creek landslide taken on July 13, 1982. The landslide, located at 40°58′N, 123°50′W, moves from north to south and terminates at the channel of Minor Creek. Boundaries of the main landslide are indicated by heavy lines on the photograph, and boundaries of less active, auxiliary landslides are indicated by lighter lines.

2. The surface slope of the landslide is broadly uniform and averages 15°. Slopes of topographic undulations depart by as much as 20° from this average, but only in very localized areas.

3. The thickness of the landslide is roughly uniform and varies from about 4.5 m to 6.5 m. There is a tendency for the landslide to be thinner nearer the toe.

4. Most shear deformation occurs in a zone about 1–2 m thick at the base of the landslide. No discrete shear plane is readily detectable in this zone. Some creep deformation occurs above the basal shear zone, but it contributes minimally to the overall downslope landslide movement. Lateral shear zones that range from several centimeters to several meters wide are present at the margins of the landslide.

5. Groundwater potentiometric levels seldom, if ever, fall below the base of Minor Creek landslide. The potentiometric levels measured near the landslide's base rise rapidly with the onset of heavy rains in November, December, and January and then remain at relatively stable, high levels until rains begin to subside in April, May, and June. Vertical potentiometric gradients persist throughout the year in the landslide, and significant water storage occurs in a thick unsaturated zone.

6. The landslide moves at slow summer creep rates of a few millimeters per month until it smoothly accelerates to rates of centimeters or decimeters per month sometime during the rainy season. The timing of the onset and later cessation of the fast-movement period corresponds reasonably well with the timing of seasonal rises and falls of potentiometric levels.

7. Movement rates during the annual fast-movement period are quite steady and exhibit few short-term fluctuations. The movement rates are not appreciably affected by potentiometric fluctuations that occur in response to individual rainstorms.

8. Long-term surficial movement rates average about 0.4 m/yr. This movement is not equally distributed from year to year; from 1973 to 1984 over half the total movement of midslope portions of the landslide occurred during two years, each of which had unusually wet Novembers.

9. Movement and deformation are not equally distributed along the landslide's length. For example, stream undercutting of the landslide's toe from 1980 to 1984 catalyzed stretching deformation and rapid motion that was not transmitted far upslope. Natural longitudinal strains as large as 1 and elevation decreases as large as 1.4 m were measured near the toe during the 1982–83 movement season. A transient mass depletion is thus inferred to have occurred at the landslide toe.

10. Nonuniform components of the landslide's motion are evidenced by zones of measurable longitudinal compression or extension. Clear morphological expression of the extension zones is normally present in the form of surface cracks and scarps. Commonly, however, there is little obvious expression of the compression zones, probably because the landslide material fails less readily in compression than in extension.

The data upon which the above statements are based are specific to Minor Creek landslide. However, based on the similarity of Minor Creek landslide to other features elsewhere (cf. Harden et al., 1978; Kelsey, 1978; Swanson and Swanston, 1977; Swanston, 1981; Wasson and Hall, 1982), it appears reasonable to propose that many of the gross behavioral characteristics of

Minor Creek landslide are representative of those of landslide complexes in general. Thus, on the basis of inferences drawn from data collected at Minor Creek landslide and deductions drawn from general physical principles, a mathematical theory of landslide complex behavior can be formulated.

Theory for unsteady, nonuniform landslide behavior

A mathematical theory for the time- and space-dependent components of downslope landslide motion is outlined here. A detailed development of the theory is presented elsewhere (Iverson, 1984). The theory is based on physical conservation laws and a postulated constitutive model for the behavior of landslide material. The theory is inherently approximate, but it is developed in a very generalized form, so it is believed to be broadly applicable to a wide variety of hillslope mass-movement styles.

Constitutive equation

The most important element of the theory is a postulated constitutive relation for the rheological behavior of landslide material. The exact nature of the rheology of landslide material is unknown and undoubtedly varies from place to place. It is essential, therefore, that the postulated constitutive relation be very generalized and capable of representing a broad range of rheological behaviors.

The constitutive relation used here represents landslide material as a nonlinear viscoplastic substance that yields in response to a critical combination of octahedral normal and shear stresses and subsequently deforms in a generalized, rate-dependent fashion. The relation is expressed mathematically as

$$S = 2\mu(D)D + \frac{k + \alpha p}{\Pi_D^{\frac{1}{2}}} D \qquad \text{for } \Pi_S^{\frac{1}{2}} > k + \alpha p \qquad (1a)$$

$$D = 0 \qquad \text{for } \Pi_S^{\frac{1}{2}} \le k + \alpha p \qquad (1b)$$

wherein

$$\mu(D) = \mu_0 \left(\frac{\Pi_D^{\frac{1}{2}}}{D_0}\right)^{(1-n)/n} \qquad (1c)$$

in which S is the deviatoric stress tensor, assumed symmetric; D is the rate-of-deformation tensor, a deviator; D_0 is the reference rate-of-deformation tensor; Π is the second principal scalar invariant of the subscripted tensor; μ is the apparent viscosity of the landslide material and is a function of D; μ_0 is the equivalent Newtonian viscosity of the landslide material and is measured when $D = D_0$; n is an index of apparent viscosity nonlinearity; k is the cohesional component of landslide material strength; α is the frictional coefficient of landslide material strength; and p is the effective confining pressure or effective octahedral normal stress. The complete phenomenological rational and derivation of (1) have been described in a previous chapter (Iverson, 1985).

The most important feature of the constitutive model represented by equations (1) is the nonlinear viscous flow element represented by (1c). By varying the value of n in (1c), equations (1) may simulate deformation styles ranging from shear-thickening (i.e., "dilatant") viscoplastic flow to a limiting case of perfectly plastic frictional slip (Fig. 14.2). Thus, if the value of n is left unspecified in a mathematical analysis of landslide motion that employs (1), deductions based on the analysis should have broad applicability.

Another important feature of equations (1) is that plastic yielding is governed by a frictional, pressure-dependent criterion. The inequalities on the right sides of (1a) and (1b) represent a viscoplastic adaptation of the plastic yield criterion proposed by Drucker and Prager (1952). The Drucker–Prager criterion accounts for linear dependence of yield strength on the effective confining pressure, and it is the simplest possible extension of the classic Mohr–Coulomb criterion to three-dimensional stress states. Thus, by direct analogy to the Mohr–Coulomb criterion, the role of porewater pressures in modifying the effective stresses and frictional strength of landslide material can be regarded as implicit in (1) (cf. Iverson, 1983).

Equations (1) imply that if landslide groundwater heads rise seasonally to sufficiently high, relatively stable levels, the landslide should first yield and then accelerate until reaching a steady speed. Figure 14.3 depicts measurements

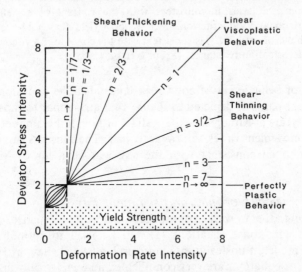

Figure 14.2 Effect of the value of n on the landslide rheology represented by equations (1). $\Pi_S^{1/2}$ and $\Pi_D^{1/2}$ are used as scalar measures of the deviator stress intensity and deformation rate intensity, respectively. The values $D_0 = \frac{1}{2}$, $k + \alpha p = 1$, and $\mu_0 = 1$ were used in constructing this plot from equations (1). Consistent units (such as meter–kilogram–second) were assumed.

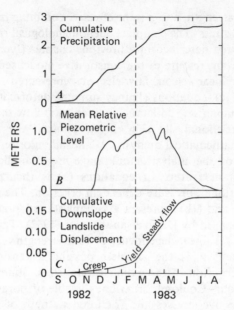

Figure 14.3 Relationship between seasonal hydrologic conditions and displacement across a lateral shear zone at Minor Creek landslide. Precipitation and displacement data were collected using continuously recording gauges. Piezometric data were collected weekly from two observation wells near the displacement gauge and are referenced to a datum 2.1 m below the ground surface. Landslide yielding in response to high piezometric levels is consistent with the concept of an effective-stress-dependent, frictional yield criterion. Steady, post-yield flow under conditions of sustained high piezometric levels is consistent with the concept of a viscous component of landslide deformation resistance. (From Iverson (1985).)

of this type of behavior at Minor Creek landslide. The hydrologic behavior and movement patterns depicted in Figure 14.3 are similar to those measured during most years at Minor Creek landslide, although the amount and timing of landslide movement varies from year to year. The concepts embodied by (1) thus appear to be consistent with the seasonal behavior of Minor Creek landslide.

Steady-state velocity profiles and sediment flux
By combining them with appropriate momentum-balance and mass-conservation equations, equations (1) can be used to formulate a three-dimensional model of unsteady, nonuniform landslide behavior. However the resulting tensor equations are very complicated and analytically intractable. A simpler approach employs the one-dimensional form of equation (1a), which can be written as

$$\tau_{yx} = (2D_0)^{(n-1)/n} \mu_0 \left(\frac{dV_x}{dy}\right)^{1/n} + c + \tau_{yy} \tan \phi \qquad (2)$$

in which τ_{yx} is shear stress acting in the x-direction on planes normal to the y-direction, τ_{yy} is effective normal stress (positive in compression) in the y-direction, and V_x is velocity in the x-direction. The soil cohesion c and residual friction angle ϕ are the one-dimensional surrogates for k and α in (1). Equation (2) embodies the assumption that all landslide motion is unidirectional in the x-direction and that all deformation occurs through inhomogeneous simple shear in the x–y plane (Fig. 14.4). The equation appears to be reasonably valid for many landslides, in which most motion is directly downslope and most resistance is provided by basal shear. A rigorous derivation of (2) from tensor equations of motion is provided elsewhere (Iverson, 1984).

Integration of (2) under the assumption of a geostatic hillslope stress field with slope-parallel groundwater seepage yields, after some manipulation, expressions that reflect idealized landslide dynamics for the simple case of steady, one-dimensional shear deformation on a uniformly inclined slope (cf. Keefer, 1977; Suhayda and Prior, 1978; Iverson, 1984). The most important expressions so obtained are those that represent the dimensionless basal-shear-zone velocity profile

$$\frac{V_x}{V_{x\max}} = 1 - \left(\frac{y - T}{h - T}\right)^{n+1} \tag{3}$$

the thickness of a more or less rigid body overlying the basal shear zone,

$$T = -A_2/A_1 \tag{4}$$

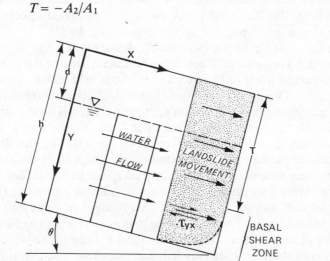

Figure 14.4 Schematic diagram and definitions for idealized, steady, one-dimensional landslide motion in the x–y plane. Downslope landslide motion is accompanied by steady, slope-parallel groundwater flow at a constant velocity relative to the solid matrix. Landslide deformation occurs as inhomogeneous simple shear in a basal shear zone of finite thickness $h - T$. The shear stress τ_{yx} in the basal shear zone is balanced by the viscous momentum flux that diffuses outward from the landslide's stable base.

and the downslope landslide sediment flux,

$$Q = w \frac{(2D_0)^{1-n} A_1^n}{\mu_0^n (n+1)} (h-T)^{n+1} \left(\frac{h-T}{n+2} - h \right) \tag{5}$$

in which w is landslide width, h is landslide thickness, and $V_{x\,max}$ is the ground surface downslope landslide velocity. A_1 and A_2 are functions of landslide material properties and geometry and are given by

$$A_1 = g \cos \theta \, [(\rho_{sat} - \rho_f) \tan \phi - \rho_{sat} \tan \theta] \tag{6a}$$

$$A_2 = c + gd \, [(\rho_{un} - \rho_{sat} + \rho_f) \cos \theta \tan \phi + (\rho_{sat} - \rho_{un}) \sin \theta] \tag{6b}$$

in which g is the magnitude of gravitational acceleration, θ is the slope angle, d is the water-table depth measured normal to the ground surface, and ρ_{un}, ρ_{sat} and ρ_f are the mean mass densities of unsaturated landslide material, saturated landslide material, and the fluid that saturates the landslide pore space, respectively.

Several significant facets of idealized landslide behavior can be assessed using (3) to (6). For example, the family of basal-shear-zone velocity profiles represented by (3) is shown in Figure 14.5A. Provided for comparison in Figure 14.5B are normalized basal-shear-zone velocity profiles inferred from inclinometer-tube measurements in Minor Creek landslide and three other landslides in the Redwood Creek basin. The measured velocity profiles fall within the range of possible profiles represented by (3).

The landslide rigid-body thickness T, defined in (4), figures prominently in (3) and (5). The value of T can either be calculated from measured values of the parameters in (4) or inferred directly from field measurements in inclinometer tubes. One special condition, in which T equals h, represents a state of limiting equilibrium in the landslide, implying that it is on the verge of motion (cf. Fig. 14.4). In fact (4), which defines T, has exactly the same meaning as the equation presented by Taylor (1948, p. 431) for the limiting stable thickness of partially saturated infinite slopes with slope-parallel seepage.

The steady-state landslide sediment flux represented by (5) depends on many parameters and can vary widely. Particularly important in affecting Q are the values of ϕ, θ, and μ_0. The value of θ normally must be at least half as large as the value of ϕ for T to be less than h and the landslide to move at all. The value of μ_0 controls to a considerable extent the rate of landslide deformation in the basal shear zone. It thus influences downslope sediment flux strongly. Typical values of μ_0 are for many landslides probably in the range 10^{11}–10^{12} N s/m^2 (cf. Savage and Chleborad, 1982; Iverson, 1984).

Approximate analysis of unsteady, nonuniform landslide behavior

The simplified, steady-motion landslide model described above provides a mathematical foundation for analyzing more realistic landslide behavior. In

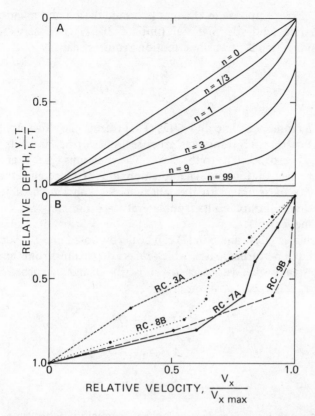

Figure 14.5 Theoretical (A) and measured (B) vertical velocity-profile shapes for basal shear zones in landslides. Velocities are normalized with respect to the ground-surface velocity, and depths are normalized with respect to the shear zone thickness. Theoretical profiles represent plots of equation (3) for different values of n. Measured profiles were inferred from repetitive inclinometer surveys of borehole casing deformation in four landslides in the Redwood Creek basin, California. Basal shear zones represented here range from 2 m to 4 m thick, and landslide velocities range from 3×10^{-10} m/s to 3×10^{-9} m/s. Hole RC-9B lies in Minor Creek landslide (Cf. Iverson, 1984).

particular, unsteady, nonuniform components of landslide motion can be anaylzed by assuming that the idealized, steady-state landslide sediment flux Q is transiently perturbed in a localized domain. The problem is then to assess the effect of the perturbed sediment flux at points far from the perturbation source. Assessment of the response of Minor Creek landslide to transient toe erosion is an example of this type of problem, and it will be considered at the end of this chapter.

The theory outlined here is based on a linear, kinematic approximation for small, unsteady, nonuniform momentum fluxes in landslides and on the principle of conservation of mass. It is mathematically similar to Nye's (1960) theory of glacier response to changes in accumulation and ablation rates. For incompressible landslide material (implied by (1)), conservation of mass during unidirectional downslope motion requires that

$$\frac{\partial Q}{\partial x} + \frac{\partial A}{\partial t} = b(x, t) \tag{7}$$

in which t represents time and $Q(x, t)$ represents the downslope volumetric sediment flux through transverse cross sections of the landslide, which have area $A(x, t)$ perpendicular to the slope. The functions $Q(x, t)$ and $A(x, t)$ are assumed to be continuous. The function $b(x, t)$ represents the volumetric lateral influx of sediment to the landslide per unit length of the landslide. Lateral sediment influx results from lateral scarp slumping and from tributary landslide movements.

The variables Q, A, and b in (7) can be decomposed into the sums of steady-state parts and perturbed parts, which reflect departures from the steady state. Denoting steady-state variables by subscript 0 and perturbed variables by subscript 1, this decomposition gives

$$Q = Q_0 + Q_1 \tag{8a}$$

$$A = A_0 + A_1 \tag{8b}$$

$$b = b_0 + b_1 \tag{8c}$$

Assuming that landslide width changes negligibly with time, A_1 can be expressed as

$$A_1 = w_0 h_1 \tag{9}$$

and by substituting (8) and (9) into (7), it can easily be shown (Iverson, 1984) that

$$\frac{\partial Q_1}{\partial x} + w_0 \frac{\partial h_1}{\partial t} = b_1 \tag{10}$$

Equation (10) is an approximate global continuity equation for landslide sediment-flux perturbations, and it is one of the equations upon which the subsequent analysis is based.

To proceed with the analysis it is necessary to know in (10) the functional form of Q_1 in relation to h_1. Such a functional form is provided by (5) if a kinematic approach is adopted. This approach requires that perturbations or departures from steady-state conditions have negligible effect on the dynamic force balance governing landslide motion. For nonuniform perturbations this requirement means that sediment-flux perturbations must be sufficiently small that attendant perturbations in slope angle, soil thickness, and water-table depth are at least an order of magnitude smaller than their steady-state values. This ensures that the style of stress in the perturbed state differs little from that in the steady state, and the behavior of arbitrarily small, unsteady, non-uniform movements can thus be studied kinematically.

Restricting attention to small perturbations, a kinematic approach is adopted; and from (5) it is deduced that Q_1 depends on nine physical parameters (n, μ_0, D_0, ϕ, c, ρ_{un}, ρ_{sat}, ρ_f, g), on landslide width, and on five kinematic variables. The kinematic dependence is represented by

$$Q_1 = Q_1(x, t, \theta_1, h_1, d_1) \tag{11}$$

This simple functional relation holds true for almost any reasonable constitutive equation that could be used to characterize rheologically the behavior of mass movements. The lack of strong dependence of (11) on the particular form of the constitutive equation is a significant advantage of this approximate theory.

Regardless of the exact functional form of (11), so long as it is continuously differentiable and single-valued, it can be expanded as a multivariable Maclaurin's series for a given x and t and then truncated to first-order terms:

$$Q_1 = h_1 \left(\frac{\partial Q_1}{\partial h_1}\right)_{h_1=0} + \theta_1 \left(\frac{\partial Q_1}{\partial \theta_1}\right)_{\theta_1=0} + d_1 \left(\frac{\partial Q_1}{\partial d_1}\right)_{d_1=0} \tag{12}$$

The approximation error due to series truncation depends on the size of the perturbations and on the nonlinearity of the function represented by (11). An analysis of the error implicit in (12) is presented elsewhere (Iverson, 1984). For theoretical purposes, however, the error of (12) can be made vanishingly small by simply reducing the perturbation size.

The partial derivatives in (12), the linearized flux equation, actually represent derivatives evaluated about the steady state (i.e., with perturbations equal to zero). Equation (12) can therefore be written as

$$Q_1 = C_0 h_1 + G_0 \theta_1 + J_0 d_1 \tag{13}$$

where

$$C_0 = \left(\frac{\partial Q}{\partial h}\right)_0 \tag{14a}$$

$$G_0 = \left(\frac{\partial Q}{\partial \theta}\right)_0 \tag{14b}$$

and

$$J_0 = \left(\frac{\partial Q}{\partial d}\right)_0 \tag{14c}$$

and the coefficients C_0, G_0, and J_0 can be explicitly evaluated by partial differentiation of (5). For reasonable values of rheological parameters and typical landslide geometries, computational experiments show that G_0 is almost universally one to two orders of magnitude larger than C_0 and J_0, which are commonly of the same order of magnitude as Q_0 (Iverson, 1984). This implies that perturbations in slope angle will have a much greater influence on landslide sediment flux than will perturbations in landslide thickness or water-table depth.

The governing equation for unsteady, nonuniform, sediment-flux perturbation behavior is derived by combining the linearized perturbation flux equation (13) with the perturbation continuity equation (10). First, it is noted that θ_1 in (13) is nearly equal to $\tan \theta_1$, because θ_1 is small by definition. It follows that a good approximation is

$$\theta_1 = -\partial h_1/\partial x \tag{15}$$

because $\tan \theta_1 = -\partial h_1/\partial x$ exactly. Inserting (15) into (13) gives

$$Q_1 = C_0 h_1 - G_0 \, \partial h_1/\partial x + J_0 d_1 \tag{16}$$

which along with (7) forms the basis for further development of the theory.

By rearranging (10), this equation can be expressed as

$$\frac{\partial h_1}{\partial t} = \frac{1}{w_0} \left(b_1 - \frac{\partial Q_1}{\partial x} \right) \tag{17}$$

Differentiation of (17) with respect to x yields

$$\frac{\partial^2 h_1}{\partial x \partial t} = \frac{1}{w_0} \left(\frac{\partial b_1}{\partial x} - \frac{dw_0}{dx} \frac{\partial h_1}{\partial t} - \frac{\partial^2 Q_1}{\partial x^2} \right) \tag{18}$$

and differentiation of (16) with respect to t yields

$$\frac{\partial Q_1}{\partial t} = C_0 \frac{\partial h_1}{\partial t} - G_0 \frac{\partial^2 h_1}{\partial x \partial t} + J_0 \frac{\partial d_1}{\partial t} \tag{19}$$

Substitution of (17) and (18) into (19) yields the linear, second-order, partial differential equation governing the one-dimensional kinematic behavior of perturbations in landslide sediment flux:

$$\frac{\partial Q_1}{\partial t} = \frac{1}{w_0} \left(C_0 + \frac{G_0}{w_0} \frac{dw_0}{dx} \right) \left(b_1 - \frac{\partial Q_1}{\partial x} \right) - \frac{G_0}{w_0} \left(\frac{\partial b_1}{\partial x} - \frac{\partial^2 Q_1}{\partial x^2} \right) + J_0 \frac{\partial d_1}{\partial t} \tag{20}$$

Equation (20) is an inhomogeneous advective–diffusive equation. The dependent variable is Q_1, and w_0, b_1, and d_1 are specified "source" or "forcing" functions which account for the mathematical inhomogeneity. The advection term (involving $\partial Q_1/\partial x$) in (20) mathematically represents undamped, down-slope propagation of sediment-flux perturbations as kinematic waves, whereas the diffusion term (involving $\partial^2 Q_1/\partial x^2$) represents damped instantaneous transfer of sediment-flux perturbations from their source to points throughout the body of the landslide.

Normalization and analysis of the advective–diffusive perturbation equation

Physical interpretation and analysis of (20) are facilitated by normalizing it with respect to steady-state landslide properties. To normalize the equation the following dimensionless parameters are introduced:

$$Q^* = \frac{Q_1}{Q_0} \qquad x^* = \frac{X}{L} \qquad t^* = \frac{tC_0}{Lw_0}$$

$$d^* = \frac{d_1}{d_0} \qquad b^* = \frac{b_1}{b_0} \qquad w^* = \frac{w_0}{L} \qquad (21)$$

$$Pe = \frac{C_0 L}{G_0} \qquad N_f = \frac{b_0 L}{Q_0} \qquad N_w = \frac{d_0 L}{Q_0}$$

In (21) L represents landslide length. One unit of dimensionless time t^* is the time required for diffusionless kinematic waves to travel the distance L (cf. Iverson, 1984). The dimensionless number Pe defined in (21) is called the landslide Peclet number by analogy to the theory of advective–dispersive solute transport. In landslides the Peclet number broadly represents the impor-tance of advective transfer relative to diffusive transfer of sediment-flux perturbations. In (21) N_f is a dimensionless number that represents the volume of sediment influx from lateral landslide margins relative to downslope, datum-state sediment flux. N_w is a dimensionless number that represents the importance of sediment-flux enhancement due to groundwater effects relative to the overall datum-state sediment flux.

Multiplication of each term in (20) by the factor Lw_0/C_0Q_0 yields the advective–diffusive perturbation equation in dimensionless or normalized form:

$$\frac{\partial Q^*}{\partial t^*} + \left(1 + \frac{1}{Pe}\frac{1}{w^*}\frac{dw^*}{dx^*}\right)\frac{\partial Q^*}{\partial x^*} - \frac{1}{Pe}\frac{\partial^2 Q^*}{\partial x^{*2}}$$

$$= N_f\left(b^* + \frac{1}{Pe}\frac{b^*}{w^*}\frac{dw^*}{dx^*} - \frac{1}{Pe}\frac{\partial b^*}{\partial x^*}\right) + N_w\frac{\partial d^*}{\partial t^*} \qquad (22)$$

The prominent role of Pe in this equation is clear. As Pe approaches infinity the equation reduces to a diffusionless kinematic-wave equation with a relatively simple inhomogeneous term. As Pe approaches zero the diffusive term gains considerable importance, and complicated elements of the inhomogeneous term assume a more significant role.

Because Pe largely determines the behavior of (22) and, hence, the behavior of sediment-flux perturbations, it is worthwhile to examine physically imposed constraints on the possible range of its values. The following relation, which expresses Pe in terms of measurable physical parameters, can be derived from (5), (14), and (21) (Iverson, 1984):

$$Pe = L \left(\frac{(\rho_{sat} - \rho_f)\tan \phi - \rho_{sat} \tan \theta}{[(h-d)(\rho_f - \rho_{sat}) - d\rho_{un}]\tan \theta \tan \phi - [(h-d)\rho_{sat} + d\rho_{un}]} \right) \left(\frac{n+2}{n+T/h} \right)$$

(23)

Trial calculations, which use the physically reasonable range of values of the parameters in (23), help identify constraints on the viable range of values of Pe. These calculations show that the Pe value is anticipated to lie in the range $0.1 \leqslant Pe \leqslant 100$ for most landslides with basal shear zones of measurable thickness (Iverson, 1984).

The style of landslide perturbation response, as indicated by solutions of (22), varies widely as a function of the value of Pe (Iverson, 1984). In cases where Pe is very small, (23) indicates that soil deformation is essentially confined to a discrete slip surface (i.e., T is nearly equal to h). Perturbations are then "felt" almost simultaneously throughout the landslide, and the whole mass responds more or less at once. The response, however, decays away from the perturbation source and is of much shorter duration than in cases where Pe is large. This type of response is consistent with the type of response expected during propagation of Rankine states of limiting equilibrium in perfectly plastic materials (e.g., Suemine, 1983). The response is affected little by viscous momentum exchange in the landslide's basal shear zone.

In contrast, for Pe values of 100, slow, wave-like response is very distinct. This is not surprising, as large values of Pe arise from physical parameter values that indicate thick basal shear zones and predominantly viscous landslide deformation styles. Diffusive effects modify perturbations somewhat even with large values of Pe, but perturbations for this Pe value travel much like simple kinematic waves at speeds that are approximated well by those of diffusionless waves. In these cases, effects of perturbations may not be felt until many tens of years later at points far downslope from the perturbation source (cf. Iverson, 1984).

Landslide response to transient toe erosion

Equation (22) can be analytically solved for several sets of boundary conditions that are relevant to landslide behavior problems (Iverson, 1984). A problem of particular interest in connection with Minor Creek landslide concerns the landslide response to a transient toe perturbation. How does the theoretical landslide response to an episode of toe erosion compare to the actual response measured during 1981–1984?

To address this problem in the simplest manner, the right-hand side of (22) is first set equal to zero. This simplification corresponds physically to a state of no lateral sediment influx or water-table fluctuation. Appropriate initial and boundary conditions to be used with (22) specify that no perturbation exists at time zero (i.e., the landslide is in equilibrium), that a finite perturbation is then enforced at the landslide toe and subsequently allowed to decay with time, and that the landslide headscarp is allowed to respond freely to downslope perturbations. Using these conditions, (22) can be solved using

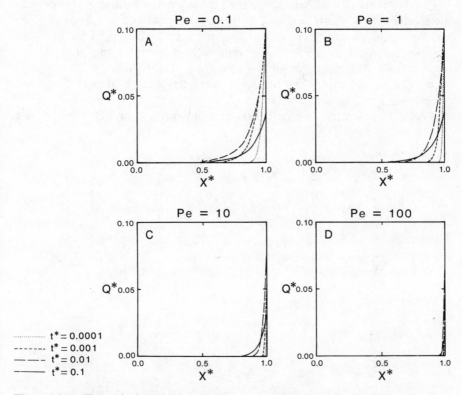

Figure 14.6 Theoretical landslide response to an episode of perturbed sediment flux at the landslide toe (i.e., at $x^* = 1.0$). The normalized sediment-flux perturbation (Q^*) has an enforced initial magnitude of 0.1 at the toe, and it subsequently decays exponentially with a half-life equal to 0.069 units of t^*. Plot A shows that where the value of Pe is small (i.e., 0.1), a perturbation can spread rapidly upslope and affect half the landslide before it decays significantly at its source. Plot D, in contrast, shows that upslope perturbation spreading is negligible where Pe is large (i.e., 100). Plots B and C show the landslide response for intermediate Pe values.

Laplace transform methods (Marino, 1978) to yield landslide sediment-flux-perturbation distributions as a function of the Peclet number, distance, and time. The solution consists of a rather complex summation of exponential functions and complementary error functions and is presented elsewhere (Iverson, 1984).

The sediment-flux-perturbation behavior represented by the solution is depicted in Figure 14.6. This figure shows that perturbations spread considerably further and faster upslope for small values of Pe than for large values. Thus, for cases where landslides are dominantly rigid, with thin basal shear zones, toe perturbations are felt more significantly upslope than in cases where landslide behavior is more viscous.

An important feature of the toe-perturbation solution for all values of Pe depicted in Figure 14.6 is that significant sediment-flux perturbations never migrate to the landslide headscarp, even though perturbations at the toe may be significant for many years. This type of response is not required by the mathematical boundary conditions, but it might be linked to the fact that the mathematical theory is strictly valid only if perturbations are small. A comparison of theoretical results to data from Minor Creek landslide shows, however, that the theory can be useful for understanding landslide responses even when perturbations are large.

Figure 14.7 depicts the distribution of sediment flux through Minor Creek

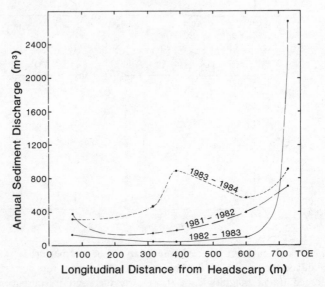

Figure 14.7 Distribution of volumetric sediment flux through Minor Creek landslide during the movement seasons of 1981–1982, 1982–1983, and 1983–1984. The plotted points mark the locations of transverse stake lines, which are instrumental in providing surface-velocity data on which sediment-flux calculations are based. Inclinometer tube data and longitudinal stake-line data were also used in constructing this plot.

landslide for the period 1981–1984. Points plotted on Figure 14.7 were derived from inclinometer measurements and longitudinal and transverse stake-line data (cf. Iverson, 1984). Owing to uncertainties that result from incomplete spatial sampling of the landslide's thickness and movement rates, the points on Figure 14.7 could be in error by as much as 50 percent. Despite this large potential for error, Figure 14.7 clearly shows a marked efflux of sediment from the landslide toe area during 1981–1984. The efflux peaked during 1982–1983, but the effects were damped strongly as the perturbation spread upslope. This style of response corresponds well with the theoretical responses depicted in Figure 14.6.

Quantitative comparisons between theory and measurements depend on the value of *Pe*. On the basis of field and laboratory measurements, the value of *Pe* for Minor Creek landslide is estimated to be about 10 (Iverson, 1984). Figure 14.6C thus shows the theoretical landslide response. The time $t^* = 1$ is equivalent to about 350 years for Minor Creek landslide (Iverson, 1984), so the curve for $t^* = 0.01$ on Figure 14.6C should roughly correspond to the situation depicted in Figure 14.7. Quantitative agreement between Figures 14.6C and 14.7 for both the timing and distance of perturbation spreading appears to be fair.

An interesting feature of Figure 14.7 that is not explained by this theory is the sediment-flux "bump" present in the central portions of Minor Creek landslide during 1983–1984. The bump could be the result of increased instability (i.e., a perturbation) generated locally in the landslide. It could, however, represent upslope propagation of the toe perturbation as a dynamic wave, which would require that stresses vary longitudinally in the landslide. Insufficient information is available at present to evaluate the relative merits of these ideas.

Conclusions

The one-dimensional theory outlined here provides a physically based framework for understanding complicated movement patterns of persistently active landslides. The chief strength of the theory lies in its generality and few constraints imposed by the constitutive equation. The major limitation of the theory is that it is strictly valid only for small departures from steady-state landslide behavior.

Qualitatively, the elements of the theory and mathematical deductions based on those elements correspond very well with field measurements. Quantitatively, the theory agrees moderately well with field measurements. A lack of perfect agreement is understandable in light of the three-dimensionality and inhomogeneity of landslides and the approximate nature of the theory.

The most important inference drawn from the theory is that a single dimensionless parameter, the landslide Peclet number, largely controls landslide response to sediment-flux perturbations. Small values of *Pe* correspond physically to landslide deformation styles that are dominantly rigid–plastic, whereas large values of Pe indicate more viscous landslide behavior. If a landslide has a small Pe value, sediment-flux perturbations spread rapidly from their source but decay with distance and time. If the Pe value is large, sediment-flux perturbations propagate slowly downslope as weakly diffusive kinematic waves. Many landslides probably exhibit a combination of both styles of perturbation response.

Acknowledgments

I am indebted to Roger Denlinger, Arvid Johnson, and Tom Pierson, who critically reviewed this manuscript, and to Robert Ziemer, who loaned unpublished inclinometer data.

References

Blackwelder, E., The Gros Ventre Slide: an active earthflow, *Bulletin of the Geological Society of America, 23*, 487–492, 1912.

Drucker, D. C., and W. Prager, Soil mechanics and plastic analysis or limit design, *Quarterly of Applied Mathematics, 10*, 157–165, 1952.

Harden, D. R., R. J. Janda and K. M. Nolan, Mass movement and storms in the drainage basin of Redwood Creek, Humboldt County, California—a progress report, *U.S. Geological Survey Open-File Report 78-486*, 1978.

Iverson, R. M., A model for creeping flow in landslides—discussion, *Bulletin of the Association of Engineering Geologists, 20*, 455–458, 1983.
Iverson, R. M., Unsteady, nonuniform landslide motion: theory and measurement, Ph.D. thesis, 303 pp., Stanford University, Stanford, California, 1984.
Iverson, R. M., A constitutive equation for mass-movement behavior, *Journal of Geology, 93*, 143–160, 1985.

Janda, R. J., K. M. Nolan and T. A. Stephens, Styles and rates of landslide movement in slump-earthflow-sculpted terrane, northwestern California (abstract), *Geological Society of America, Cordilleran Section 76th Annual Meeting, Abstracts with Programs*, 113, 1980.

Keefer, D. K., Earthflow, Ph.D. thesis, 317 pp., Stanford University, Stanford, California, 1977.
Keefer, D. K., and A. M. Johnson, Earthflows: morphology, mobilization and movement, *U.S. Geological Survey Professional Paper 1264*, 1983.
Kelsey, H. M., Earthflows in Franciscan melange, Van Duzen River basin, California, *Geology, 6*, 361–364, 1978.

Marino, M. A., Flow against dispersion in nonadsorbing porous media, *Journal of Hydrology, 37*, 149–158, 1978.

Nolan, K. M., and R. J. Janda, Sediment discharge and movement of two earthflows in Franciscan terrane, northwestern California, in Geomorphic processes and aquatic habitat in the Redwood Creek drainage basin, *U.S. Geological Survey Professional Paper*, in preparation, 1985.

Nye, J. F., The response of glaciers and ice-sheets to seasonal and climatic changes, *Proceedings of the Royal Society (London), Series A, 256*, 559–584, 1960.

Savage, W. Z., and A. F. Chleborad, A model for creeping flow in landslides, *Bulletin of the Association of Engineering Geologists, 19*, 333–338, 1982.

Suhayda, J. N., and D. B. Prior, Explanation of submarine landslide morphology by stability analysis and rheological models, *Proceedings of the Tenth Annual Offshore Technology Conference, Houston*, 1075–1082, 1978.

Suemine, A., Observational study on landslide mechanism in the area of crystalline schist (Part I)—an example of propagation of Rankine state, *Bulletin of the Disaster Prevention Research Institute, Kyoto University, Japan, 23*, 105–127, 1983.

Swanson, F. J., and D. N. Swanston, Complex mass-movement terrains in the western Cascade Range, Oregon, *Geological Society of America Reviews in Engineering Geology, 3*, 113–124, 1977.

Swanston, D. N., Creep and earthflow erosion from undisturbed and mangement-impacted slopes in the Coast and Cascade Range of the Pacific Northwest, U.S.A., in Erosion and Sediment Transport in Pacific Rim Steeplands, *International Association of Hydrological Sciences Publication 132*, 76–94, 1981.

Taylor, D. W., *Fundamentals of Soil Mechanics*, 700 pp., Wiley, New York, 1948.

Wasson, R. J., and G. Hall, A long record of mudslide movement at Waerenga-O-Kuri, New Zealand, *Zeitschrift für Geomorphologie, 26*, 73–85, 1982.

15

The morphology and mechanics of large-scale slope movement, with particular reference to southwest British Columbia

Michael J. Bovis

Abstract

Large, slow-moving landslides in the Interior Plateau of British Columbia, hitherto referred to as earthflows, display distinctive features such as discrete lobes, lateral deposits, flow bifurcation, plug-flow, and irregular accumulation fans, which are commonly observed on much smaller-scale mudslides and debris flows. The possible origins of the features are examined by considering material properties and the mechanics of deformation. Surface displacements show systematic variation related to convergence and divergence of the moving mass. Extending zones, analogous to those on glaciers, are associated with acceleration of the moving mass and attainment of the active Rankine state. Compressing zones, in areas of reduced gradient, show morphological and subsurface evidence of slow thrusting along shear planes, which are at residual strength. Field moisture contents are close to the plastic limit, even to depths of 10 m. This produces a rigid debris mass, with most of the downslope displacement occurring at well-defined boundary shear zones. Vertical velocity profiles show only slight internal deformation down to depths of 14 m. These deformation properties suggest the term "earthslide" is more appropriate than earthflow. The latter term should be applied to undrained failures in sensitive soils that exhibit true flow.

Introduction

Large, slow-moving landslides, hitherto referred to as earthflows, have left a distinctive imprint on the terrain of the Interior Plateau region of British Columbia. Elongated masses of hummocky debris, some totaling more than 30 million cubic meters, occupy minor tributary valleys incised into the steep

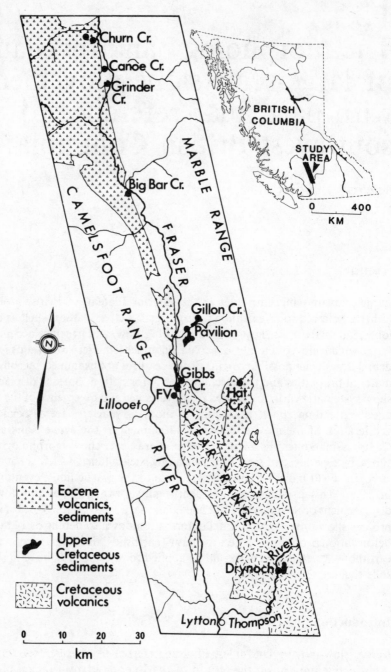

Figure 15.1 Study area, showing earthflow locations and simplified geology.

margins of the plateau (Fig. 15.1). The morphology of the deposits indicates several phases of movement, in some cases extending over at least a 6500 year period. Apart from preserving valuable information on long-term slope movement, the deposits exhibit distinctive morphological features that provide clues to the character of movement. This chapter examines the origin of the typical macro- and microscale forms. There is not an extensive literature on the morphology of large earthflows. In British Columbia two fairly detailed geotechnical studies have been conducted, prompted by the actual or potential interference of earthflow movement with engineering works (VanDine, 1980; Rawlings, 1984). Both give a detailed treatment of shear strength, texture and plasticity, groundwater regime, and geologic conditions, but largely ignore slope morphology.

The main morphological elements of earthflows within the western part of

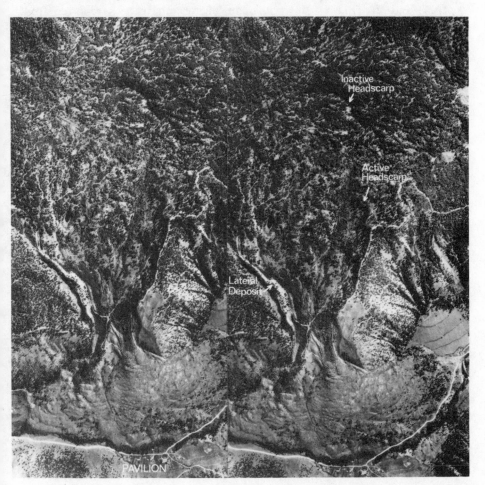

Figure 15.2 Stereopair of Pavilion earthflow (BC 7788 : 242–243).

the Interior Plateau are all present on the Pavilion earthflow, which has also been the focus of detailed investigations (Fig. 15.2). These morphological elements are: (1) arcuate headscarps, occurring as broadly concentric sets, (2) linear ridge-like deposits along the margins of both the actively moving, and the apparently inactive parts of the earthflow complex, (3) discrete lobe forms, which terminate in steep-fronted snouts, (4) irregular shaped, closed depressions, some containing ephemeral and perennial ponds, (5) lateral boundary shear surfaces, which define the longitudinal boundaries of actively moving masses within the earthflow complex, (6) large-scale bifurcation and rejoining of the moving flow mass, and (7) a pronounced downslope elongation of the earthflow complex and of the actively moving masses in particular.

Similar morphological features have been noted on other large earthflows, notably by Keefer (1977), Swanson and Swanston (1977), Gil and Kotarba (1977) and VanDine (1980). Analogous features have been reported from much smaller "mudslides" by Hutchinson (1970) and Hutchinson et al. (1974), and from debris flows by Johnson (1970). This suggests rheological similarities over a considerable range of scale, from tens, to hundreds of cubic meters of material in the case of debris flows, through to tens of millions of cubic meters in the largest earthflows. Over this same range, material type and movement rate change from saturated slurries, moving at the order of meters per second, through to rigid earth masses of approximately zero liquidity index, moving at the order of meters per year. It appears that similar morphological features can arise over a wide range of material strengths and strain rates. With the exceptions of Johnson (1970) and Keefer (1977), few authors have addressed this apparent scale-independent effect in the morphology of earth-matrix mass movements.

Some authors have attempted to explain earthflow morphology by means of physical and mathematical models; notable examples that are reviewed briefly here are those of Brückl and Scheidegger (1973), and Keefer (1977). The smaller, single-lobed features, such as the Beltinge mudslide described by Hutchinson (1970), appear to be the most tractable. An acceptable agreement between the observed and the predicted longitudinal profile form was obtained by Brückl and Scheidegger, at least for the toe area at Beltinge. From this analysis and that carried out on an earthflow described by Crozier (1969), they concluded that the active Rankine state was most probable in earthflow tongues. It is worth pointing out, however, that upward sloping slip lines, suggesting longitudinal earth pressures approaching the passive Rankine state, are clearly evident in Hutchinson's profile of the Beltinge mudslide. Keefer's analysis deals with an entire earthflow in longitudinal section from headscarp region to the toe. Three elements are identified. The first element is an upper tension zone, depicted as providing an active thrust to a second element, the main earthflow mass, which is assumed to move *en bloc*. The third element forms the toe of the earthflow, and mobilizes a passive thrust directed in an upslope direction against the lower boundary of the central element. In contrast to the infinite slope, limit-equilibrium analysis that is often applied to elongated slope failures (Skempton and Hutchinson, 1969; VanDine, 1980), Keefer's analysis takes "end effects" into account.

As might be expected, many earthflows when examined in detail, appear to deviate somewhat from both of these idealized modes of deformation. The principal complicating factors, at least in a number of British Columbian earthflows, derive from (1) irregularity of the underlying topography that the earthflow is overriding, (2) shear-strength contrasts in the materials comprising the earthflow, and (3) significant lateral constraints on earthflow movement provided by either bedrock topography or by massive lateral deposits produced by earlier phases of earthflow movement. The last factor is a notable complication because it implies that earthflows are likely to modify their boundary conditions over time. Given the morphological complexity of earthflows with multiple phases of movement, such as occurs at Pavilion and at most other earthflows in the region, a complete quantitative analysis of deformation seems unlikely. The existence of significant lateral constraints to movement means that the assumptions of plane stress analysis are not met, and the data requirements of a three-dimensional analysis are beyond the reach of this study. Consequently, the approach followed will use conventional limit-equilibrium analysis, earth pressure theory, and rheology to account for specific elements of the morphology, rather than attempt an explanation of the overall form from first principles. Preliminary data on the deformation of earthflows in British Columbia are compared with the results of earthflows in other mountain areas. It is suggested that the term *earthslide* is more appropriate to describe the deformation of these relatively rigid debris masses.

The study area

A notable concentration of large earthflows occurs within the Camelsfoot and Clear Ranges, which form part of the rugged, western boundary of the Interior Plateau (Fig. 15.1). Three principal topographic units are identified. The most extensive is a dissected, plateau surface developed at 1000 to 1300 m above sea level, which forms the uplifted, western margin of an earlier, mid-Tertiary erosion surface. Rising above the dissected plateau, the second unit comprises the deeply dissected mountain blocks of the Camelsfoot, Marble, and Clear Ranges, in which isolated summits attain 2000 m. The third element consists of deeply incised trunk-stream and tributary drainage lines, a legacy of the late-Tertiary uplift of the Western Cordillera. Local relief along the steep-sided Fraser and Thompson Valleys increases in a downstream direction and ranges between 500 and 1500 m. The deeply incised valleys lie within the semiarid zone of ponderosa pine and bunchgrass that receives an average annual precipitation of 250 mm. The undulating plateau surface receives about 450 mm and supports a Douglas fir forest. This merges upwards into

Engelmann spruce with subalpine fir above 1250 m, with small areas of alpine tundra above 1900 m.

The study area is underlain by volcanic, intrusive and sedimentary rocks, ranging in age from Pennsylvanian to Mio-Pliocene. As earthflows are confined to Cretaceous and early Tertiary rocks, only these units are identified on the location map (Fig. 15.1). Many of the contacts with earlier rocks are fault-defined and follow the dominant northwest-to-southeast regional trend. In the lower Cretaceous, earthflows are confined to portions of the Spences Bridge Volcanics (predominantly andesites) in which interbedded sediments are important. The large earthflows at Drynoch and Pavilion are developed in upper Cretaceous sediments containing bentonitic shales, sandstone, and conglomerate. Earthflows are common within the Eocene dacite-to-rhyolite volcanics at Gibbs Creek–Fountain Valley, in the Hat Creek Graben, and along the Fraser Valley upstream of Big Bar Creek, particularly where thick interbedded sediments occur. The sediments comprise bentonitic claystones and siltstones, conglomerate, sandstone, and coal.

The entire study area has been glaciated, probably several times, by Cordilleran ice sheets. However, virtually all of the drift deposits visible today accumulated during either the Fraser Glaciation, which ended about 10 000 years ago, or during the short, but geomorphically active, "paraglacial" period (Ryder, 1971) that occupied the first millennium of postglacial time. Most of the upland surfaces are mantled by basal till capped with variable amounts of ablation deposits. Large amounts of till have been moved downslope by mass movement but, with few exceptions, glacial deposits have simply been rafted by deformation of underlying clay-rich materials derived from Cretaceous and Tertiary sediments. Along the major valleys, thick accumulations of late-glacial to early postglacial drift occur, totaling well over 200 m in places. Glacio-lacustrine silts and outwash sands and gravels are usually overlain by a thick sequence of debris flow deposits, the latter dating from the paraglacial period. The entire valley fill has been sculpted into stream-cut terraces by the Fraser and Thompson rivers. Many earthflows have overridden or breached these materials during their prolonged, postglacial descent of the steep plateau margins.

Earthflow morphology

Planimetric form
The earthflows in the Fraser Valley study area are highly elongated, sometimes sinuous, zones of hummocky topography up to 6 km in length and 2 km in width. In plan form, they are either compound (multiple-lobed) or simple (single-lobed). Compound flows possess several feeder or tributary masses, which have converged over time to a central depression (Figs. 15.2 and 15.3). The identity of individual feeder lobes is often preserved on account of shear furrows developed by differential movement between converging flow masses.

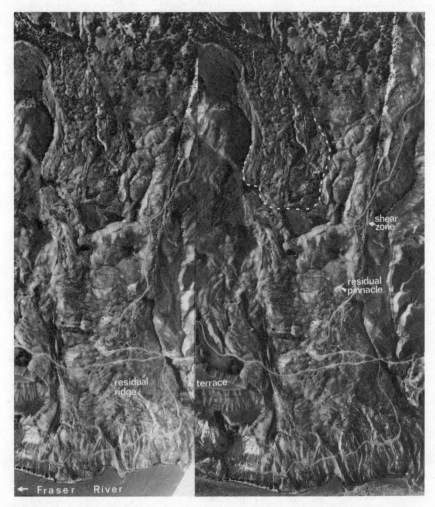

Figure 15.3 Stereopair of Churn Creek earthflow (BC 7714:85–86). Note ridges and depressions created by compressional failure of flow mass. The broken white line identifies the front of the lobe.

Most of the feeder lobes can be traced upslope to well-defined, arcuate headscarps (Fig. 15.4), the product of prolonged, slow retrogression. Headscarps are produced mainly by rotational slumping rather than lateral spreading, as shown by numerous instances of uphill-tilted trees.

Divergent motion of earthflow lobes is also fairly common and is well developed at Pavilion (Figs. 15.4 and 15.5), Big Bar Creek, Grinder Creek, and Hat Creek earthflows. In the latter two cases, bifurcation is caused by bedrock at or close to the surface. At Pavilion and Big Bar Creek it is apparently due to relatively resistant drift deposits that have higher shear strength than materials comprising the advancing earthflow mass.

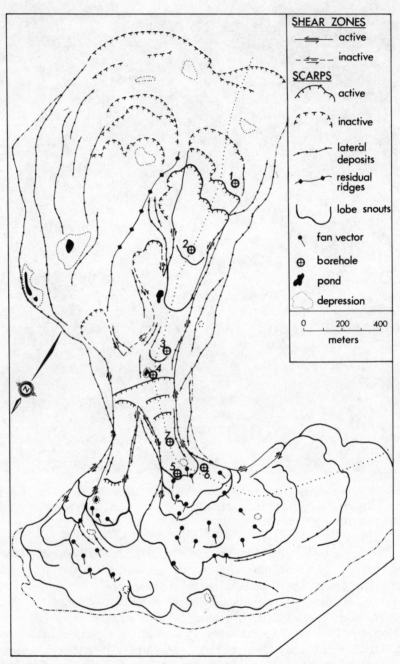

Figure 15.4 Morphological features of Pavilion earthflow (compare with Fig. 15.2 and 15.5). The dotted longitudinal line shows the location of the profile in Figure 15.8.

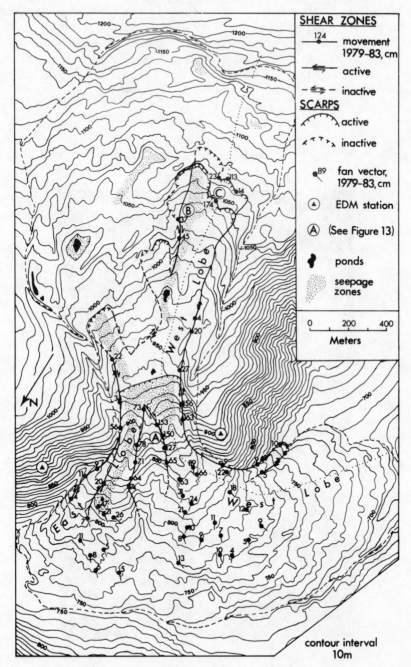

Figure 15.5 Displacement at Pavilion earthflow, 1979–1983.

Large fans of earthflow deposits have accumulated wherever postglacial stream erosion has been modest. The largest fans occur at Pavilion, Gibbs Creek, and Fountain Valley (Figs. 15.1 and 15.2). The large, poorly drained area directly east of the Pavilion fan shows the effects of partial blockage of the Pavilion Creek drainage over a prolonged period. All of the remaining flows, with the exceptions of Gillon Creek and Canoe Creek, are periodically eroded by fluvial action and therefore have negligible fan development.

Lateral deposits

Prominent linear ridges or lateral deposits occur close to the lateral margins of active or formerly active earthflow masses. They appear to be analogous to

Figure 15.6 Lateral deposit section at elevation 895 m on the margin of the east lobe at Pavilion (Fig. 15.5). Section height is 3.5 m. Light tones are reworked till and dark tones red–brown diamicton from Cretaceous sediments. Note vertical and lateral accretion since the Mazama tephra fall.

debris flow levees, though they are considerably larger, attaining a relative relief of 20 m in some cases. At Pavilion where downslope movement has been measured, there is a clear association between lateral deposit formation and active shear zones along the flow boundaries (Fig. 15.4). This allows other roughly parallel sets of linear ridges, stranded up to 40 m above the level of the presently active flow, to be interpreted as the boundaries of earlier, somewhat larger, flow masses. Most of the large earthflows show this tendency to progressive lowering of the actively moving surface over time as material has been steadily discharged from the source region. Of similar form, though of different origin from lateral deposits, are residual ridges (Figs. 15.3 and 15.4). These are remnant features analogous to glacial aretes, produced by upslope expansion of headscarps, whereas lateral deposits are produced by both lateral and vertical accretion at the earthflow margins.

Roadcuts through lateral deposits usually show a complex stratigraphy. The section at Pavilion earthflow (Fig. 15.6) clearly shows vertical accretion since the Mazama tephra fall (6600 B.P.) in addition to a lateral growth of the

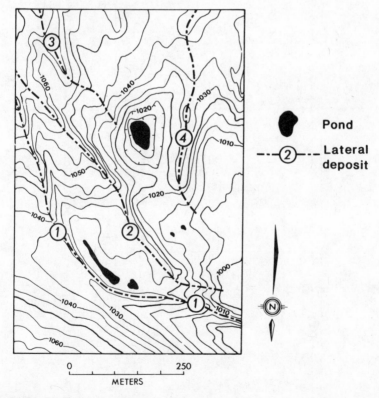

Figure 15.7 Topographic detail of depressions in the upper, east part of Pavilion earthflow (compare with Fig. 15.5). Depressions are due to accretion of successively younger lateral deposits (1 = oldest).

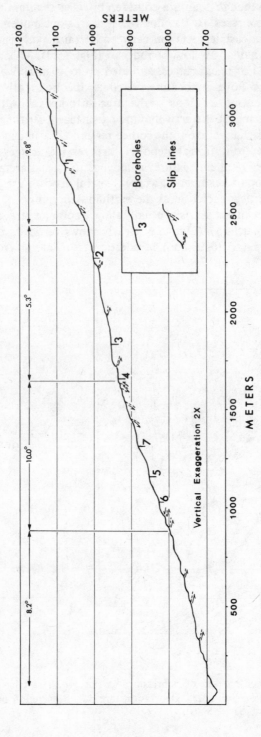

Figure 15.8   Longitudinal profile of Pavilion earthflow (2 × vertical exaggeration). See Figures 15.4 and 15.5 for location of section line.

deposit. Lateral deposits are particularly well-developed where flows have discharged through topographic constrictions, as at Pavilion (Fig. 15.2 and 15.5). Lateral deposits are largely absent or are weakly developed in areas where earthflow motion is notably divergent, such as on the depositional fans (Fig. 15.4). Linear ridges on the Pavilion fan have been produced by differential movement between the lobate masses which comprise the fan.

Echelon arrangement of lateral deposits produced by successively younger phases of earthflow movement has produced a complex ridged topography, notably at Pavilion and Churn Creek (Figs. 15.2 and 15.3). Depressions are thereby created, many containing ephemeral or perennial ponds (Fig. 15.7).

Earthflow lobes

The actively moving parts of large earthflows usually consist of several apparently overlapping lobes, each of which terminates in a steep frontal face or snout. These are depicted in detail in Figure 15.4 and produce an irregular longitudinal profile (Fig. 15.8). The most rapidly advancing toes are generally steeper than stationary lobes and have loose, dilated soil spilling down their frontal faces. This suggests that a certain amount of vertical mixing of earthflow material occurs as lobes are slowly thrust over earlier deposits. The rate of advance of individual lobes is sufficiently slow (Fig. 15.5) for the basal shear surfaces to be concealed by colluvial activity along the lobe fronts. Some lobes have reversed slopes and closed depressions, apparently produced by upthrusting of the frontal part of the lobe. Fresh, slickensided blocks of local bedrock material are often found near these apparently upthrust areas. The pattern of contemporary deformation in these areas is compatible with compressing flow and is discussed later under "Earthflow deformation."

The main flow mass upslope of a given lobe front may show morphological evidence of either extending or compressing flow analogous to glacier motion. At Pavilion numerous transverse tension zones exist upslope of borehole 7, where the longitudinal gradient increases abruptly probably due to bedrock control (Fig. 15.8). Upslope of borehole 3 numerous, small-scale undulations, mounds, and closed depressions may signify compressing flow in a zone of substantially reduced gradient (Fig. 15.4). Compressing flow also occurs where active lobes discharge onto the earthflow accumulation fans.

The transverse profile of lobes is generally convex where flow is laterally unconfined and concave where it is constricted by earlier lateral deposits or local bedrock topography. Concavity is due in part to an upward buckling of the rigid, near-surface edges of the flow mass, and in part to a more rapid movement of the medial part of the lobe.

Material properties

Two distinctive types of material are found in earthflow deposits in the study area. The first consists of a grey to brown matrix silty unit, representing the upland basal till. Cobbles and boulders from the Coast Mountains to the west of the study area are incorporated to produce a distinctive, diamicton deposit. Although these materials are considered to have played no active part in earthflow initiation other than contributing to local shear stress and influencing near-surface hydraulic conductivity, large amounts of till have become lodged along the margins of many earthflows during the earliest phases of slope movement. It is appropriate to consider the mechanical properties of this material, as it is evident that lateral deposits of relatively high strength till have partly controlled the pattern of more recent slope movements. This is especially the case at Pavilion earthflow (Fig. 15.2).

The second material type consists of a cohesive, clay-rich material ranging in color from red to grey, depending on geologic age, and derived from the breakdown of poorly lithified shales, claystones, and siltstones of Cretaceous or Eocene age. Large blocks of more resistant, stratigraphically adjacent rocks, such as volcanic tuff, lava, sandstone, and conglomerate, have been incorporated into the clay matrix to produce a diamicton material that is similar in general appearance to the glacial diamicton but possesses quite different geotechnical properties. Most of the data presented in this section are derived from Pavilion earthflow were a detailed geotechnical study has been carried out. Materials from other earthflows in the region are discussed briefly.

The plasticity chart (Fig. 15.9) shows the distinct separation between samples derived from glacial till and Cretaceous materials in the Pavilion area. Included also are samples from the much smaller Gillon Creek earthflow located 3 km northeast of Pavilion. Pavilion samples were drawn from near-surface levels exposed in road cuts, from river bluffs along the toe of the earthflow, and from seven boreholes drilled to depths of 9.5 to 16.5 m. Gillon

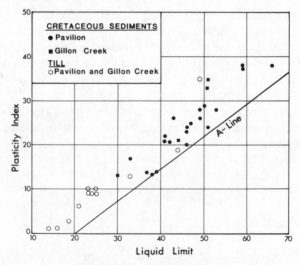

Figure 15.9 Plasticity chart.

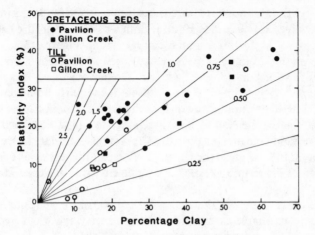

Figure 15.10   Activity chart.

Creek earthflow is developed in similar glacial and Cretaceous materials. The clear separation of glacial and Cretaceous materials is also seen in the activity diagram (Fig. 15.10). Activity is defined here as the ratio of plasticity index to weight percentage of clay ($< 2 \, \mu m$). X-ray diffraction analyses generally showed montmorillonite as the first-ranked mineral in this fraction, which accounts for the fairly high activity ratios of 1.0 to 1.5 in samples derived from Cretaceous sediments. Values lower than about 0.75 are usually classed as inactive, and most of the glacial till samples fall in this region.

Shear strength envelopes derived from consolidated, drained direct-shear tests with multiple reversals are given in Figure 15.11 and refer to residual

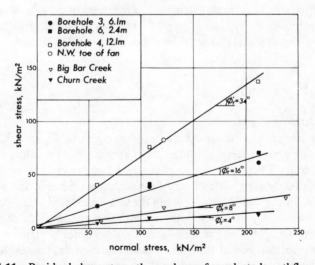

Figure 15.11   Residual shear-strength envelopes for selected earthflow materials.

parameters. In that no pore pressure measurements are made during this type of test, a slow feed rate of 4.8×10^{-2}mm/min was used to ensure fully drained conditions. Two important points emerge from Figure 15.11. First, an appreciable shear strength contrast exists between the Pavilion glacial till ($c' = 0$, $\phi_r' = 34°$), and the Pavilion Cretaceous clay-rich diamicton ($c' = 7.0$ kN/m^2, $\phi_r' = 16°$). The second point is the fairly homogeneous shear strength of different samples drawn from the same formation. In Figure 15.11 the full circles show the results of a three-stage test on a disturbed sample of red, cohesive sandy clay, typical of the Cretaceous material, from a depth of 6.1 m in borehole 3, whereas the full squares depict the test results for material from a depth of 2.4 m in borehole 6. A single envelope has been fitted to both samples.

It is generally recognized that cohesion tends to zero at residual strength. Based on this reasonable assumption, a second envelope was forced through the origin yielding revised parameters of $c' = 0$, $\phi_r' = 18°$. This adjustment does not materially change the shear strength contrast between glacial and nonglacial materials in this area. The peak strength parameters from these same remolded samples were $c' = 9.6$ kN/m^2, $\phi_r' = 20°$. In Figure 15.11 the open squares refer to a till sample recovered from a depth of 12.2 m in borehole 4, and the open circle to till exposed in the river bluff at the extreme northwest toe of the earthflow fan (Fig. 15.5). These test results again indicate reasonable homogeneity of shear strength.

Strength envelopes of samples taken from river bluff exposures at the toes of the Big Bar Creek and Churn Creek flows (Fig. 15.1) are also given in Figure 15.11. In both cases $c' = 0$, with ϕ_r' values of 7.6° and 4.3°, respectively. These much lower values were both obtained from Eocene samples rich in montmorillonite. Although the Churn Creek value is very low, and may have been biased downward by pore pressure rise during the test, the Big Bar Creek value is comparable with back-calculated parameters of $c' = 0$, $\phi_r' = 6.8°$ and laboratory values of $c' = 0$, $\phi_r' = 8.2–9.0°$ recently reported from the Eocene claystones of the Hat Creek earthflow (Rawlings, 1984).

Earthflow deformation

Surface displacements

A four-year record of displacements is summarized in Figure 15.5. Lateral shear-zone movements were obtained from strain nets straddling the clearly defined boundary shear planes. Displacement vectors on the earthflow fan were determined by electronic distance measurement. Shear-zone stations on opposite sides of a given lobe at roughly the same elevation show comparable displacements over the four-year period, with the exception of the east lobe (Fig. 15.5), where figures on the west shear zone are notably higher. Both lobes below the bifurcation point show acceleration as material passes through the steeper gradient, confined zones followed by pronounced deceleration as both lobes diverge at the apexes of their respective fans. Information from the upper flow area, although fragmentary, shows comparable totals on both sides of the flow at the level of borehole 4. Above this level, displacements on the west lobe steadily decrease, then increase again above the level of borehole 2. This suggests that the lobe front directly downslope of borehole 2 is advancing at a faster rate than the material between this point and borehole 3. Displacements at the headscarp are variable, reflecting differential movement between several slumped blocks. The figure at site C (Fig. 15.5) refers to displacement over the two-year period of 1982–84.

Displacement vectors on the earthflow fans are broadly consistent with the configuration of advancing lobe fronts and show a general downslope decrease in velocity consistent with the conservation of mass in a divergent flow. Measurements in the vicinity of boreholes 5 and 6 at the apex of the west lobe fan suggest that the lobe front directly downslope of borehole 5 is advancing across the lobe containing borehole 6 (compare Figs. 15.4 and 15.5).

The typical horizontal velocity profile (Fig. 15.12) was obtained by resurvey of pegs arranged normal to the flow 10 m downslope of borehole 7. A similar profile was obtained from the displacement of fence posts on the east lobe at an elevation of 830 m. The horizontal profile shows plug-flow, similar to that reported for much smaller-scale debris flows (Johnson, 1970), and mudslides (Hutchinson, 1970). Most of the profiles reported from other large earthflows by Crandell and Varnes (1961), Keefer (1977), and Kelsey (1978) show that the bulk of surface displacement also occurs at lateral shear zones rather than within the moving mass. In this respect, earthflow movement is substantially different from glacier flow.

Displacement–time curves are shown in Figure 15.13 for three slope movement stations identified by upper-case letters in Figure 15.5. Site A is representative of the most rapidly moving part of the west lobe over the period 1979–84. Seasonal acceleration associated with changes in effective stress produced by snowmelt recharge of groundwater is clearly visible in 1981 and is less clear in 1983. A similar pattern of movement is seen at site B in the upper

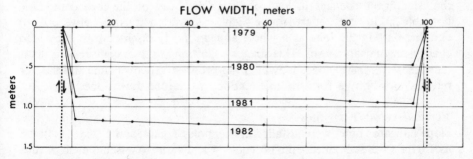

Figure 15.12 Horizontal velocity profile 10 m downslope of borehole 7. See Figure 15.4 for location. Broken lines are lateral shear zones.

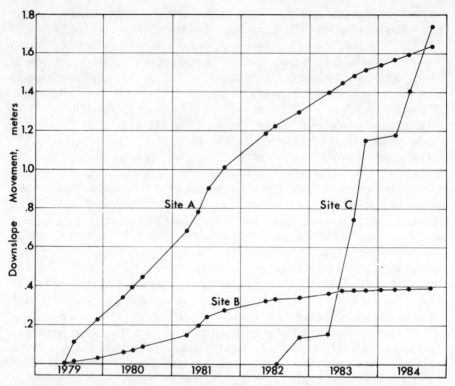

Figure 15.13 Cumulative displacement, 1979–1984 for selected Pavilion sites. See Figure 15.5 for locations.

part of the west lobe. Since the middle of 1983 negligible movement has occurred at this site, suggesting a "stick–slip" type of response. Site C shows a pronounced, seasonal regime of movement that is well correlated with the period of highest piezometric levels over most of the earthflow (April–August). Although this movement refers only to a mass that extends about 30 m downslope of the headscarp, the two-year record at site C suggests that significant changes in load occur at the head of the lobe containing borehole 2. However, apart from a slight acceleration at site B between April and June 1981, the lobe as a whole appears fairly insensitive to such load changes in the short term. This suggests that longitudinal compressive stress from these surcharge loads may build up progressively over time and then be released slowly by a forward acceleration of the lobe downslope.

Vertical velocity profiles

Slope indicator pipes were installed at boreholes 1, 2, 4, and 5 (Fig. 15.4). No useful data were obtained from boreholes 1 and 4 due to misalignment of tube sections. It should be stated at the outset that the data on vertical velocity profiles are preliminary only. They are adequate to assess the probable nature

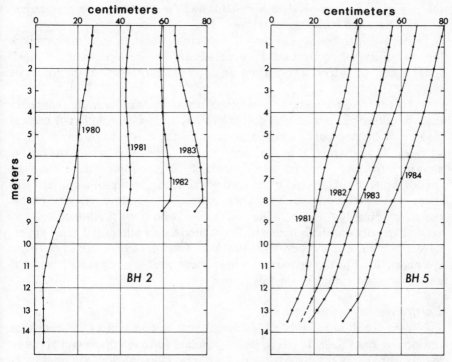

Figure 15.14 Vertical velocity profiles, Pavilion earthflow. See Figure 15.4 for borehole locations.

of earthflow deformation and also serve as a basis for comparison with other published profiles. Data from boreholes 2 and 5 are shown in Figure 15.14. Borehole 2 shows a shear zone at about 8.5 m and a pronounced development of upslope tilt over a four-year period. The inclinometer–piezometer pipe could still be plumbed to 14 m with a weighted line in October 1984. This profile is consistent with a rotational motion of the lobe containing this borehole along an elongated, curved failure zone that probably extends from the lobe front below borehole 2 to the headscarp at station C. Although the factor of safety along the entire shear plane may be close to unity, a limit equilibrium analysis at borehole 2 yielded a minimum factor of safety of 1.25, assuming full saturation to the surface and groundwater flow parallel to the surface. The following values were used in the analysis: bulk unit weight $\gamma = 19.1$ kN/m^3, slope angle $\beta = 8°$, and residual shear strength parameters $c_r' = 0$, $\phi_r' = 18°$. The groundwater assumption is reasonable, given extensive discharge along the east side of the lobe (Fig. 15.5). This is active almost year-round and produces a thick, surface icing in winter. However, the maximum recorded piezometric level at borehole 2, referred to the base of the slope indicator pipe at 15.8 m depth, is 6.3 m below local ground level. This is not compatible with groundwater discharge at the surface close by, and suggests the existence of a second, deeper groundwater flow system below the perched system associated with the shear zone.

The above analysis suggests that in the vicinity of borehole 2, the earthflow lobe is being pushed forward by materials further upslope having a much lower factor of safety. The pattern of displacements at site B supports this idea.

At borehole 5 an increase in downslope inclination has occurred, though no discrete failure plane is identified. The profile suggests distribution of vertical shear over a thick zone, which may point to the existence of several shear surfaces. The existence of a shear zone or plane below the base of the tube is shown by the fact that total displacement at the tube top is greater than the the computed displacement at the base of the tube. Both profiles have a ten-times horizontal exaggeration. When replotted to scale a velocity profile typical of plug-flow is seen. The conclusion from both the horizontal and vertical velocity profiles is that earthflow materials are fairly rigid in the uppermost 10–15 m. This is consistent with field water contents close to or below the plastic limit in the uppermost 5 m in many areas, with values somewhat above the plastic limit below this level.

Earth pressure considerations

Published longitudinal profiles of earthflows and mudslides indicate that the moving mass tends to adjust its longitudinal form to correspond approximately with that of the underlying more resistant material, which in some cases is bedrock. At Pavilion a break in slope occurs at approximately 930 m on both the east and west lobes above the bifurcation point and is considered to be caused by a series of resistant conglomerate layers, which crop out at a comparable elevation just west of the earthflow (Fig. 15.5).

Similar changes in longitudinal gradient and the attendant changes in longitudinal stress are well documented for glacier ice (Nye, 1952). The slip-line fields associated with extending and compressing flow have their analogs in active and passive Rankine zones in soils. Morphological evidence of active zones are seen in the headscarp areas of most of the British Columbian earthflows and show up as tension scarps at Pavilion (Fig. 15.4). A second zone occurs at Pavilion just above the bifurcation point, where a fairly abrupt change in longitudinal gradient occurs. These masses, in which the active Rankine state has clearly been attained, supply a lateral thrust to soil masses further downslope. This thrust may be sufficient to produce downslope displacements in masses where the local factor of safety is greater than unity. The existence of passive failure zones is difficult to prove on morphological grounds, though the presence of bulbous toe areas and surface undulations suggests that the passive resistance may locally have been exceeded. However, the soil masses subjected to lateral compression, for example at the apex of the accumulation fan at Pavilion, have already experienced considerable downslope translation. All shear surfaces within the mass will therefore be at residual strength. The mass above the shear plane, or between shear planes, will still retain a higher frictional strength and some cohesion. This is shown by the horizontal velocity profile. Using peak strength parameters of $c' = 19 \text{ kN/m}^2$ and $\phi' = 20°$, the passive resistance determined graphically from Mohr circle constructions is estimated to be about five time greater than the residual shearing resistance. This means that a soil mass residing in a zone where the local factor of safety is greater than unity (for example, borehole 2) will tend to be pushed forward along pre-existing, residual strength shear planes as a relatively rigid mass. The slope indicator data confirm this tendency.

Discussion and conclusions

Earthflows are a significant type of mass movement in the western part of the Interior Plateau of British Columbia. During Holocene time, hundreds of millions of cubic meters of material have moved into the adjacent valleys from the steep plateau margins. In terms of total slope movement, they may prove to be more significant than catastrophic rockslides in the nearby Coast and Cascade Mountains. Contemporary displacements are largely explained by seasonal changes in effective stress as snowmelt-induced fluctuations in piezometric level occur. This is the case at Pavilion and has been noted elsewhere in the region by VanDine (1980) and Rawlings (1984). Materials in the near-surface zone are close to the plastic limit. This gives rise to a distinctive plug-flow, with most of the downslope movement occurring at well-defined, boundary shear surfaces. Longitudinal changes in movement rate are associated with extending flow zones, analogous to those found on glaciers, in which the soil attains the active Rankine state. Flow divergence, or deceleration in lower gradient areas, produces compressing flow. Movement in these areas is thought to occur by slow thrusting along a series of shear planes along which residual strength is mobilized.

Gross rheological similarities exist between large earthflows and smaller-scale mudslides and debris flows. Features common to all three include discrete lobes with bulbous snouts, lateral deposits, and plugflow. The macroscopic impression of slow, viscous flow in earthflows, obtained from aerial photographs, is not borne out by detailed observation of slope movements. Surface and subsurface measurements suggest that sliding is the dominant mode of movement, indicating that the term earthflow is not appropriate. Where field water contents are close to the plastic limit the term earthslide is deemed more appropriate. The term mudslide would then be reserved for much smaller-scale features with liquidity indices significantly greater than zero and showing a tendency to undrained loading effects, as previously documented by Hutchinson and Bhandari (1971) and Hutchinson et al. (1974). Although localized, undrained failures may occur in the headscarp regions of large earthslides, the bulk of the shear deformation probably involves drained failure. Annual displacements of between 0.41 and 0.01 m/yr translate to shear strain rates of the order 10^{-3} to 10^{-4} mm/min, and are probably slow enough to ensure drained conditions. A completely different mode of failure and deformation is seen in quick-clay materials in eastern Canada (Mitchell and Markell, 1974) and elsewhere. Undrained failure and liquefaction of highly sensitive soils occurs, producing considerable internal deformation once the moving mass has discharged from the failure zone. It is suggested that the term earthflow should be reserved for these features, even though they are morphologically similar to mudslides and earthslides.

Acknowledgments

This work was supported by a grant from the Natural Sciences and Engineering Research Council, Ottawa, Canada. Thanks are due to the faculty and staff of Geotechnical Engineering, University of Alberta, for access to laboratory facilities and to Dr M. Roberts, Simon Fraser University, for the loan of a trailer-mounted drilling rig and for field instruction on its operation.

References

Brückl, E., and A. E. Scheidegger, Application of the theory of plasticity to slow mud flows, *Géotechnique, 23,* 101–107, 1973.

Crozier, M., Earthflow occurrence during high intensity rainfall in Eastern Otago (New Zealand), *Engineering Geology, 3,* 325–334, 1969.

Crandell, D. R., and D. J. Varnes, Movement of Slumgullion earthflow, near Lake City, Colorado, *U.S. Geological Survey Professional Paper 424B,* 136–139, 1961.

Gil, E., and A. Kotarba, Model of slide slope evolution in flysch mountains (an example drawn from the Polish Carpathians), *Catena, 4,* 233–248, 1977.

Hutchinson, J. N., A coastal mudflow on the London Clay cliffs at Beltinge, North Kent, *Géotechnique, 20,* 412–438, 1970.

Hutchinson, J. N., and R. K. Bhandari, Undrained loading, a fundamental mechanism of mudflows and other mass movements, *Géotechnique, 21,* 353–358, 1971.

Hutchinson, J. N., D. B. Prior, and N. Stevens, Potentially dangerous surges in an Antrim mudslide, *Quarterly Journal of Engineering Geology, 7,* 363–376, 1974.

Johnson, A., *Physical Processes in Geology,* 577 pp., Freeman, Cooper, San Francisco, 1970.

Keefer, D. K., Earthflow, Ph.D. thesis, 317 pp., Stanford University, 1977.

Kelsey, H. M., Earthflows in Franciscan melange, VanDuzen River basin, California, *Geology, 6,* 361–364, 1978.

Mitchell, R. J., and A. R. Markell, Flowsliding in sensitive soils, *Canadian Geotechnical Journal, 11,* 11–31, 1974.

Nye, J. F., The mechanics of glacier flow, *Journal of Glaciology, 2,* 82–93, 1952.

Rawlings, G. E., Active slide in bentonitic clays, Upper Hat Creek, British Columbia, *Proceedings of the 4th International Symposium on Landslides, 1,* 503–510, 1984.

Ryder, J. M., The stratigraphy and morphology of para-glacial alluvial fans in south-central British Columbia, *Canadian Journal of Earth Sciences, 8,* 279–298, 1971.

Skempton, A. W., and J. N. Hutchinson, Stability of natural slopes and embankment foundations *Proceedings of the 7th International Conference on Soil Mechanics and Foundation Engineering, State-of-the-Art Volume,* 291–340, 1969.

Swanson, F. J., and D. N. Swanston, Complex mass-movement terrains in the western Cascade Range, Oregon, *Reviews in Engineering Geology, 3,* 113–124, 1977.

VanDine, D. F., Engineering geology and geotechnical study of Drynoch landslide, British Columbia, *Geological Survey of Canada Paper 79-31,* 1980.

16
Processes leading to landslides in clay slopes: a review

R. J. Chandler

Abstract

A classification of landslide causitive processes is proposed in terms of external effects that result in increased shear stresses within the potential landslide mass, and internal effects that result in decreased shear strengths. These processes are reviewed in the context of stiff clays, with particular reference to the internal effects.

The more important causes of strength loss within slopes are due to either (1) decreasing effective stresses or (2) reduction in the soil strength parameters (c', ϕ') of the slope material. Processes leading to (1) include post-erosion swelling and rising groundwater within the slope, whereas those leading to (2) include weathering, shear-rate effects, and progressive failure.

The relation between laboratory measured strengths and strengths inferred from landslide stability analyses is discussed, and it is concluded that for low-plasticity clays the agreement between the two strengths is close. For plastic clays it is suggested that slope processes (perhaps particularly progressive failure) reduce the field strength well below that measured in the laboratory.

Introduction

A primary objective of research on landslides is the prediction not only of the likelihood that a landslide may take place at a given location but also of such factors as its spatial extent and rate and magnitude of displacement. The necessary stages in achieving these objectives are to identify and quantify those processes or phenomena that result in the occurrence of landslides. Once the processes involved are understood, then they may be quantified in the context of the particular site under investigation, and hence the landslide potential may be assessed.

So far as landslides in clays are concerned, considerable progress has been made in this direction, based primarily on field studies of landslides in comparatively uniform clay soils. Comparison of the failure conditions in these slopes with laboratory measurements of the strength of the clay of which they are composed adds considerably to our understanding of the mechanisms involved.

The purpose of this chapter is to review the state of knowledge on landslides in stiff clays, the geological material for which our understanding of landslide processes is greatest. Most of what follows has been obtained from observations of man-made excavated slopes. The basic principles, however, apply equally to either naturally formed or man-made slopes, and it is hoped that the advances that have evolved in an engineering context will also be of geomorphological value.

Definitions

A landslide occurs when the average shear stresses due to gravity (and earthquake forces if relevant) acting on any potential failure surface within an inclined mass of particulate (i.e., soil) or rock material exceed the stresses that the material along that surface is capable of resisting. The resisting stresses result from the strength of the material in shear.

This definition of a landslide provides the basis of a classification of landslide inducing processes as either those that result in an increase in the gravitational/earthquake stresses imposed on the slope or those that cause a reduction in the material shear strength available within the slope. In either event there is a reduction in the ratio of shear strength to shear stress; if this ratio, the factor of safety, falls below unity, the slope fails. This classification is shown in Table 16.1 and used as the basis for subsequent discussion.

Effective stress
The two-phase nature of particulate materials with fluid-filled pores results in the mechanical behavior of the material being controlled by the effective

Table 16.1 A classification of processes leading to the occurrence of a landslide.

External factors	Internal factors
Increase in shear stress due to	Decrease in shear strength due to
(a) slope being steepened by	(a) reduced effective stresses caused by
(i) erosion/excavation at foot	(i) swelling following erosion
(ii) loading at crest	(ii) general rise in water table
(b) earthquake forces	(b) reduced parameters c', ϕ' caused by
(c) drawdown of external water level	(i) weathering
	(ii) shear-rate effects
	(iii) progressive failure

stresses. The effective stress σ' is defined as

$$\sigma' = \sigma - u \tag{1}$$

where σ is the total stress and u the pore-fluid pressure. Working in effective stresses has the advantage that the effects of seepage forces are automatically taken into account and need not be explicitly considered (e.g., Taylor, 1948). The reader is referred to standard soil mechanics texts for further discussion of this principle.

The shear strength of the soil may then be written as

$$\tau = c' + (\sigma - u) \tan \phi' \tag{2}$$

as shown in Figure 16.1. The term c' is the cohesion intercept or cohesive component of strength, and ϕ' the angle of friction or angle of shearing resistance.

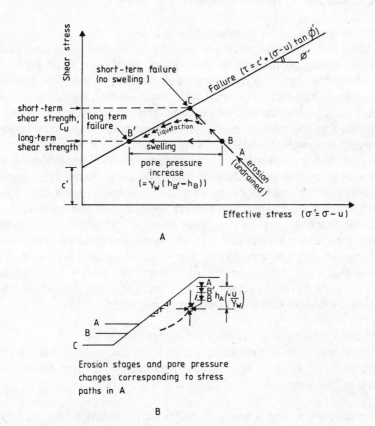

Figure 16.1 (A) Strength envelope and stress paths to failure of typical soil elements in a first-time clay landslide. (B) Key diagram showing the stages in erosion and the corresponding pore pressures.

The prime indicates that the parameters are in terms of effective stress. Unless stated otherwise, subsequent discussion is in terms of effective stress.

First-time landslides

The distinction between a landslide that occurs for the first time and one that is a reactivation of a previous landslide movement is important. With the latter the reactivation will occur (at least in part) on a pre-existing sliding surface that is at or close to the "residual strength." In clay soils the residual strength is the consequence of the reorientation of the platy clay minerals to a position parallel to the sliding surface. In this orientation they provide the minimum resistance to displacement of the landslide, and the corresponding strength may be very considerably less than the peak strength that applies when shearing occurs for the first time.

Classification of landslide-causative processes

Processes involving an increase of shear stress

With these definitions one may now refer to Table 16.1. The main causative factors involving an increase in shear stress within the slope generally act externally to the slope. These are processes that steepen the slope, the removal of the restraining effect of external water, and loading, including earthquake forces. Although internal loading resulting from the introduction of water into a slope falls in the third category, this will generally cause only a marginal increase in unit weight and is unlikely to increase the shear stresses significantly. Consideration of the environment of the slope should enable reasonable quantification of these factors to be made. Sarma (1980) considers the analysis of the stability of slopes affected by earthquakes and also has a useful bibliography. Examples of natural slopes at reservoir margins that have failed by landsliding following drawdown of the reservoir level, thus increasing the shear stresses acting on the slope, are given by Schuster (1979) and Schuster and Embree (1980). There will also be pore-water pressure changes associated with reservoir drawdown, as considered subsequently. Such events are a classic engineering problem, the analysis of which has been discussed by Bishop (1954) and Morgenstern (1963).

Shear strength loss due to reduced effective stresses

Turning now to processes operating within the slope, we find that our concern is primarily with factors that operate to reduce the strength of the material comprising the slope.

Swelling and short- and long-term conditions. In clays where the soil mass comprising the slope has a low permeability, there are two specific situations for which the stability of the slope must be considered (Fig. 16.1). Following erosion at the toe of the slope, there will have been increases in the shear stresses in the slope. Although these stress changes are accompanied by a reduction in pore-water pressure (which with stiff clays at shallow depths may become negative), the low permeability of the clay ensures that there is initially little pore-water migration in the slope, and hence negligible change in the available strength. This is the first of the situations and is known as the "short term" or "undrained" case. It applies where the erosion rate is high relative to the rate of pore-water migration, as shown in Figure 16.1A by path A–B–C. This path illustrates the sequence of stress changes that a typical soil element undergoes on the sliding surface of the landslide, and corresponds to the erosion sequence or falling external water level A–B–C shown in Figure 16.1B. Undrained failure of the slope occurs as erosion reaches C at an undrained shear strength c_u.

This stress path represents the behavior of a stiff clay slope; a similar path will be followed if undrained failure results from debris emplacement at the crest of the slope. With soft or, more particularly, with sensitive clays and materials subject to liquefaction, the corresponding pore pressures are higher, so that the failure line will be reached by stress path A–B–C at a point to the left of point C shown in Figure 16.1A. The undrained strength will be correspondingly lower. With such soils the pore pressures may continue to increase after failure and hence the strength will continue to decrease, with the stress path moving closer to the origin. This is the basic mechanism of flow slides, though with granular debris it may be necessary to invoke entrapped air rather than water as the pore fluid.

The second case is the "long-term" one and is also shown in Figure 16.1. If erosion does not continue until undrained failure occurs at C, but ceases at B, then failure may still occur, but only after a time delay. As described above, following erosion, the pore pressures in the slope are initially reduced. The magnitude of this change is shown by the fall in the piezometric head at the soil element, from h_A to h_B. Subsequently, the piezometric head will rise, but only slowly, to $h_{B'}$, the final value. This time-dependent pore pressure increase causes a reduction in the effective stress of $(h_{B'} - h_B)\gamma_w$. If this reduction is of sufficient magnitude, failure will occur when the stress path from B reaches the failure line at B'. Since no further erosion occurs below B, shear stresses remain constant beyond B, and path BB' is horizontal.

Thus, in summary, unloading the slope by erosion causes the pore pressures in the slope to fall, temporarily maintaining the available strength of the clay. Subsequent migration of water into the slope reduces the effective stresses and hence the strength of clay, and may result in a landslide many years after the original erosion phase.

The distinction between the short- and long-term situation is discussed in detail by Bishop and Bjerrum (1960), who classify the short term as being a situation where the pore pressure is *dependent* on the stress changes, and the long term as one where the pore pressures are *independent* of any previous stress changes.

The process by which long-term pore pressures are attained is that of "swelling," and as shown in Figure 16.1, it results (after a time delay) in a reduced soil strength being available in the long term compared with the short term. Note that the strength parameters c' and ϕ' remain unchanged. As a consequence of their low permeability, short-term conditions can apply only to clay soils. Slopes composed of more freely draining granular soils or jointed rock masses will have a groundwater regime that responds rapidly to erosion; thus at all erosion stages, the slopes will be in the long-term state.

Field evidence for delayed swelling   The time delay involved in the swelling processes still requires detailed study, though an analysis has been carried out by Eigenbrod (1975). By analogy with the process of consolidation (the process of reduction of excess positive pore pressures, leading to a decrease in water content and increase in strength), some of the factors controlling the rate of swelling are (1) the coefficient of permeability of the clay k, (2) the coefficient of volume expansion m_s, and (3) the square of the drainage path length d^2. Items (1) and (2) may be combined as the coefficient of swelling, c_s ($= km_s^{-1}\gamma_w^{-1}$). Hence, the rate of swelling is directly proportional to c_s and inversely proportional to d^2.

Even in an engineering context the rate of swelling in clay slopes following excavation is often comparatively fast. Chandler (1984a) concluded that in most late Pleistocene and Recent clays the combination of relatively high values of c_s (≥ 10 m^2/yr) and short path lengths (< 5 m) to pervious horizons (which often occur within the slope) results in the swelling process typically taking less than 18 months. In contrast, with overconsolidated clays or shales for which $c_s \leq 1$ m^2/yr and which are comparatively free from more permeable horizons, swelling periods may be lengthy. Slope failures delayed by up to 50 years have been observed in 10-m deep excavated slopes in the Eocene London Clay (Skempton, 1977; Vaughan and Walbancke, 1972).

These time periods suggest that with high erosion rates the swelling process may be of geomorphological significance in stiff clays. This will be the case particularly where a slope of comparatively high relief is being eroded so that drainage path lengths are long, as occurs at the London Clay coastal cliffs off the Isle of Sheppey, Kent (Bromhead and Dixon, 1984). Here the cliffs are about 40 m high and are being eroded at average rates of up to 4 m/yr (Hutchinson, 1973). Piezometers installed in the cliffs show pore pressures depressed by as much as 15 m below the eventual values were erosion to cease. Similarly, low pore pressures have been reported from Folkestone Warren, Kent (Hutchinson et al. 1980), where Chalk overlies the heavily over-consolidated Gault Clay, and where erosion protection works were completed some 25 years previously. These cliffs are about 160 m high. These examples show that if erosion ceases (or is prevented), slow swelling will cause reduction in strength and promote the continuation of landslipping for many years.

Groundwater level rise.   A number of situations other than swelling can cause pore pressures within a slope to rise, leading to a loss of strength. These include the removal of vegetation, irrigation, seasonal or other variations in the normal rainfall pattern, and an increase in external water levels, such as occurs when a reservoir is impounded.

This last effect is important, but may be neglected on the assumption that rising external water levels imply reduced shear stresses and hence greater stability of the slope. However, this is not necessarily true if there is a concomitant rise in the groundwater level within the slope, and if the frictional component of the soil strength is dominant. These conditions apply to slopes of granular soils or fractured rocks where the water table is initially low. Although the rising external water increases the stability of the slope, the increasing pore pressures within the slope reduce the effective stress and hence the strength. For a given slope, potential failure surface, and strength parameters, there will be an external water level at which the internal loss of strength is greater than the stabilizing effect of the external water, and the factor of safety will reach a minimum.

This is the "critical pool" level phenomenon, examples of which are given by Jones et al. (1961) and Schuster (1979). They describe landslides that occurred in the Pliestocene glaciofluvial deposits composing the slopes surrounding Roosevelt Lake, the reservoir impounded by the Grand Coulee Dam, Washington. Of some 500 landslides during the period 1941 to 1953, 49 percent occurred during first filling of the reservoir in 1941 and 1942, doubtless the consequence of the critical pool level phenomenon.

Local migration of pore water leading to local swelling and loss of strength can also occur to cause landslides, but there are few well documented examples. Ward et al. (1955) report the case of a newly constructed embankment where lateral migration of pore water from beneath the bank along a peat layer resulted in failure. It is possible, too, that falls of debris onto mudslides (Hutchinson and Bhandari, 1971) result in pore-water migration toward the mudslide basal surface, thus adding further to the instability caused by debris loading.

Shear strength loss due to reduced strength parameters

The previously discussed landslide-inducing processes all involve a change in strength resulting from reduced effective stresses, but no change in the effective strength parameters c' and ϕ'. The other category of mechanisms by which soil strength may be lost is that where the average strength parameters operating on the sliding surface are reduced. Such mechanisms include weathering, rate of shear effects, and progressive failure.

Weathering and shear strength.   In overconsolidated clays and clay—shales, weathering can conveniently (though perhaps rather simplistically) be defined as those processes that in saturated materials lead to an increase in water content in excess of that required for swelling. (This definition is, of course, not correct for soft clays where weathering, assisted by desiccation, results in the development of a stiff surface layer.)

Because in the absence of cementation an increase in water content implies a reduction in strength, weathering is a strength-reducing process. As such, its effect on clays and shales has been examined by a number of workers (e.g., Chandler, 1969, 1972; Herstus, 1971; Spears and Taylor, 1972; Cripps and Taylor, 1981) both in terms of undrained strength and effective stress. The general conclusion that emerges is that the effect of weathering is to remove in part the effects of the previous consolidation history. Thus so far as undrained strength is concerned, weathering results in a change in the relationship between strength and water content, as shown for different weathering zones in the Upper Lias (Chandler, 1972) (Fig. 16.2). The data refer to unconsolidated, undrained triaxial tests; unfortunately corresponding relationships in terms of effective stress are not available, but the general pattern indicated in Figure 16.3 can be inferred, with the process of weathering resulting in a reduction of both c' and ϕ'. Experimental data on Miocene clay (Herstus, 1971), Triassic marl (Chandler, 1969), and Carboniferous shale (Spears and Taylor, 1972) all demonstrate that weathering primarily results in a reduction in c', with only a modest reduction in ϕ'. This is consistent with the hypothesis that weathering modifies the effects of overconsolidation.

Shear-rate effects.   Laboratory strength tests show that if the rate at which soil specimens are sheared is reduced, then there is a corresponding reduction in the soil strength. Such effects are usually regarded as rheological or time-

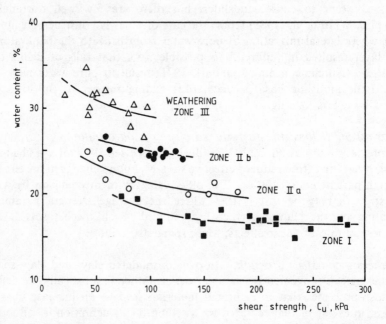

Figure 16.2 Relations between strength and water content for different weathering zones in Upper Lias. Weathering increases from zone I to zone III.

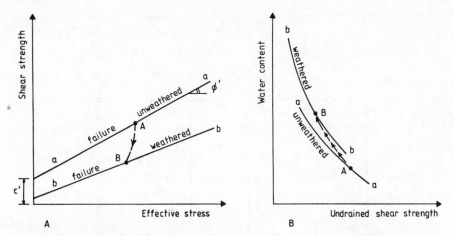

Figure 16.3 Idealized effects of weathering on the strength of clays and clay–shales: (A) in terms of effective stress, and (B) in terms of water content. Weathering from A to B increases the water content and decreases the available strength.

dependent. Their examination requires individual experiments to last for considerable time periods; hence research on the subject has been limited, particularly regarding the consequences for effective stress parameters.

Bishop and Henkel (1957) report for times to failure up to 14 days a loss of strength of 3.5 percent per log cycle of time for remolded Weald Clay. Tavenas et al. (1978) obtained a strength loss of about 7 percent per log cycle for the extra-sensitive St. Albans Clay, with tests lasting up to 34 days. Tests on undisturbed, unweathered Oxford Clay by Petley (1984) taking up to 80 days suggest a much higher rate of strength loss, up to 35 percent per log cycle. There is a suggestion in most of these data that the proportional reduction in strength falls with log time. This is a subject in which further work is required.

Progressive failure. Geological materials show widely varying load–displacement relationships. Such behavior is often brittle (Fig. 16.4, curve 1),

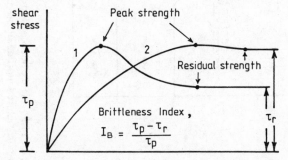

Figure 16.4 Stress–strain curves and brittleness. For curve 1 $I_B \simeq 0.4$ and for curve 2 $I_B \simeq 0$.

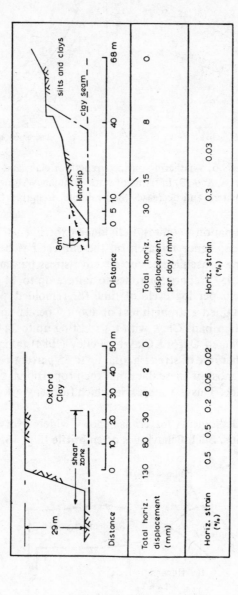

Figure 16.5 Field evidence of progressive failure of slopes: (A) pre-failure displacements and strains at Saxon Pit, Cambridgeshire, and (B) post-failure landslide displacements and strains, Seattle, Washington.

though low-plasticity clays usually show low brittleness (Fig. 16.4, curve 2). In a slope composed of brittle material, landsliding by progressive failure may occur if prefailure movements along a potential sliding surface are non-uniform. Thus, when the slope fails some points on the sliding surface will be at peak strength, while elsewhere local failure will have occurred and the strength will tend to the residual. The average strength along the sliding surface is thus intermediate between peak and residual.

By far the best field data concerning progressive failure are observations made by Burland et al. (1977) of the deformations of the side slopes of a brick-clay pit (Saxon Pit) in Cambridgeshire (Fig. 16.5A). The excavation was 29 m deep in Oxford Clay, a horizontally bedded Jurassic clay–shale. The clay has both high strength and high brittleness, with a brittleness index of 0.6. The slope face was continually excavated, and the reduction in horizontal stress resulted in movement of the face toward the excavation. Instrumentation showed that most of the horizontal strain occurred at a point 20 to 30 m back in the slope; thus a shear zone or sliding surface was developing as shown in Figure 16.5A.

As a result of this displacement, the shear zone would have been at residual strength over most of its length. Had a slope failure later occurred, the presence of the low-strength shear zone would have resulted in an average strength along the sliding surface considerably below the laboratory peak value.

Even after a landslide has occurred, nonuniform displacements may still occur within the landslip mass. Evidence for this is provided by a case record from Seattle, Washington (Fig. 16.5B) (Wilson, 1970), where excavation for a freeway caused a landslide in heavily overconsolidated, horizontally bedded silts and clays, whose shear surface coincided with a particularly plastic clay layer. This geological setting is not dissimilar to that at Saxon Pit. Observations of movement were not commenced until after the landslide occurred; even so, progressive movements at quite a high rate were observed. These movements strongly suggest that the landslide originally developed as a consequence of progressive failure.

Back-analysis of clay landslides

Our present knowledge of the behavior of slopes with respect to the strength of the soil is based on field observations of pore pressures and to a lesser extent displacements within both stable and unstable slopes, and also on comparison of laboratory measurements of strengths (usually on few small samples) with that obtained from limit equilibrium analysis of actual landslides. The latter method is known as "back-analysis," and the strength obtained from a back-analysis will be referred to as the "field strength."

Terzaghi (1936) observed that it was frequently found that laboratory strengths overestimated the field strength. The discrepancy between these two strengths is conveniently quantified by the strength ratio τ_{field}/τ_{lab}. This ratio may be considerably less than unity for plastic clays. Because such clays are often fissured, Terzaghi hypothesized that a softening or weathering process was involved, with water entering the clay along the fissures. In contrast, low-plasticity clays are often free from fissures ("intact"), and yield strength ratios are close to unity.

Since the publication of Terzaghi's paper, many back-analyses of clay landslides have been published that confirm the somewhat different behavior of intact and fissured clays. More recently, however, evidence has been produced from laboratory work on fundamental aspects of the shear strength of clays (reviewed by Chandler (1984a)) which suggests that plasticity and perhaps to a lesser extent particle size distribution have a significant influence on the stress–strain behavior. Thus clays of low plasticity inherently tend to be soils that require significant displacement to peak strength and exhibit low brittleness. Conversely, high-plasticity clays are both brittle and often reach peak strength with only small displacements. Thus, if clay brittleness is a function of plasticity, then progressive failure, which inevitably reduces the strength ratio τ_{field}/τ_{lab}, is also a function of plasticity. It follows that differences in behavior of intact and fissured clays may be more appropriately attributed to the plasticity and brittleness of the clay than to the presence or absence of discontinuities. Indeed, it has been shown experimentally (High et al., 1979) that even if discontinuities in low-plasticity, low-brittleness clay exist, they can have no influence on strength. An important consequence of this is that both undisturbed and remolded samples (provided this is done at constant water content) show similar strengths.

Low-plasticity clay landslides

There are comparatively few cases of landslides in slopes composed of low-plasticity clay, and further studies are required, particularly for the more heavily overconsolidated clays. One case is the Lodalen landslide (Sevaldson, 1956; Janbu, 1973), which occurred near Oslo, Norway, in October 1954. When the slope failed, five years had elapsed since the slope was excavated by widening an existing excavation in a soft clay with an average plasticity index of 18 percent. Stability analyses, in which the laboratory strength ($c' = 10$ kPa, $\phi' = 27°$) was used, gave factors of safety in the range 1.05 (Sevaldson) to 0.96 (Janbu). Thus the strength ratio was approximately unity.

A similar record for a stiff, low-plasticity clay is the Selset landslide in Yorkshire (Skempton and Brown, 1961), which took place at some unknown date in a glacial till having a plasticity index of 13 percent. Again, the use of laboratory strength parameters ($c' = 8.5$ kPa, $\phi' = 32°$) gave a factor of safety (and hence strength ratio) of unity. The stability of many other low-plasticity till slopes in northern England is consistent with the Selset record: not only is a substantial value of ϕ' implied, but in addition the c' term is significantly greater than zero (Chandler, 1984a). Thus the present evidence supports the use of careful laboratory strength determination for the assessment of the stability of slopes composed of either soft or stiff low-plasticity clays. Also implied by the good agreement between laboratory and field strengths is that weathering, shear-rate effects, and progressive failure are unimportant in these materials.

High-plasticity clay landslides

A comprehensive series of data concerning the field strengths exhibited by a number of landslides in two stiff plastic clays was published some years ago (Chandler and Skempton, 1974), but more recently laboratory measurements of the "bulk" strength of both clays have been made using triaxial test specimens up to 260 mm in diameter (Chandler, 1984*b*). These laboratory test data are compared with the results of back-analyses of the field failures in Figure 16.6. The two clays are London Clay and Upper Lias Clay, the latter of Lower Jurassic age. Their plasticity indices are 52 percent and 32 percent respectively.

The term "bulk" strength used above refers to the fact that the test results are thought to be the lowest reasonable estimate of the strength of an element of the landslip mass, the tests having been made on highly weathered and fissured

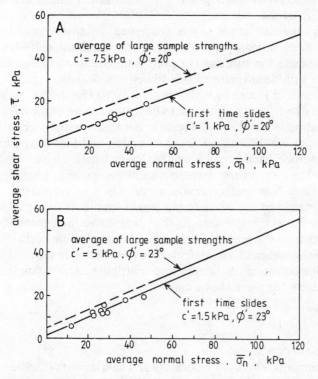

Figure 16.6 Comparison of slope failures and large sample strengths: (A) London Clay, and (B) Upper Lias.

specimens of the clays. If small samples are used for the laboratory measurement of the strength of plastic clays, then it is generally recognized that the bulk strength will be overestimated. In particular, the value of c' obtained is likely to be excessive, apparently because small samples do not contain a representative network of fissures. The evidence, though limited, suggests that for these clays the value of ϕ' obtained is independent of sample size, but that c' varies inversely with sample size, only becoming constant when the sample is large enough to contain a representative fissure network. Confirmation of this is provided by direct measurement of the joint and/or fissure strength of London Clay by Skempton and Petley (1967), and Santa Barbara Clay by Calabresi and Manfredini (1973). Both pairs of authors concluded that although ϕ' was little changed in comparison with that obtained in the absence of joints/fissures, c' was reduced to zero along a discontinuity.

Further factors must be considered in comparing the bulk strength with the field strength. Of these, the more important are the rate at which laboratory samples are sheared and the effects of strength anisotropy. In the case of the London and Lias Clays, rates of compression were in the range 5 to 15 mm/day. This compares with movements of 0.5 to 3 mm/day observed prior to failure in landslides (Skempton and Hutchinson, 1969; Mitchell and Eden, 1972), so even these large-sample tests may marginally overestimate the bulk strength of the clay.

As far as strength anisotropy is concerned, relative strengths of samples tested at different orientations (Skempton and Hutchinson, 1969) suggest that for most landslides in stiff clay the effect is unimportant, though this may not be the case with deep rotational landslides in soft clays.

It can be seen in Figure 16.6, that in spite of the fact that the measured bulk strengths for both clays are thought to be representative of the most weathered and fissured material, the field strengths are significantly lower. The average strength ratios are 0.7 and 0.8 for the London and Lias Clays, respectively. All the landslide back-analyses are for cases where the pore pressures are known, so the cause of the strength ratio being less than unity is probably a slope process such as weathering, shear-rate effects, or progressive failure. The weathered nature of the laboratory strength specimens would appear to rule out weathering as a major factor. That progressive failure does occur in the field is clear from the previous discussion, though the role of shear-rate effects remains problematical. Consequently, progressive failure seems likely to prove the most important of the factors that contribute to the strength ratio being less than unity for these plastic clays.

Other case records

Further comparison between laboratory and field strengths can be made on the basis of other landslide case records in the literature. Preliminary results of this survey are shown in Figure 16.7, where strength ratios are plotted against plasticity index for long-term failures of both soft Late- or Post-Glacial clays and also stiff, overconsolidated clays.

The laboratory strengths reported are all from tests on 38 mm-diameter specimens. Thus these strengths cannot be compared to the large-sample tests on London and Lias Clay, and will presumably overestimate the bulk strength of the slope material in many instances due to sample size, shear-rate, and anisotropic effects. With the softer clays, sampling disturbance may lead to an underestimate of the laboratory strength.

In spite of all these drawbacks, Figure 16.7 is useful as it shows (1) that laboratory small-sample strengths are often at least twice that observed in the field; (2) that typically the strength ratio is greater for soft clays, suggesting that overconsolidation leads to lower strength ratios; and (3) that there is a tendency in both soft and stiff clays for the strength ratio to fall with increasing plasticity. Figure 16.7 also confirms that at plasticity indices below 15

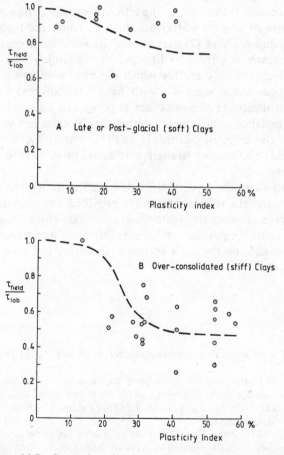

Figure 16.7 Strength ratios from back-analyses of landslides.

to 20 percent, strength ratios approach unity, and laboratory strength data may consequently be used directly to assess slope stability. Thus at higher plasticities the internal processes of weathering, shear-rate effects, and progressive failure appear to become more important; their combined effects on the field strength of slopes of more plastic clay must be considered.

Conclusions

A classification of processes that lead to the occurrence of landslides are given in Table 16.1, providing a basic check list of the factors to be considered in assessing the landslide potential of any particular site. External causative factors generally lead to increases in shear stresses within the slope and can probably be assessed with reasonable accuracy from environmental studies. Internal causative factors lead to a reduction in the shear strength of the material composing the slope. These internal factors may be further divided into those processes (1) that reduce the effective stresses (rising pore pressures or a general rise of ground water) but do not change the material strength parameters c' and ϕ', and (2) that reduce the material strength parameters (weathering, shear-rate effects, and progressive failure).

Those relatively plastic clays for which the most comprehensive landslide studies have been made seem likely to have been affected by progressive failure, though shear-rate effects cannot at present be discounted. Thus even the use of large laboratory samples, tested slowly, does not enable the field strength of the clay to be established. If small laboratory specimens are tested, the ratio of field to laboratory strength may be as low as 0.27, though typically values are higher.

With low-plasticity clay (plasticity index < 20 percent), the ratio of field to laboratory strength is close to unity, provided that the disturbance of the softer clays is minimized. Weathering, shear-rate effects, and progressive failure are much less important, and reasonably accurate assessments of slope stability are possible on the basis of small laboratory test specimens.

References

Bishop, A. W., The use of the pore-pressure coefficients in practice, *Géotechnique, 4*, 148–152, 1954.

Bishop, A. W., and L. Bjerrum, The relevance of the triaxial test to the solution of stability problems, *Proceedings of the Conference on the Shear Strength of Soils*, 437–501, 1960.

Bishop, A. W., and D. J. Henkel, *The measurement of soil properties in the triaxial test*, 228 pp., Edward Arnold, London, 1957.

Bromhead, E. N., and N. Dixon, Pore water pressure observations in the London Clay cliffs at the Isle of Sheppey, England, *Proceedings of the 4th International Symposium on Landslides, 1*, 385–390, 1984.

Burland, J. B., T. I. Longworth, and J. F. A. Moore, A study of ground movement and progressive failure caused by a deep excavation in Oxford Clay, *Géotechnique, 27,* 557–591, 1977.

Calabresi, G., and G. Manfredini, Shear strength characteristics of the jointed clay of S. Barbara, *Géotechnique, 23,* 233–244, 1973.

Chandler, R. J., The effect of weathering on the shear strength properties of Keuper marl, *Géotechnique, 19,* 321–334, 1969.

Chandler, R. J., Lias clay: weathering processes and their effect on shear strength, *Géotechnique, 22,* 403–431, 1972.

Chandler, R. J., Recent European experience of landslides in overconsolidated clays and soft rocks, *Proceedings of the 4th International Symposium on Landslides, 1,* 61–81, 1984a.

Changler, R. J., Delayed failure and observed strength of first-time slides in stiff clay: a review, *Proceedings of the 4th International Symposium on Landslides, 2,* 19–25, 1984b.

Chandler, R. J., and A. W. Skempton, The design of permanent cutting slopes in stiff fissured clays, *Géotechnique, 24,* 457–464, 1974.

Cripps, J. C., and R. K. Taylor, The engineering properties of mudrocks, *Quarterly Journal of Engineering Geology, 14,* 324–364, 1981.

Eigenbrod, K. D., Analysis of the pore-pressure changes following the excavation of a slope, *Canadian Geotechnical Journal, 12,* 429–440, 1975.

Herstus, J., Influence of weathering on effective values of shear strength of Miocene clay, *Proceedings of the 4th European Conference on Soil Mechanics and Foundation Engineering,* 135–142, 1971.

Hight, D. W., M. K. El-Ghamrawy, and A. Gens, Some results from a laboratory study of a sandy clay and its implications regarding its in situ behaviour, *Proceedings of the 2nd International Conference on Behaviour of Offshore Structures, 2,* 133–150, 1979.

Hutchinson, J. N., The response of London Clay cliffs to differing rates of toe erosion, *Geologia Applicata e Idrogeologia, 8,* 221–239, 1973.

Hutchinson, J. N., and R. Bhandari, Undrained loading, a fundamental mechanism of mudflows and other mass movements, *Géotechnique, 21,* 353–358, 1971.

Hutchinson, J. N., E. N. Bromhead, and J. Lupini, Additional observations on the Folkestone Warren landslides, *Quarterly Journal of Engineering Geology, 13,* 1–31, 1980.

Janbu, N., Slope stability computations, in *Embankment–Dam Engineering* (Casagrande Volume), edited by R. C. Hirschfeld and S. J. Poulos, pp. 47–86, John Wiley, New York, 1973.

Jones, F. O., D. R. Embody, and W. L. Peterson, Landslides along the Columbia River Valley, north eastern Washington, *U.S. Geological Survey Professional Paper, 367,* 1961.

Mitchell, R. J., and W. J. Eden, Measurements of slope displacements, *Canadian Journal of Earth Sciences, 9,* 1001–1013, 1972.

Morgenstern, N., Stability charts for earth slopes during rapid drawdown, *Géotechnique, 13,* 121–131, 1963.

Petley, D. J., Shear strength of over-consolidated fissured clay, *Proceedings of the 4th International Symposium on Landslides, 2,* 167–172, 1984.

Sarma, S. K., A simplified method for the earthquake resistant design of earth dams, *Proceedings of the Conference on Dams and Earthquake,* 155–160, 1980.

Schuster, R. L., Reservoir-induced landslides, *Bulletin of the International Association of Engineering Geology, 20,* 8–15, 1979.

Schuster, R. L., and G. F. Embree, Landslides caused by rapid draining of Teton Reservoir, Idaho, *Proceedings of the 18th Annual Symposium of Engineering Geology,* 1980.

Sevaldson, R. A., The slide in Lodalen, October 6th, 1954, *Géotechnique, 6*, 167–182, 1956.

Skempton, A. W., Slope stability of cuttings in brown London Clay, *Proceedings of the 9th International Conference on Soil Mechanics and Foundation Engineering, 3*, 261–270, 1977.

Skempton, A. W., and J. D. Brown, A landslide in boulder clay at Selset, Yorkshire, *Géotechnique, 11*, 280–293, 1961.

Skempton, A. W., and J. N. Hutchinson, The stability of natural slopes and embankment foundations, *Proceedings of the 7th International Conference on Soil Mechanics and Foundation Engineering, State-of-the-Art Volume*, 291–340, 1969.

Skempton, A. W., and D. J. Petley, The strength along structural discontinuities in stiff clays, *Proceedings of the Geotechnical Conference, 2*, 29–46, 1967.

Spears, D. A., and R. K. Taylor, The influence of weathering on the composition and engineering properties of in situ Coal Measures rocks, *International Journal Rock Mechanics and Mining Science, 9*, 729–756, 1972.

Tavenas, F., S. Leroueil, P. LaRochelle, and M. Roy, Creep behaviour of an undisturbed, lightly over-consolidated clay, *Canadian Geotechnical Journal, 15*, 402–423, 1978.

Taylor, D. W., *Fundamentals of soil mechanics*, 700 pp., John Wiley, New York, 1948.

Terzaghi, K., Stability of slopes of natural clay, *Proceedings of the 1st International Conference on Soil Mechanics and Foundation Engineering, 1*, 161–165, 1936.

Vaughan, P. R., and H. J. Walbancke, Pore pressure changes and the delayed failure of cutting slopes in over-consolidated clay, *Géotechnique, 23*, 531–539, 1973.

Ward, W. H., A. Penman, and R. E. Gibson, Stability of a bank on a thin peat layer, *Géotechnique, 5*, 154–163, 1955.

Wilson, S. D., Observational data on ground movements related to slope stability, *Journal of the Soil Mechanics Division, Proceedings of the American Society of Civil Engineers, 96*, 1521–1544, 1970.

17
Hollows, colluvium, and landslides in soil-mantled landscapes

William E. Dietrich, Cathy J. Wilson, and Steven L. Reneau

Abstract

Field measurements from 23 sites in California and from five other widely distributed locations on soil-mantled hillslopes reveal a strong inverse relation between maximum drainage area of unchannelized basins and their average hollow gradients. The area–slope product was found to equal about 4000 m^2. Hollow length was also negatively correlated to average hollow gradient. These observations indicate that the number of sources and, consequently, drainage density increase with slope. The conservation of mass equation for a tipped triangular trough and a slope-dependent transport law are used to develop an expression for the rate of colluvium accumulation in hollows. Maximum colluvium depth is found to increase by the one-half power of time owing to the vertically increasing cross-sectional area of the trough. Rate of accumulation is proportional to side-slope gradient and to the difference between the side-slope and hollow gradient. The ratio of hollow to side-slope gradient is typically about 0.8. Models of shallow subsurface flow and deeper ground water flow are used to predict, via the Mohr–Coulomb failure criterion, the relation of hollow length to its gradient. It is not yet known which flow path plays the dominant role in controlling the position of the channel head, but both models predict that angle of internal friction ϕ' of the colluvium or weathered bedrock strongly influences the size and slope of the unchannelized basins and that basin size must rapidly decline above a gradient of 0.7 tan ϕ'.

Introduction

As a consequence of the seminal work by Hack and Goodlett (1960) and the numerous field studies that their study stimulated, it is now becoming widely recognized that small unchannelized valleys (Fig. 17.1) dominate the drainage

Figure 17.1 View of unchannelized basins on hillslopes in Marin County, California. Note tributaries or subhollows within larger unchannelized basins.

area of most soil-mantled hillslopes and greatly influence runoff and sediment transport processes. The valleys focus shallow subsurface storm flow or throughflow, which may produce high pore pressures and saturation overland flow, in the downslope ends of the valleys. The spoon-shaped surface topography also forces the convergence of soil material transported downslope by creep and biogenic transport processes, and consequently the valleys tend to develop a thick mantle of colluvium (e.g., Hack and Goodlett, 1960; Pierson, 1977; Dietrich and Dunne, 1978; Marron, 1982; Lehre, 1982; Woodruff, 1971; Reneau et al., 1984). Landslides most commonly occur in these topographic convergent zones (e.g. Pierson, 1977; Tsukamoto et al., 1982; Shimokawa, 1984; Dietrich and Dunne, 1978; Woodruff, 1971; Hack and Goodlett, 1960; Lehre, 1982; Iida and Okunishi, 1983) and the thick deposits tend to produce large rapid debris flows that scour to bedrock the first- and second-order channels that lie downslope (Pierson, 1977; Dietrich and Dunne, 1978; Swanston and Swanson, 1976). Debris flows from thick colluvium in unchannelized valleys have caused considerable destruction of property and loss of life (Brown, 1984; Reneau et al., 1984; Woodruff, 1971; Tsukamoto et al., 1982) and play a major role as sediment sources in catchment-level sediment budgets (Dietrich and Dunne, 1978; Lehre, 1982; Swanson et al., 1982; Pierson, 1977).

A qualitative model is fairly well established for how the unchannelized valleys work. In essence, the valleys are sites that experience infrequent, but recurrent rapid evacuation of colluvium, primarily by landsliding, followed by periods of slow colluvium accumulation (Hack and Goodlett, 1960; Woodruff, 1971; Pierson, 1977; Calver, 1978; Dietrich and Dunne, 1978; Kirkby, 1978; Humphrey, 1982; Lehre, 1982; Shimokawa, 1984). The convergent subsurface flow in the spoon-shaped bedrock geometry leads to highest pore pressures in the lower portion of the valley. These pore pressures may be further enhanced by the discharge of deeper ground water in the hollow. Under a certain combination of colluvium strength and antecedent moisture conditions, storm precipitation can produce failure in the colluvium. Weathering and soil development may produce permeability and strength boundaries such that failures occur within the colluvium (Reneau et al., 1984; Tsukamoto and Kusakobe, 1984). The landslide may only partially evacuate the colluvial mantle, after which sheetwash and gullying may continue to erode the exposed slide scar. In some cases, gullying may be the primary cause of colluvium removal. Revegetation on the scar contributes significantly to reestablishing accumulation in the hollow (Lehre, 1982; Shimokawa, 1984).

The frequency of flushing events in hollows is not yet well documented. Shimokawa (1984) used dendrochronology and soil thickness measurements on landslide scars of various ages in steep, forested hillslopes of Japan to estimate recurrence interval of landsliding in hollows and adjacent slopes. As Iida and Okunishi (1983) have also proposed, Shimokawa suggested that recurrence of landsliding is largely controlled by the rate of "top-soil" accumulation because landsliding is thought to be very likely once the soil reaches a critical depth. Estimated recurrence for landslides at a site varied with bedrock type and slope gradient from 12 to 30 years on 50–70° slopes developed on ash and pumice to about 1000 years on 20–40° slopes developed on granite. In the Pacific Northwest of the U.S.A., dendrochronology has been useful in giving estimates of the rate of deposition in hollows in the early phase of post-landslide accumulation (Dietrich and Dunne, 1978), but it appears that for the vast majority of sites the recurrence of major flushing is much greater than can be ascertained from this method. Palynology (Dietrich and Dorn, 1984) and radiocarbon dating of thick colluvial deposits in hollows in the California Coast Ranges and Tehachapi Mountains (Marron, 1982; Reneau et al., 1984 and in preparation) suggest that the frequency of erosion of colluvium to the bedrock surface is of the order of 10 000 years. Six of nine sites dated, however, yield basal ages between 11 000 and 14 500 B.P. (Reneau et al., in preparation), and it may be that these hollows record a period of more frequent, intense storms. Much work is still needed to sort out the significance of climatic fluctuations. We see both landsliding and net accumulation under current climatic conditions, hence climatic change is not required to induce either net accumulation or erosion in individual hollows. A similar conclusion was reached by Hack and Goodlett (1960) for hollows in the Appalachian Mountains, and by Gray and Gardner (1977) for colluvial deposits on hillslopes of the Appalachian Plateau.

There is now a need to develop quantitative, physically based models for the hydrology, sediment transport, weathering, and instability of the colluvium and link these models to predict the geometry of unchannelized basins. Some valuable first steps have been taken in modeling subsurface flow in small unchannelized bedrock valleys using finite-element methods (Humphrey, 1982) and contributing-area concepts (Humphrey, 1982; Iida, 1984; Kirkby, 1978). Iida and Okunishi (1983) treated the stability problem as being entirely dependent on the depth of colluvium, and modeled the lowering rate of the bedrock surface as a function of the rate of colluvium accumulation and bedrock weathering, which in turn control the recurrence interval of landsliding.

In this chapter, we focus on the end-member problem of what controls the size and shape of the largest unchannelized basins on hillslopes—that is, those basins large enough to support channelized flow at their downslope ends. In essence we ask: How much drainage area in a basin does it take to initiate a channel? We first present field data on area, length, and slope relationships of such basins and then analyze the shape of the basins and the accumulation of colluvium in the hollow. Finally, through the use of simple flow models and slope stability calculations we propose a quantitative explanation for the observed field relationships.

Basin geometry

Other than the qualitative impression that the colluvium-mantled bedrock valleys appear to be typically spoon-shaped with concave contours and concave-up downslope axes, little is known about the geometry of these features. To describe the large-scale, three-dimensional properties of slopes, Hack and Goodlett (1960) introduced the now widely used terminology of *nose* for convex contours, *side slope* for straight contours, and *hollow* for concave contours. Typically all three elements make up the basin upslope of the head of the channel. Marcus (1980) proposed a morphologic classification of first-order basins based in part on the planform shape of the unchannelized valley in the first-order basins. Tsukamoto et al. (1982) computed the areal proportions of convergent, divergent, or planar slopes in Japanese catchments. On average, convergent slopes make up 60 percent of the entire basin area. Tsukamoto et al. propose calling convergent slope areas zero-order basins and state that these are sites of ephemeral streams during heavy storms.

Four measures of unchannelized basin geometry that are fairly simple to obtain and appear to be quite useful are (1) the drainage area of the basin upslope of the channelway, (2) the maximum horizontal length of the axis of the basin, (3) the hollow and side-slope gradients, and (4) the basin amplitude (Fig. 17.2). The first three measures involve careful field inspection to locate the farthest downslope extent of the unchannelized, colluvial portion of the hollow (axial region of the basin). The average hollow gradient, which will be used in the analysis to follow, is computed as the elevation difference between ridge crest and channel head divided by the horizontal distance. In general, the hollows that are direct upslope extensions of the first-order channels rather than tributaries to them will be the largest unchannelized basins. After

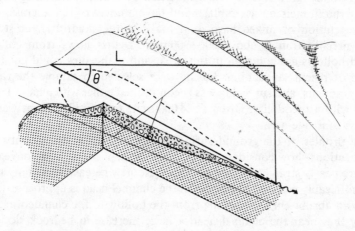

Figure 17.2 Illustration of the geometric properties of unchannelized basins. L and θ are the horizontal length and average gradient of the hollow, respectively. Basin amplitude is the distance from the bedrock to the horizontal line between the noses along a line normal to the bedrock surface in the hollow axis. The broken line encloses the depositional zone in the hollow.

erosional events, the channel head may temporarily extend well up the hollow, as described by Hack and Goodlett (1960) and many other subsequent authors; hence for the purpose of defining the maximum unchannelized basin geometry, the hollows without significant evidence of recent flushing should be used. Alternatively, basins can be randomly sampled without regard to recent flushing and trends in geometry can be established using the largest unchannelized basins encountered. The fourth measure, basin amplitude, quantifies how deeply the hollow is incised relative to adjacent interfluves. Amplitude can be computed from a detailed topographic map and information on thickness of the colluvium in the hollow. A straight line can be drawn across the basin connecting points of equal elevation on adjacent drainage divides on the noses. The maximum difference in elevation along this line between the nose and the bedrock surface of the hollow multiplied by the cosine of the interfluve slope is a measure of basin amplitude. The thickness of the colluvium in the hollow can be determined by drilling, by examining exposure in landslide scars, by using geophysical methods, such as seismic refraction, or by employing some combination of these techniques. As amplitude will vary along the basin, for purposes of comparison, maximum amplitude is used.

Observations

Results of field surveys in northern California and comparison with sites elsewhere raise several intriguing questions regarding basin geometry and controls on channel inception. Basins in the rounded, soil-mantled hillslopes just north of San Francisco in Marin County were selected to obtain a broad range in hollow gradient. Sites are underlain by chert, greenstone, and sandstone of the Franciscan assemblage and have a cover of dense grass, coastal scrub vegetation or mixed hardwood (oak, laurel, madrone) forest. Mean annual precipitation is about 600–900 mm. Debris flows from colluvium-mantled hollows are common in this area, and many sites could not be used because of recent landsliding. Basins were selected to sample the range of hollow gradients, and surveying was accomplished either with a hand-level and tape or with a theodolite. Often 1 : 24 000-scale topographic maps were used to define drainage area.

Slope profiles of the ground surface of hollows in Marin County and at other locations show considerable variation. Except for a short convex ridge, many are nearly straight or concave throughout, whereas others only become either noticeably convex or concave as the channel head is approached. Often there is an abrupt change in slope from the hollow to the channelway, and it is likely that near the channel head a large increase in bedrock slope leads to a convex profile in the colluvial fill, whereas a strong decrease in bedrock slope results in a concave colluvial slope profile. A large change in slope from hollow to channelway probably significantly affects the groundwater flow and consequent pore pressures develop in the downstream end of the colluvial deposit. For three cases where a strong steepening of slope was observed, the hollow and local channelway slopes were averaged and data were noted separately (Fig. 17.3). It is not yet clear whether this averaging procedure is the best representation of field relationships. In two gentle profiles with gradients less than 10°, discontinuous gullies were present 50–70 m upslope from the continuous channel head, and in one case, winter storm flow discharged from a large pipe in the upper part of the discontinuous gully. Regardless of surface slope, the longitudinal profile of the underlying bedrock surface in hollows appears generally to be concave-up (Reneau et al., 1984; Dietrich and Dorn, 1984; Pierson, 1977; Iida and Okunishi, 1983).

Drainage area of surveyed unchannelized basins in Marin County increases in proportion to the 1.2 power of the total hollow length (Fig. 17.3A). This low rate of increase in area with hollow length implies that small and large basins differ more in their length than in their width. As Hack and Goodlet (1960) originally noted, however, area increases quite rapidly with distance down the hollow within individual basins.

Data from field sites in Marin County reveal an inverse relationship between drainage area upslope of the channel head and average hollow gradient (Fig. 17.3B). The most accurate field data give area proportional to the −1.04 power of hollow gradient. Area is proportional to the −0.99 power of hollow

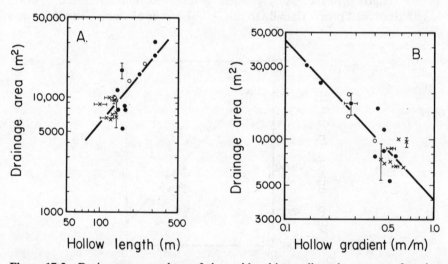

Figure 17.3 Drainage area upslope of channel head in small catchments as a function of (A) horizontal length of hollow and (B) hollow gradient for hillslopes in Marin County, immediately north of San Francisco, California. Open circles represent sites in which slope strongly steepened downstream of the channel head and the average of upslope and downslope gradients on either side of the channel head was used (unpublished data from K. Whipple and P. Templet, University of California, Berkeley, 1984). The crosses represent forested sites where landslide scars are common and make it more difficult to define precisely the channel head. Error bars denote range of uncertainty. The full line in (A) represents the least-squares regression equation $A = 22.9 L^{1.21}$, $r = 0.79$, $n = 20$.

gradient for the entire 20-point data set. Hence the area–slope product is essentially a constant and for the nine data points represented by full circles in Figure 17.3B the constant is 4043 m². If all available data for Marin County are used in conjunction with eight other data points from gravelly soils in northern California, Oregon, Virginia, and Japan to define the area–slope correlation, essentially the same result is obtained (Fig. 17.4). In this case, area decreases in proportion to the – 1.05 power of hollow gradient and the area—slope product equals 4121 m². There are, of course, many smaller unchannelized basins that intersect as tributaries to channels or other hollows (Fig. 17.1) before they capture sufficient drainage area to produce a channel. Strong geologic controls may force ground water toward the surface, which can generate sufficient pore pressure and consequent instability to initiate and maintain a channel in basins smaller or less steep than expected from Figure 17.4. With more data, it may be possible to detect differences due to geology, climate, or vegetation. This small sample suggests, however, that in general the area–slope product necessary to maintain a channel head at the end of an unchannelized basin in well-vegetated, coarse-textured colluvium on a soil-mantled hillslope is a constant and equals about 4000 m².

Data from the sites used in Figure 17.4 also show a strong decrease in total hollow length with increasing gradient (Fig. 17.5). For all 28 data points, length decreased proportionally to the – 0.67 power of slope. For example, a

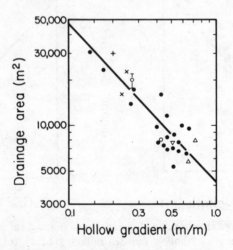

Figure 17.4 Relationship between drainage area and hollow gradient for basins upslope of channel heads in Marin County, California (●), Clear Lake, California (+ , Dietrich and Dorn, 1984), coastal northern California (× , W. Trush, University of California, Berkeley, unpublished data, 1984), Virginia (○, without bar, Hack and Goodlett, 1960; Ō, with bar, W. Dietrich, unpublished data, 1984), Japan (▽ , Tanaka, 1982), and central coastal Oregon (△ , Pierson, 1977, personal communication, 1984). The full line represents the regression equation $A = S^{-1.05}$, $r = -0.885$, $n = 28$.

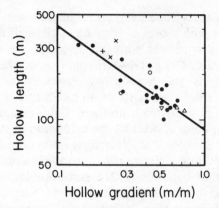

Figure 17.5 Relationship of hollow length to average gradient for unchannelized basins. Symbols are the same as in Figure 17.4. The regression equation is $L = 87.7S^{-0.67}$, $r \doteq -0.823$, $n = 28$.

hollow with a $10°$ slope should be about 281 m long, whereas a steeper, $30°$ hollow should be only about 127 m long. Because the hollow gradient equals the total elevation drop from ridge to channel head divided by horizontal length of the basin, the regression equation indicates that hollow length decreases as the -2.0 power of elevation drop.

The length–slope observation has important implications for drainage density. The area required to initiate a channel head is also the area necessary to produce a source for an exterior link. Hence, the maximum number of sources in a basin equals the basin drainage area divided by the observed area required to initiate a channel. Clearly, the actual number of sources will be much less because most of a drainage basin drains to the sides rather than to the heads of channels. Nonetheless, the inverse relationship between area and slope shown in Fig. 17.4 would suggest that the number of sources in a basin is directly correlated with slope. Intuitively, one would expect that the greater the number of sources in a drainage basin, the greater the drainage density. This intuition is confirmed if certain simple assumptions are made (R. Shreve, personal communication, 1985). If we assume that in a large drainage basin the number of exterior links, which is the same as the number of sources, is equal to the number of interior links, and we denote the number of sources *per unit area* by N, then the drainage density D is

$$D = \frac{2N\bar{l}_L}{2N\bar{a}_L} \tag{1}$$

for a given mean link length $\bar{l}_L$ and link area $\bar{a}_L$. Because $\bar{a}_L = 1/(2N)$ and $\bar{l}_L^2 = K\bar{a}_L$ (see discussion in Abrahams (1984)), where K is a constant, the drainage density can also be expressed as

$$D = (2KN)^{0.5} \tag{2}$$

Thus our empirical area–slope relationship implies that steeper basins of a given drainage area with more sources, have higher drainage density than gentler basins of the same area with fewer sources. These preliminary data support the inference by Dunne (1980) that in regions where subsurface flow, and consequent piping and landsliding, play a major role in landscape evolution, drainage density is higher on steeper slopes.

Limited data (four sites) from northern California on basin amplitude suggest that the amplitude divided by the half-width of the basins is roughly equal to the hollow gradient. For example, a 100-m wide basin on a 9° slope would have 8 m elevation difference between the interfluve and the hollow, whereas a 60-m wide basin on a 27° slope would have 15 m elevation difference. Although unchannelized basins tend to be narrower on steeper slopes, the steeper gradient of the hollow results in greater hollow amplitude.

Because the gradients down the interfluves, down the hollow, and from the interfluve to the hollow are similar, it is reasonable to suggest that the gradients are controlled by the same mass-wasting processes. We cannot as yet predict these gradients.

Sediment transport

On well-vegetated, soil-mantled hillslopes, soil is transferred downslope by the mass movement processes of landslides, creep, and biogenic transport. Landsliding is rare on noses or side slopes to hollows. On hillslopes where soil properties and biologic activity are relatively uniform, rates of creep and biogenic transport are probably primarily dependent on slope gradient. Hence, in unchannelized basins, the focusing of colluvium toward the hollow, which is generally less steep than the side slopes, results in net deposition and a thickening of colluvium over time.

Landslide processes involving the colluvial deposits in the hollow appear to vary with texture. Instability in clay-rich colluvium tends to produce slow-moving earthflows (e.g., Keefer and Johnson, 1984): in a single season total displacement of the center of mass of the colluvial fill may be only a few meters or tens of meters. Movement is periodic and the intervals of inactivity can be sufficiently long that the landslide morphology (such as the crown scarp and the lobate snout) may be indistinct. In contrast, landslides in coarse-textured colluvium in hollows, which are the focus of this paper, typically initiate as a slide but quickly liquefy, producing a rapidly flowing debris flow (e.g., Pierson, 1977). Liquefaction of colluvium may be caused by the sudden slide-induced increased shear that affects an undrained loading condition. Recent work by Kramer (1985) has demonstrated that loose fine sand will liquefy when subjected to undrained shear, and he has suggested that coarser debris will act similarly. Typically only a portion of the accumulated colluvium in the hollow is discharged with the debris flow. Subsequent gullying and slumping may erode a large portion of the remaining colluvium before net accumulation is initiated.

In the following discussion, we examine two simple models for colluvium accumulation in hollows. There are several reasons to build such models. At least in the early phases of deposition in a hollow after a flushing event, increasing colluvium thickness should lead to less stable deposits because of the progressively reduced effectiveness of the apparent cohesion provided by roots (Dietrich and Dunne, 1978). Hence, an important linkage may exist between rate of infilling and frequency of instability (see also Pierson, 1977, p. 133). As mentioned above, Iida and Okunishi (1983) proposed a deterministic model that predicts failure of colluvium once it reaches a critical thickness. The colluvium was assumed to be saturated, and they stated that intense saturation-producing storms in Japan are sufficiently frequent for this to be a reasonable assumption. On the steep slopes of the coastal mountains in the western United States, however, colluvium in many areas has reached thicknesses much greater than that which can be easily saturated, perhaps due to an extended period in the Holocene with a low frequency of intense rainstorms (Reneau et al., in preparation). Colluvium thickness in hollows ranges from less than 1 m to over 10 m, but typical maximum depth is 4–5 m (Reneau et al., 1984). If rates of deposition can be predicted, some of the observed variations may be explicable quantitatively in terms of basin geometry and time since flushing. Finally, develoment of a sediment transport model will provide, some insight into the form of the transport law (*sensu* Kirkby, 1971; Smith and Bretheron, 1972) controlling soil transport and the morphology of the nose and side slopes contributing to the hollow accumulation. This sediment transport model should allow us to explore the adjustments in slope, soil thickness, and transport rates for different bedrock.

A conical basin

The most convenient mathematical representation of an unchannelized basin is that of a slice of a cone (Fig. 17.6). As Carson and Kirkby (1972, p. 392) have shown, the mass balance equation for colluvium transport in a cylindrical

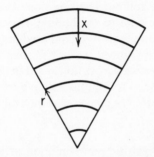

Figure 17.6 Coordinate system for conical-shaped basins.

coordinate system (Fig. 17.6) in which the mass transport law is

$$q_S = -a \frac{\partial z}{\partial x} \qquad (3)$$

can be written as

$$-a \frac{\partial^2 z}{\partial x^2} - \frac{a}{r} \frac{\partial z}{\partial x} = -\frac{\partial z}{\partial t} \qquad (4)$$

The first term in (4) is the diffusion component in which a is the transport coefficient, z is the elevation above an arbitrary datum, and x is the distance downslope from the ridge crest (Fig. 17.6). The second term arises from the convergence caused by contour concavity and increases as the distance from the basin outlet r decreases. The last term is the local elevation change of the colluvial surface with respect to time t. Equation (4) can be solved numerically for a variety of upslope and downslope boundary conditions. If profile concavity is small then the diffusion component is negligible and (4) yields

$$h \approx \frac{a}{r} St \qquad (5)$$

where h is the thickness of the accumulated colluvium and S is the constant slope. Depth of colluvium increases downslope linearly with time. For a given amount of time since onset of deposition, thickness is proportional to slope. More complex results, involving significant contribution from the diffusion component, can be obtained if profile concavity is allowed and the effects of discharge at the end of the basin are permitted to propagate upslope and alter the surface slope. This seems unwarranted, however, because the basin geometry, although mathematically convenient, does not appear to represent natural basins.

Inspection of unchannelized basins in the field or on sufficiently detailed topographic maps (1- to 2-m contour interval) confirms Hack and Goodlett's (1960) observation that such basins are best described as having convex contours (nose) along a narrow drainage divide and straight contours (side slopes) between the nose and the hollow. The side-slope element, which is the transportation surface of colluvium to the hollow, is not well represented by the coordinate system shown in Figure 17.6 and used in (4) and (5). This geometry also does not yield the pronounced 'U' or 'V' shaped cross-sections typically revealed in road-cut exposures across hollows. Finally, unchannelized basins, although spoon-shaped, tend to be long relative to their widths and to have subparallel rather than strongly convergent interfluves.

A trough-shaped basin

We propose that a more realistic representation of an unchannelized basin is that of a tipped triangular trough. As illustrated in Figure 17.7A, sediment

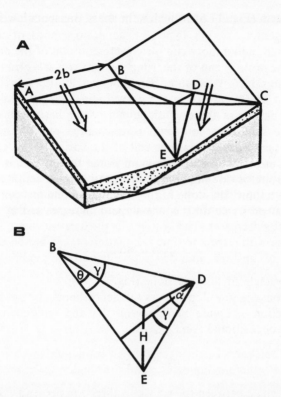

Figure 17.7 (A) Diagram of an unchannelized basin represented as a tipped triangular trough. The horizontal plane ABC intersects the surfaces of the colluvium-mantled hollow and side slopes along contours AB and BC, respectively. Note that convergent transport, represented by arrows, leads to accumulation of thick deposits of colluvium. (B) Geometric relationships of hollow gradient θ side-slope gradient α and the angle γ formed between a line parallel to the side slope and perpendicular to the depositional zone in the hollow.

transport from straight contour side slopes intersects the less steep hollow at an acute angle. The trough is composed of a source area of side slopes, where the soils are thin, and a depositional zone of thicker colluvium in the hollow. With increasing depth of colluvium, the cross-sectional area increases rapidly and produces a distinct 'V' shaped deposit. Profile concavity of the trough axis is not considered in this geometric model, although the model can be modified to include this feature. The nose is reduced to a line, but this should have negligible effect on deposition of colluvium in the hollow. Perhaps more importantly, creep and biogenic transport processes will tend to make the slope transition from the side slope to the hollow less abrupt, probably leading to straight contours on accumulated colluvium along the hollow boundary. This effect will not be considered here, as it is viewed as nonessential to understanding the form of deposition over time.

To model the deposition of colluvium in the tipped trough we make several simplifying assumptions. (1) Typically the upslope ends of unchannelized basins are curved, and a more complete representation of the basin would be to include at the upslope end of the trough a short conical form similar to that depicted in Figure 17.6. We assume here that contribution of colluvium to the upslope end of the trough from the curved end is exactly balanced by the colluvium discharged at the downslope end of the basin. This implies that although some side-slope colluvium is discharged out of the basin, it is replaced with that from the upslope end of the basin. In this case, then, the change in volume with time of colluvium in the hollow equals the discharge of colluvium from the side slopes into the depositional zone. (2) As the deposit thickens in the hollow, the slope of the colluvial fill remains constant. (3) The downslope transport rate on the side slope q_S is expressed by equation (3), rewritten here as

$$q_S = a \tan \alpha \tag{6}$$

where α is the angle of the side slope (Fig. 17.7B).

Sediment from the side slope enters the depositional zone at an angle (Fig. 17.7A). Inspection of Figure 17.7B reveals that the component of transport normal to the depositional zone q_{Sn}, is

$$q_{Sn} = q_S \cos \gamma \tag{7}$$

where γ is the angle between the transport direction perpendicular to the side slope contours and the transport perpendicular to the depositional zone, which is assumed to lie parallel to the axis of the trough. Analysis of Figure 17.7B indicates that

$$\sin \gamma = (H/\sin \alpha)(H/\sin \theta)^{-1} \tag{8}$$

where H is the vertical height shown in Figure 17.7B and α and θ are the slopes of the side slope and hollow, respectively. Substitution of (8) and (6) into (7) yields

$$q_{Sn} = a \tan \alpha \left(1 - \frac{\sin^2 \theta}{\sin^2 \alpha}\right)^{1/2} \tag{9}$$

which can be simplifed to

$$q_{Sn} = a \left[\frac{\cos^2 \theta}{\cos^2 \alpha} - 1\right]^{1/2} \tag{10}$$

The total transport Q_{Sn} into the hollow of length l is then

$$Q_{Sn} = a2l \left[\frac{\cos^2 \theta}{\cos^2 \alpha} - 1\right]^{1/2} \tag{11}$$

Note than when the trough is horizontal, $\theta = 0$ and $Q_{Sn} = 2al \tan \alpha$, as required by the geometry of the basin. The conservation of mass equation for a depositional fill of volume V is

$$Q_{Sn} = \frac{dV}{dt} \qquad (12)$$

The volume of the deposit at any time t is simply the cross-sectional area A times the length l of the deposit. The cross-sectional area is the half-width b times the maximum depth h of the fill measured perpendicularly to the bedrock axis in the hollow

$$A = bh \qquad (13)$$

Again, analysis of the geometric relationships expressed in Figure 17.7B yields the following

$$b = h[\cos \theta (\tan^2 \alpha - \tan^2 \theta)^{\frac{1}{2}}]^{-1} \qquad (14)$$

where h is measured perpendicular to the bedrock axis of the hollow, which is inclined at slope θ. Combining (13) and (14) gives the equation for the volume of the fill with the width term eliminated

$$V = h^2[\cos \theta (\tan^2 \alpha - \tan^2 \theta)^{\frac{1}{2}}]^{-1}l \qquad (15)$$

The mass balance equation (12) can now be written using (11) and (15) as

$$a2l\left[\frac{\cos^2 \theta}{\cos^2 \alpha} - 1\right]^{\frac{1}{2}} = \frac{d}{dt}[h^2(\cos \theta (\tan^2 \alpha - \tan^2 \theta)^{\frac{1}{2}})^{-1}l] \qquad (16)$$

which gives

$$a \cos \theta (\tan^2 \alpha - \tan^2 \theta)^{\frac{1}{2}}\left[\frac{\cos^2 \theta}{\cos^2 \alpha} - 1\right]^{\frac{1}{2}} dt = h\, dh$$

Performing the integration and using the boundary conditions that $h = 0$ at $t = 0$ leads to

$$h = \left[a2 \cos \theta (\tan^2 \alpha - \tan^2 \theta)^{\frac{1}{2}}\left[\frac{\cos^2 \theta}{\cos^2 \alpha} - 1\right]^{\frac{1}{2}}t\right]^{\frac{1}{2}} \qquad (17)$$

Inspection of available detailed topographic maps, including those reported by Hack (1965), indicates that the ratio of $\tan \theta$ to $\tan \alpha$ is typically about 0.8. This ratio probably ranges from 0.5 to 1.0. Substitution of possible slope combinations in $[(\cos^2 \theta/\cos^2 \alpha) - 1]^{\frac{1}{4}}$ shows that this term differs by less than 20 percent from $(\tan^2 \alpha - \tan^2 \theta)^{1/4}$; hence, for simplicity, (17) can be reduced to

$$h = [a2 \cos \theta (\tan^2 \alpha - \tan^2 \theta)t]^{1/2} \tag{18}$$

and for $S_S = \tan \alpha$ and $S_h = \tan \theta$, it becomes

$$h = [a2 \cos \theta (S_S^2 - S_h^2)t]^{1/2} \tag{19}$$

Equation (19) shows that the depth increases in proportion to the 0.5 power of time after the onset of net deposition; hence the rate of increase in depth diminishes with time. This results from the increasing cross-sectional width with increasing depth in the triangular trough. In Figure 17.8, (19) is plotted for two ratios of hollow to side-slope gradient. For a given ratio, the maximum depth of colluvium at a given time is greater for steeper side slopes. Equation (19) also shows that the gradient difference between the hollow and side slope strongly influences accumulation rates. For example, the change in S_h/S_S from 0.8 to 0.5 depicted in Figure 17.8 results in a 45 percent increase in depth of colluvium.

For a given site, if the time since net deposition began can be determined then the coefficient a in the transport equation can be evaluated. Equation (19) can then be used to define quantitatively the thickness variations of colluvium with respect to time and slope. Reneau et al. (1984) report a radiocarbon date of 12 900 B.P. from charcoal in basal colluvium in a small, grass-covered basin with a maximum thickness of approximately 450 cm. From a detailed topographic map, the gradients of the hollow and side slope (which includes the edge of the depositional zone) were found to be 17° and 21°, respectively. Hence the ratio of the tangents of these two gradients is about 0.8. Equation

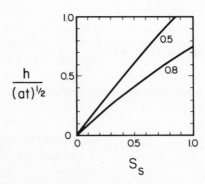

Figure 17.8 Relationship of the dimensionless ratio soil thickness h divided by the square root of the product of transport coefficient a and time t to side-slope gradient S_S for two different ratios (0.5 and 0.8) of hollow to side-slope gradients in a tipped triangular trough.

(19) can be rewritten to solve for a and, using the above values, yields $a = 152 \ cm^3/cm \ yr$. This value is relatively large as it is equivalent to the entire 30 cm thick soil column on the side slope moving downslope at the speed of 5 cm/yr. Biogenic transport caused by burrowing animals is quite apparent on the grass-covered hillslopes, but is probably not strong enough to cause this high transport rate (see discussion in Lehre (1982)). We have used a maximum value of colluvium depth rather than an average colluvium depth in the hollow axis of 400 cm, but this is a small correction. A large possible source of error in evaluating a is the assumption that the hollow was free of colluvium at 12 900 B.P. Reworking of colluvium not removed by a major erosional event should increase the apparent accumulation rates.

The simple sediment transport model based on the tipped triangular trough provides a quantitative tool for understanding the accumulation of colluvium in hollows. Extensive radiocarbon dating through vertical sections in hollows will allow direct testing of the form of the transport equation and evaluation of the transport coefficient. Such dates for different basins will also allow us to test the findings that the basin geometry strongly influences deposition rates in the hollow. Our first crude test points to a need to examine carefully the exposures of colluvium in hollows in order to identify stratigraphic or pedologic evidence for partial evacuation of colluvium and subsequent reworking. It also demonstrates the need for direct measurement of colluvium transport rates on side slopes.

Runoff and slope stability

The slope length or catchment area required to initiate a channel head must be one that provides sufficiently frequent runoff and high pore pressures that colluvium cannot accumulate further downslope. The instability that leads to flushing of colluvium in the hollow and temporary extension of the channel head upslope is not necessarily the same as that required to maintain the channel head in its farthest downslope position. Flushing of the colluvium is infrequent, occurring at intervals of hundreds to thousands of years, whereas the channelway appears to be the site where channelized flow and at least some alluvial transport regularly occur. The transition from the hollow to the channelway varies between basins. In some cases the channel gradually develops over a 5–10 m length of valley floor, but more often the channel head is an abrupt, steep wall in colluvium more than a meter in height. In this latter case we have seen abundant subsurface flow issuing from the base of the wall.

The specific mechanisms of channel-head maintenance are not yet well understood. In the California basins, the bed of the channel just downslope from the hollow is usually bedrock. Hence the channel is a seepage face draining the groundwater from the upslope unchannelized basin. As Dunne (1980) described, high seepage forces due to excessive pore pressures generated in the bedrock or colluvium may develop at channel heads and erode the colluvium. On very steep basins, partial or complete saturation of the colluvium without excessive pore pressures, can be sufficient to cause landslides. This can occur frequently enough to maintain a channel head (Humphrey, 1982). Because of the dense vegetation cover in the downslope ends of hollows it seems unlikely that saturation overland flow could provide sufficient boundary shear stress to initiate a channel. Transport by the channelized flow at the channel head, however, should play some role in channelway maintenance, particularly in fine-grained colluvium where shallow flow is capable of significant transport. The distance downslope to the channel head also should be influenced by climate and thus vary with climatic change, although it is not obvious what precipitation characteristic should be most important.

In the following, we hypothesize that slope stability controls the location of the channel head and thereby directly or indirectly controls the size and geometry of the upslope unchannelized basin. We examine how shallow subsurface flow in the colluvial mantle, and excessive pore pressures generated from groundwater flow in the bedrock, may be responsible for the empirical relationships presented in the first part of this chapter. Throughout the following analysis we make the reasonable initial assumption that hollows have straight longitudinal profiles.

Shallow subsurface flow

If we represent the stability conditions leading to formation of a channel head with the infinite slope model of the Mohr–Coulomb failure law (e.g., Carson and Kirkby, 1972, p. 152), we can write for failure of colluvium of thickness z on a bedrock surface inclined at slope θ

$$\rho_s gz \sin \theta \cos \theta = c' + (\rho_s gz \cos^2 \theta - \rho_w gh \cos^2 \theta)\tan \phi' \qquad (20)$$

where ρ_s and ρ_w are the bulk density of colluvium and water, respectively, g is the gravitational acceleration, c' is the effective cohesion, ϕ' is the angle of internal friction, and h is the vertical height of saturated soil above the failure plane. If we can compute the saturation height h as a function of contributing area properties, then substitution into (20) will permit calculation of the length or area and hollow gradient that provide the saturation height requisite for slope failure.

Independently, Humphrey (1982) and Iida (1984) developed simple, analytic expressions for the depth of saturation caused by shallow subsurface flow during a specified rainstorm of constant intensity R_0 on a hillslope of simple geometry with isotropic hydraulic conductivity in the colluvium and an underlying impermeable bedrock. Under these conditions, the discharge at time t of subsurface flow per unit contour length $Q(t)$ is equal to the rainfall rate R_0 times the contributing area $a(t)$ per unit contour length at time t:

$$Q(t) = R_0 a(t) \qquad (21)$$

This is equivalent to Iida's equation (8) and Humphrey's equation (3.39). The discharge rate can be related to slope and soil properties via Darcy's law. The specific flux q for flow parallel to a bedrock boundary inclined with slope θ is

$$q = K_s \sin \theta \tag{22}$$

where K_s is the saturated hydraulic conductivity. The horizontal velocity component V_s is equal to $q \cos \theta$; thus

$$V_s = K_s \sin \theta \cos \theta \tag{23}$$

The discharge per unit contour length for a depth of saturation $h(t)$ becomes

$$Q(t) = h(t)K_s \sin \theta \cos \theta \tag{24}$$

Combining (24) and (21) and solving for $h(t)$ gives

$$h(t) = \frac{R_0\, a(t)}{K_s \sin \theta \cos \theta} \tag{25}$$

which is equivalent to Iida's equation (9) and Humphrey's equation (3.41). The maximum depth of saturation will be achieved when all of the catchment upslope of the channel head is contributing. Assuming uniform width, hence no flow convergence, (25) reduces to

$$h = \frac{R_0 L}{K_s \sin \theta \cos \theta} \tag{26}$$

The effect of flow convergence is treated somewhat differently by Iida and Humphrey, but both made use of a cylindrical coordinate system. Our previous analysis argues against this approximation and instead suggests that unchannelized basins are well represented by a tipped triangular trough. In this case the simplest approximation is to assume that shallow subsurface flow leaves the basin through the colluvium in the hollow; hence the ratio c of basin width to hollow width ($2b$) is a measure of the increased catchment area per unit contour length. Basin width is usually much larger than hollow width. As Hack (1965) pointed out, the basin to hollow width ratio may vary with hollow slope. Thus in (25) $a(t)$ equals cL (where L is basin length) for maximum saturation depth, which suggests that the consequence of flow convergence in a typical unchannelized basin is to increase the contributing area per unit length by several-fold, perhaps typically about five times for steep basins. The magnitude of the increase is inversely proportional to the ratio of hollow to basin width because it is assumed that flow lines are perpendicular to surface contours in the hollow. With these limitations in mind we can rewrite (26) to account for flow convergence:

$$h = \frac{cR_0L}{K_s \sin\theta \cos\theta} \tag{27}$$

The equation shows that for a given rainfall the depth of saturation increases with gentler slopes, larger hydraulic conductivity, smaller hollow widths, and greater hollow lengths, all of which is physically realistic.

Slope stability
The linkage between basin length, strength properties and hillslope hydrology can be achieved by substituting (27) into (20) and solving for cohesionless soils with basin length L:

$$L = \frac{\rho_s K z}{c\rho_w R_0} \sin^2\theta \left[\frac{1}{\tan\theta} - \frac{1}{\tan\phi'}\right] \tag{28}$$

Because of the assumption used in deriving (28), it is only valid for hydrostatic pore pressures where the water depth h does not exceed the soil depth; hence $(\rho_s - \rho_w)\rho_s^{-1} \tan\phi' < \tan\theta \le \tan\phi'$. The upper boundary $\tan\phi'$ defines those slopes that would be unstable under dry conditions. As expected, for a given slope and angle of internal friction, the basin length should be greater for soils with high hydraulic conductivity relative to the rainfall rate and for basins with weak convergent topography. Within the range of applicable $\tan\theta$, L at first declines relatively slowly with hollow gradient, but above $0.7 \tan\phi'$, it declines rapidly to zero at $\tan\theta = \tan\phi'$. Humphrey (1982, p. 122) obtained similar results, although for much shorter slope lengths, for a basin he called an "elongate wedge" in which colluvium mantles a bedrock step. Kirkby (1978, p. 359) has also sketched qualitatively a decreasing distance to divide with increasing slope for a hypothetical hillslope.

The role of climate is expressed in the steady rainfall rate R_0 in (28). The soil depth z is of order 0.1 to 1.0 and for typical unchannelized basins with $c \sim 5$ it appears that the ratio K_s/R_0 should be about 10^3. An assumption used in (28) is that the entire basin is contributing runoff, and for a given R_0 the time required for water from the upslope end of a hollow to reach the channel head is the hollow length divided by the horizontal velocity component of subsurface flow V_s. According to (23), $L/V_s = L(K_s \sin\theta \cos\theta)^{-1}$, which is of the order of 10^6 to 10^7 or 10 to 100 days. This implies that the location of the channel head depends on the cumulative rainfall over a long period and the greater the rainfall over this characteristic period, the shorter the unchannelized basin length. Shorter basins imply higher drainage density; thus (28) agrees with observations that drainage density generally increases with greater precipitation in humid areas (see Dunne, 1980; Abrahams, 1984).

In order to compare the functional relation given in (28) to the field data, we have matched the equation with the field observation that a basin of

$\tan \theta = 0.518$ (27.4°) should be about 140 m long. This allows us to treat all the constants in (28) as one, evaluate it under the assumption that $\phi' = 46°$ (Reneau et al., 1984), and use this constant to depict (28) in Figure 17.9 (broken curve). The regression line from Figure 17.4 is shown for values less than $0.5 \tan \phi'$. At values close to $0.5 \tan \phi'$, (28) predicts a length–gradient relationship close to the empirical one. The available field data (eight points) do not indicate whether (28) or the empirical line derived mostly from less steep basins provides a better fit. The rapid decline in length with increasing slope shows that hillslopes close to $\theta = \phi'$ are much less stable than hillslopes only a few degrees less steep. This suggests that slopes with θ approaching ϕ' would be rare; and, in fact, the steepest hollow with a colluvial fill (which has recently failed) that we have found in California is $0.75 \tan \phi'$ (38°), and few occur above $0.7 \tan \phi'$. In Oregon and Washington, a survey of road-cuts exposing colluvium-mantled hollows yielded only six hollows with thick deposits steeper than 40° in the 47 hollows observed (Dietrich et al., 1982, Fig. 3A). If the structure of (28) is correct it also suggests that drainage density may rapidly increase on steeper hillslopes. This may explain in part why side slopes, which are steeper than the hollows, often support smaller subhollows (Fig. 17.1). As a final comment, (28) indicates that colluvium strength, as controlled by the angle of internal friction ϕ' strongly influences the required basin length. The form of the broken curve in Figure 17.9 remains unaltered, but

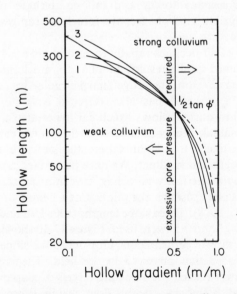

Figure 17.9 Factors controlling the relation of length to hollow gradient in unchannelized basins. The full straight line is the empirical fit to the data. Curves represent results for (29) where $\psi = CL^j(\tan \theta)^k$ and j and k vary with curve number: (1) $j = 1.5$, $k = 0.2$, (2) $j = 1.0$, $k = 0.0$, (3) $j = 1.0$, $k = 0.2$. The broken curve is the calculated relation based on shallow subsurface flow for $\phi' = 46°$.

it is shifted either right or left with corresponding changes in $0.5 \tan \phi'$. For example, if $\phi' = 41°$ rather than $46°$, at $\tan \theta = 0.577$ $(30°)$ the length of the basin would, according to (28), decline by 25 percent, and for steeper slopes the difference in lengths would be much larger.

Groundwater flow

Many of the basins used to define the area, length, and slope relationships presented here have hollow gradients much less than that necessary for instability in the colluvium at saturation. Not only do the high values of angle of internal friction for the coarse textured colluvium suggest this, but saturation overland flow (*sensu* Dunne, 1978) is commonly observed in the downslope ends of the unchannelized basins (e.g., Lehre, 1982; Pierson, 1977; and our own observations in California). Except for the case where a fine-textured soil of low hydraulic conductivity overlies a coarser soil of high conductivity (e.g., Rogers and Selby, 1980) it is probably rare for excessive pore pressures to develop as a result of shallow subsurface flow. High pore pressures and pressure gradients, however, can develop near the surface in the deep groundwater flow system drained at the channel head (Dunne, 1980). These high pore pressures may contribute to the headward advance of a gully at the downslope end of the basin. As in the previous section, then, we propose that the channel head occurs where frequent excessive pore pressure causes erosion of colluvium and prevents burial of the channel. This instability depends both on seepage forces and on the strength of the colluvium. Neglecting the contribution from seepage forces, we can rewrite (20) in terms of pressure head ψ:

$$\psi = \frac{\rho_s z \sin \theta \cos \theta}{\rho_w} \left[\frac{1}{\tan \theta} - \frac{1}{\tan \phi'} \right] \tag{29}$$

Unlike the shallow surface flow, which can be treated as hydrostatic and largely two-dimensional, deeper groundwater flow is not easily approximated algebraically. Intuitively we might expect the pressure head to increase with both slope length and gradient. We have performed a preliminary test of this relationship by numerically modeling flow in a saturated, straight, two-dimensional hillslope using the integrated finite-difference method (Narasimham et al., 1978). Three slope lengths (100, 150, and 200 m) and three gradients (0.125, 0.25 and 0.5) were used. Figure 17.10 shows the equipotential lines (to which the flow is perpendicular) for the 200-m slope of 0.25 gradient, a geometry similar to that observed in the field. The high excessive pore pressures in the lower end of the slope are largely a consequence of the deep downslope vertical boundary on the flow region (lower left wall). This boundary condition is appropriate for the case of the hillslope facing a valley with a slope of comparable size and gradient on the opposite side. For this boundary condition, Figure 17.10B shows the variation in pressure head at 2 m below the surface for a site 5 m upslope from the downslope boundary and

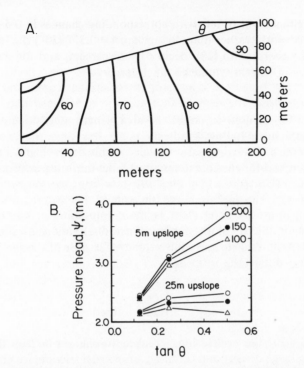

Figure 17.10 Results of numerical modeling: (A) equipotential lines (in meters) in a two-dimensional straight hillslope under complete saturation with a hydraulic conductivity of 10^{-5} cm/s and a surface gradient of 0.25; and (B) changes in pressure head with gradient for slope lengths of 100, 150, and 200 m at two positions 5 m and 25 m upslope.

25 m upslope from the boundary on the three slope lengths and gradients. As anticipated, pressure head ψ varies with both slope length and gradient, with gradient being more important upslope from the boundary. A two orders of magnitude increase in hydraulic conductivity does not significantly alter these pressure values. The actual pressures are strongly dependent on the flow region boundary conditions, but for a variety of geometries the above dependence on slope length and gradient was still observed. Because of the simplification of the natural system represented by the model, it seems unrealistic to attempt to use an equation for ψ as a function of L and $\tan \theta$ that is derived from the numerical model to explain the field relationships. Instead these results primarily suggest that ψ increases with both variables.

Slope stability

Trial and error analysis of substituting values for the exponents in the equation $\psi = CL^j(\tan \theta)^k$, solving for L, and comparing with Figure 17.4 and 17.10 by matching the solution with field data at $0.5 \tan \phi'$ suggests that slope length has a stronger influence than gradient on pressure head (i.e., the exponent j

may be greater than 1.0 and the exponent k less than 0.3). Also, all trials yielded convex-up curves rather than the linear relations fit to the data. L increases very slowly with low values of tan θ, and at high tan θ it increases very rapidly. The predominance of slope length over gradient may be physically correct because L is a surrogate for drainage area, and strong flow convergence towards the seepage face at the channel head may be a major contributor to the requisite pressure head. Further numerical modeling must be done to test this hypothesis.

The trial and error analysis also revealed one other problem. Projection of various solutions for tan θ greater than 0.5 tan ϕ' in all cases gave a curve quite similar to the model based on subsurface flow, but the maximum slope lengths were less. This finding raises the issue of whether shallow subsurface flow or deeper groundwater flow is more important in controlling pore pressures responsible for slope stability. Field observations on pore pressures and flow paths should help solve this problem. Because of the similarity of the results, largely due to the form of

$$\sin \theta \cos \theta \left[\frac{1}{\tan \theta} - \frac{1}{\tan \phi'} \right]$$

in (28) and (29), the conclusions reached in the subsurface flow section regarding drainage density and the angle of internal friction are not altered by this analysis.

An extreme case of "strong" colluvium that under this model would require large excessive pore pressures may occur at the sites studied by Mills (1981) in which massive boulders of quartzite are deposited in relatively low gradient valleys formed in shale. The high angle of internal friction and permeability both prevent the development of instabilities due to high pore pressures and, as Mills proposed, the basins may tend to migrate laterally. In contrast, Hack (1965) presented maps of two hollows in the Martinsburg Shale of Virginia and commented on the fine texture of the hollow colluvium. Based on the position of channel heads shown on the maps, drainage areas were calculated and compared with the regression in Figure 17.4. The areas were 0.63 and 0.72 of those predicted. The fine texture of the colluvium would result in a low angle of internal friction, which should require less drainage area to maintain a channel. The proposed model can be improved by accounting for the seepage forces that undermine channel head walls and, in so doing, bring about slope failure. Although it is possible to write a simple force balance between the weight of the colluvium and the outward-directed drag force caused by seepage (e.g., Dunne, 1980), it is difficult at this point to use it in a model for channel initiation. It is not yet clear whether the channel head tends to be located where the seepage force just equals the weight of the colluvium, or where it exceeds it by some large amount. Also it is difficult to estimate the outward component of the local pressure gradient because it is strongly dependent on the geometry of the flow region. Hopefully future field studies will reveal the most appropriate boundary conditions to use in a model that incorporates seepage forces.

Conclusion

Pieces of a quantitative model for the evolution, geometry, and mechanics of unchannelized basins on soil-mantled hillslopes are beginning to come together. An essential first step is to develop quantitative measures of basin characteristics. Initial field observations suggest that there exists a strong inverse relationship between the drainage area necessary to initiate a channel head and the average hollow gradient of the basin. A similar relationship occurs between hollow length and gradient. These data suggest that steeper slopes should have more exterior link sources, resulting in a greater drainage density. Basin form, consisting of noses, side slopes, and a hollow appears to be well represented by the geometry of a tipped triangular trough and typically the ratio of hollow slope to side slope is about 0.8. Mass balance calculations reveal that rate of thickening of colluvium in the hollow diminishes with time due to increasing cross-sectional area of the hollow, and that basins with steeper side slopes and more convergent topography have high rates of colluvium accumulation.

We have made an initial attempt to explain the field relationships by developing simple approximations for saturation flow depth and pressure head as a function of basin geometry and by using these approximations in the Mohr–Coulomb failure law to predict the hollow length–gradient relationship observed in the field. This analysis has probably raised more questions than it has answered. We have not yet firmly established from quantitative field observation what processes control the position of the channel head at the end of the hollow. Our models suggest that the geometry of low gradient basins is controlled by the development of significant excessive pore pressures and that changes in length and commensurate drainage area are more important than gradient in controlling the pressure head at the downstream sections of basins. Both the subsurface flow and the groundwater flow models, when used in the Mohr–Coulomb failure law, predict for slopes greater than $0.7 \tan \phi'$ a rapid decline in the slope length necessary to initiate a channel head, consequently, we rarely find basins this steep. Despite the limitations of these many simplistic models, they strongly suggest that area–slope and length–slope relations vary significantly with the strength of the colluvium, particularly the angle of internal friction. They also suggest that the functional relationship for the length–slope curve should be convex-up rather than log-linear. An improvement on the proposed model can be made when seepage forces are properly included.

We have not tackled, by any means, all the major problems concerning unchannelized basin geometry. We do not know what controls the width of these basins. For a given width, it is not yet known what determines either the ratio of hollow slope to side slope or the basin amplitude. Hollow bedrock profiles appear to be concave, but we have insufficient data to attempt generalizations regarding geometry. The cause of the axial concavity is not readily observed either. The mechanism of Calver (1977), whereby bedrock scouring varies in proportion to the frequency of high runoff events which in turn tend to diminish upslope through the hollow, may be applicable, although in this case the curvature might instead be related to the frequency of landslide-induced exposure of the bedrock surface. In some cases, the concavity may be the result of a curved failure plane within weathered bedrock, as illustrated by Foxx (1984). Finally, as Hack (1965) has emphasized, in many unchannelized basins the principal hollow may have several tributaries or subhollows. These features also dissect the side slopes to major streams. Perhaps, as we have suggested, they arise because the side slopes have steeper gradients than the hollows.

Acknowledgments

J. T. Hack has contributed significantly to this chapter by taking the senior author on a field trip to his study area in Virginia and by freely giving of his ideas and time in several lengthy conversations. H. Mills and C. Hupp also kindly showed the senior author their study areas in Virginia. Valuable conversations were also held with R. Shreve, T. Pierson, N. Humphrey, T. Dunne, C. Wahrhaftig, L. Leopold, M. Power, and R. Dorn. L. Reid helped substantially in analyzing the geometric relationships in a tipped triangular trough. T. Narasimhan gave generously of his time in the development of the integrated finite-difference groundwater flow model. Field observations contributed by W. Trush, K. Whipple, P. Templet, D. Rogers, K. Vincent, J. Sowers, and C. Meade and field assistance given by M. Power, L. Collins, and S. Raugust were essential. L. Leopold, P. Whiting, A. Abrahams, L. Collins, and T. Dunne, made several useful comments on an earlier draft of this manuscript. In response to a telephone conversation, R. Shreve derived the relationship between sources and drainage density. Funding for our research was provided in part by the National Science Foundation (Grant EAR-84-16775).

References

Abrahams, A. D., Channel networks: a geomorphological perspective, *Water Resources Research*, *20*, 161–188, 1984.

Brown, W. R. III, *Overview and summary of a conference on debris flows, landslides and floods in the San Francisco Bay region*, 83 pp., National Academy of Science, Washington D.C., 1984.

Calver, A., Modelling drainage headwater development, *Earth Surface Processes, 3*, 233–241, 1978.

Carson, M. A., and M. J. Kirkby, *Hillslope Form and Process*, 475 pp., Cambridge University Press, Cambridge, 1972.

Dietrich, W. E., and R. Dorn, Significance of thick deposits of colluvium on hillslopes: a case study involving the use of pollen analysis in the coastal mountains of northern California, *Journal of Geology, 92*, 147–158, 1984.

Dietrich, W. E., T. Dunne, N. F. Humphrey, and L. M. Reid, Construction of sediment budgets for drainage basins, in *Sediment Budgets and Routing in Forested Drainage Basins*, edited by F. J. Swanson, R. J. Janda, T. Dunne, and D. N. Swanston, pp. 5–23, U.S. Department of Agriculture, Forest Service General Technical Report PNW-141, Pacific Northwest Forest and Range Experiment Station, Portland, Oregon, 1982.

Dietrich, W. E., and T. Dunne, Sediment budget for a small catchment in mountainous terrain, *Zeitschrift für Geomorphologie Supplement Band, 29*, 191–206, 1978.

Dunne, T. Field studies of hillslope flow processes, in *Hillslope Hydrology*, edited by M. J. Kirkby, pp. 227–293, Wiley, Chichester, 1978.

Dunne, T., Formation and controls of channel networks, *Progress in Physical Geography, 4*, 211–239, 1980.

Foxx, M., Slope failures in the Felton Quadrangle, 1981–83, and analysis of the factors that control slope failure susceptibility of the Monterey Formation, M.S. thesis, 139 pp., University of California, Santa Cruz, 1984.

Gray, R. E., and G. D. Gardner, Processes of colluvial slope development at McMechen, West Virginia, *Bulletin of the International Association of Engineering Geologists, 16*, 29–32, 1977.

Hack, J. T., Geomorphology of the Shenandoah Valley, Virginia and West Virginia and origin of the residual ore deposits, *U.S. Geological Survey Professional Paper 484*, 1965.

Hack, J. T., and J. C. Goodlett, Geomorphology and forest ecology of a mountain region in the central Appalachians, *U.S. Geological Survey Professional Paper 347*, 1960.

Humphrey, N. F., Pore pressures in debris failure initiation, M.S. thesis, 169 pp., University of Washington, Seattle, 1982.

Iida, T., A hydrological model of estimation of the topographic effect on the saturated throughflow, *Transactions of the Japanese Geomorphological Union, 5*(1), 1–12, 1984.

Iida, T., and K. Okunishi, Development of hillslopes due to landslides, *Zeitschrift für Geomorphologie Supplement Band, 46*, 67–77, 1983.

Keefer, D. K., and A. Johnson, Earth flow morphology, mobilization, and movement, *U.S. Geological Survey Professional Paper 1264*, 1983.

Kirkby, M. J., Hillslope process–response models based on the continuity equation, *Institute of British Geographers Special Publication 3*, 15–30, 1971.

Kirkby, M. J., Implications for sediment transport, in *Hillslope Hydrology*, edited by M. J. Kirkby, pp. 325–363, Wiley, Chichester, 1978.

Kramer, J., Non-seismic liquefaction in fine grained sand, Ph.D. thesis, University of California, Berkeley, 1985.

Lehre, A. K., Sediment mobilization and production from a small mountain catchment: Lone Tree Creek, Marin County, California, Ph.D. thesis, 375 pp., University of California, Berkeley, 1982.

Marcus, A., First-order drainage basin morphology–definition and distribution, *Earth Surface Processes, 5*, 389–398, 1980.

Marron, D. C., Hillslope evolution and the genesis of colluvium in Redwood National Park, northwestern California: the use of soil development in their analysis, Ph.D. thesis, 187 pp., University of California, Berkeley, 1982.

Mills, H. H., Boulder deposits and the retreat of mountain slopes, or "gully gravure" revisited, *Journal of Geology*, *89*, 649–660, 1981.

Narasimhan, T. N., T. A. Witherspoon, and A. L. Edwards, Numerical model for saturated-unsaturated flow in deformable porous media, II: the algorithm, *Water Resources Research*, *14*, 255–261, 1978.

Pierson, T. C., Factors controlling debris-flow initiation on forested hillslopes in the Oregon Coast Range, Ph.D. thesis, 166 pp., University of Washington, Seattle, 1977.

Reneau, S. L., W. E. Dietrich, C. J. Wilson, and J. D. Rogers, Colluvial deposits and associated landslides in the northern San Francisco Bay area, California, USA, *Proceedings of the 4th International Symposium on Landslides*, pp. 425–430, 1984.

Rogers, J. W., and M. J. Selby, Mechanisms of shallow translational landsliding during summer rainstorms: North Island, New Zealand, *Geografiska Annaler, 62A*, 11–21, 1980.

Shimokawa, E., A natural recovery process of vegetation on landslide scars and landslide periodicity in forested drainage basins, paper presented at the *Symposium on the Effects on Forest Land Use on Erosion and Slope Stability*, Honolulu, Hawaii, May, 1984.

Smith, T. R., and F. P. Bretherton, Stability and conservation of mass in drainage basin evolution, *Water Resources Research, 8*, 1506–1529, 1972.

Swanson, F. J., R. L. Fredricksen, and F. M. McCorison, Material transfer in a western Oregon forested watershed, in *Analysis of Coniferous Forest Ecosystems*, edited by R. L. Edmonds, pp. 233–266, Dowden, Hutchinson, and Ross, Stroudsburg, Pa., 1982.

Swanston, D. N., and F. J. Swanson, Timber harvesting, mass erosion, and steepland forest geomorphology in the Pacific Northwest, in *Geomorphology and Engineering*, edited by D. R. Coates, pp. 199–221, Dowden, Hutchinson, and Ross, Stroudsburg, Pa., 1976.

Tanaka, T., The role of subsurface water exfiltration in soil erosion processes, *International Association of Hydrological Scientists Publication, 137*, 73–80, 1982.

Tsukamoto, Y., and O. Kusakobe, Vegetative influences on debris slide occurrences on steep slopes in Japan. Paper presented at the *Symposium on the Effects of Forest Land Use on Erosion and Slope Stability*, Honolulu, Hawaii, May, 1984.

Tsukamoto, Y., T. Ohat, and H. Noguchi, Hydrological and geomorphological studies of debris slides on forested hillslopes in Japan, *International Association of Hydrological Scientists Publication, 137*, 89–98, 1982.

Woodruff, J. F., Debris avalanche as an erosional agent in the Appalachian Mountains, *Journal of Geography, 70*, 399–406, 1971.

18
Relative slope-stability mapping and land-use planning in the San Francisco Bay region, California

Tor H. Nilsen

Abstract

Slope-stability studies in the 1970s by the U.S. Geological Survey resulted in the publication of estimates of landslide damage, maps of estimated landslide abundance, slope maps prepared by photomechanical processes, photointerpretive maps of landslide, colluvial, and other surficial deposits, and maps of relative slope stability for the San Francisco Bay region. These studies documented that landslides were very abundant, caused significant numbers of deaths, and resulted in damage amounting to millions of dollars per year. Regional slope-stability maps prepared from bedrock geological maps, slope maps, and photointerpretive landslide inventory maps have been used by the Bay region community for many land-use planning decisions. Development in upland parts of the region that are underlain by unstable to marginally stable slopes generally cannot be undertaken at present without careful evaluation of slope stability based on published data and well-designed site investigations.

Introduction

The U.S. Geological Survey (USGS) in the 1970s completed a multidisciplinary study of the geological aspects of the San Francisco Bay region in northern California (Fig. 18.1). The study, entitled "San Francisco Bay Region Environment and Resources Study" (SFBRS), was undertaken cooperatively with the U.S. Department of Housing and Urban Development (HUD), the California Division of Mines and Geology (CDMG), the Association of Bay Area Governments (ABAG), various county and city governments, and a number of private consulting firms. The study was initiated in 1970 and most of the work was completed by 1975. The overall scope, organization, and

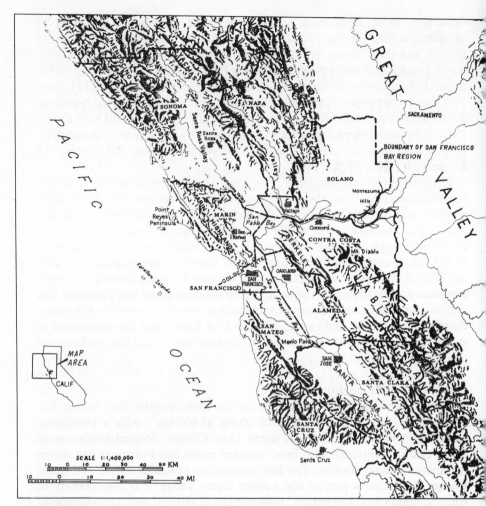

Figure 18.1 Index map of the San Francisco Bay region showing general physiography, location of counties, and principal cities.

planning of the study were outlined in the program design (U.S. Geological Survey and U.S. Department of Housing and Urban Design, 1971).

The study focused on both the geologic framework of the region and the various geologic hazards that affect land-use planning in the region. Among the more important aspects summarized in the reports are seismic safety and land-use planning (Borcherdt, 1975; Brabb, 1979); the geology of flatland deposits and their significance to land-use planning (Helley et al., 1979); flooding and land-use planning of flood-prone areas (Limerinos et al., 1973; Waananen et al., 1977); various aspects of the geology, geologic history, and use of San Francisco Bay (McCulloch et al., 1970; Atwater et al., 1977;

Conomos, 1979; Kockelman et al., 1982); the liquefaction potential of sediments around the bay (Youd, 1973; Youd et al., 1975); sedimentation and erosion and their impact on land-use planning (Brown and Jackson, 1973); waste disposal and ground-water pollution (Hines, 1973; Perkins et al., 1977); mineral resources and their impact on land-use planning (Bailey and Harden, 1975); quantitative land-capability analysis (Laird et al., 1979); and geological guidelines and principles for land-use decisionmakers (Brown and Kockelman, 1983). The applications and use of these studies by the Bay area community, particularly land-use planners and decisionmakers, was documented by Kockelman (1975, 1976, 1979).

The purpose of this chapter is to summarize the types of studies that were conducted on landslides, the use of these studies in the preparation of slope-stability maps for land-use planning, and the usefulness of this work in the Bay region. My emphasis herein will be on regional studies because the objectives of SFBRS were to produce regional slope-stability maps at a scale of 1 : 125 000. Many second-generation studies, spawned from original work during SFBRS, have been completed during the past 10 years, leading to a more complete understanding of the nature of landsliding in the San Francisco Bay region. A series of very destructive landslides, which occurred in the region during January 1982 and caused a number of deaths and the destruction of many homes (Brown, 1984), led to a major re-evaluation of the usefulness of the previous studies.

Geographic and geologic setting

The San Francisco Bay region includes the nine counties that border San Francisco Bay, a total land area of about 18 000 km^2, with a population exceeding 5 million (Fig. 18.1). Santa Cruz County, located southwest of Santa Clara County, has also been included in the San Francisco Bay region for some regional studies. The Bay region lies primarily within the Coast Ranges but includes part of the western Great Valley. The great variety of climatic conditions, vegetation, topographic situations, and geologic conditions makes the area attractive for future growth. Population has largely been confined to the flatlands surrounding San Francisco Bay and its adjacent waterways, the city of San Francisco, and some of the larger inland valleys. At present, however, development is spreading rapidly into adjacent hillside areas, where landsliding has become an increasing problem.

The geology of the region is complex. Many different types of rocks (Schlocker, 1968, 1971) and numerous active faults (Brown, 1970) are present, and the structural and tectonic history of the region has been complicated. The following characteristics of the San Francisco Bay region contribute to widespread landsliding: (1) steep, irregular slopes; (2) abundant and seasonally intense rainfall; (3) extensive human activity, including logging, grading, and cutting of slopes; (4) an abundance of rock types that are susceptible to sliding, including extensively crushed and fractured Franciscan melange complexes and poorly consolidated upper Tertiary to Holocene sediments; (5) thick

Figure 18.2 Landslides on natural slopes in the San Francisco Bay region: (A) large slump block, labeled 'A', east of San Gregorio, San Mateo County, (photograph by E. E. Brabb, 1971); and (B) debris flow near Dublin, Alameda County, 1971.

unconsolidated colluvial deposits and thick weathered zones on steep slopes; (6) abundant expansible clay soils; and (7) frequent damaging earthquakes.

More than 90 percent of the precipitation in the Bay region occurs during the winter months of November to April, chiefly as rain except for some snow in the higher mountains. However, the seasonal and monthly rainfall varies greatly from year to year. Most landslides occur during the winter months, commonly during or shortly after major storms. Orographic effects are pronounced, with greater precipitation in mountainous areas, and rainfall also generally decreases eastward across the region. The coincidence of major storms, extensive logging or development on steep slopes, and major earthquakes could trigger major episodes of landsliding in the region.

Previous work
Most pre-SFBRS studies focused primarily on mapping bedrock units rather than on slope stability. Although consulting engineering geologists had examined the slope stability characteristics of many small parcels of land in detail, little of this work had been published, and few regional studies had been undertaken. Some of the earlier studies, which were concerned with landslide features and slope-stability characteristics in the region and provided much necessary data and insight for the SFBRS studies, included those by Kachadoorian (1956), Schlocker et al. (1958), Bonilla (1960a, b, 1971), Radbruch (1957, 1969), Radbruch and Weiler (1963), Kojan (1968), Harding (1969), Clague (1969), Pampeyan (1970), Rogers (1971), and Waltz (1971). Within most of the region, however, information regarding the distribution of landslide deposits was very meager at the time SFBRS began. To attain the goal of a regional slope stability map for the entire San Francisco Bay region, much new data and mapping were needed, including the mapping of landslide deposits, bedrock geology, active faults, slopes, recent landslides, other surficial deposits, and the engineering properties of bedrock and soil units.

A wide variety of different types and sizes of landslides are generated in the Bay region each year. The landslides vary in size, shape, geometry, thickness, rate and style of movement, and type of materials involved. The more rapidly moving landslides pose the greatest hazard to life because they can destroy buildings or damage roads with little warning. Slowly moving landslides can cause large amounts of damage but are usually less life-threatening. Landslides in the Bay region can generally be divided by style of movement into slides, slumps, falls, flows, and combinations thereof. Slumps and flows are particularly common and can be easily recognized in natural settings (Fig. 18.2). Slumps, flows and slides generated on man-modified slopes can cause considerable damage to man-made structures (Fig. 18.3).

Estimated costs of landslide damage

Before SFBRS, it was known that the yearly cost of landslides in the Bay region was great, but no reliable figures were available. This type of infor-

Figure 18.3 Landslides on slopes modified by man that have caused damage to man-made structures: (A) landslide damage from slumping, U.S. Interstate Highway 80 near Pinole, Contra Costa County (photograph by Norman Prime, 1969); (B) landslide damage from debris flow, Oakland, Alameda County (photograph from Oakland Tribune, 1958); (C) landslide damage from a small rock slide near Pleasanton, Alameda Country, 1971—sliding has taken place along the dipping bedding planes of the exposed sandstone.

mation is useful not only in pointing out the necessity of slope-stability studies prior to construction but also in informing various local governmental agencies of the costs that might ultimately accrue from ignoring geologic problems during development.

Costs related to landsliding are difficult to ascertain, primarily because few public agencies keep accurate records about landslide damage. Data on the costs of damage to structures, re-evaluation of land or housing for tax purposes, and expenditures for road, sidewalk, sewer, and railroad repairs are available from some public agencies, but much damage goes unreported, and many other costs are impossible to determine. Typical indeterminable costs are those for litigation, salaries of firemen and policemen, and detours.

The total damage in the United States caused by all ground-failure hazards, which include landslides, subsidence, and expansive soils, is greater than the losses from all other natural hazards combined. From 1925 to 1975, for example, $75 billion in damage resulted from landslides and subsidence compared to a total of $20 billion in damage from floods, hurricanes, tornadoes and earthquakes (Committee on Ground Failure Hazards, 1975).

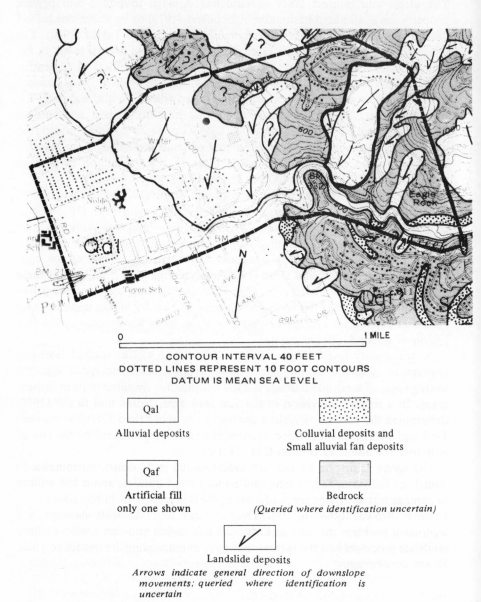

CONTOUR INTERVAL 40 FEET
DOTTED LINES REPRESENT 10 FOOT CONTOURS
DATUM IS MEAN SEA LEVEL

| Qal |
Alluvial deposits

Colluvial deposits and
Small alluvial fan deposits

| Qaf |
Artificial fill
only one shown

Bedrock
(Queried where identification uncertain)

Landslide deposits
*Arrows indicate general direction of downslope
movements; queried where identification is
uncertain*

Figure 18.4 Map showing preliminary photointerpretation of landslide and other surficial deposits in the San Jose Highlands area of northeastern San Jose, Santa Clara County (from Nilsen and Brabb, 1972). The most severely damaged area is marked by the heavily outlined landslide deposit.

The direct and indirect costs of landslide damage to public and private property were estimated nationally by Schuster (1978) to be about one billion dollars per year, a figure also determined by Fleming and Taylor (1980). The Committee on Ground Failure Hazards (1985) estimated annual losses in the United States from landslides to be 1 to 2 billion dollars and 25 to 50 deaths. Brabb (1984) estimated that the *minimum* costs in the United States between 1973 and 1983 was $200 million annually, based chiefly on cost estimates for state roads and private property; California had the highest amount, an average of about $100 million annually. CDMG previously had estimated that without improvement in reducing the cost of landslide damage, the state of California could expect an average of more than $330 million in damage annually from 1970 to 2000 (Alfors et al., 1973).

In the San Francisco Bay region, Taylor and Brabb (1972) determined a cost of $25 million for the rainy season of 1968–69 and Taylor et al. (1975) a cost of about $10 million for the rainy season of 1972–73. These estimates were determined from data compiled by governmental and private agencies, dividing costs into public, private, and miscellaneous categories for each county. Nilsen, Taylor, and Dean (1976) analyzed the data from both of these rainy seasons and calculated costs per capita, per dwelling unit, and per urban square mile for each of the counties. Fleming and Taylor (1980) updated these data and derived an average annual cost from landslide damage in the Bay region of about $1.30 per capita.

At the county level, Nilsen, Taylor, and Brabb (1976) studied landslide damage in Alameda County from 1940 to 1971, and concluded that the average cost of landslide damage is $100 per year for dwelling units in hillside areas. In a single subdivision in the San Jose area, Nilsen and Brabb (1972) determined the costs of landslide damage to be more than $760 000 between 1968 and 1971; this subdivision eventually had to be purchased by the city of San Jose for about $2 million (Fig. 18.4).

The severe rainstorm of January 1982 resulted in 24 deaths attributable to landslides in Marin, San Mateo, and Santa Cruz Counties; about $66 million in damage was estimated for landslides in the Bay region, including Santa Cruz County (Brown, 1984, p. 25). Thus is clear that landslide damage is a significant problem for the San Francisco Bay region and that studies of both landslide processes and the factors that control landsliding are needed to guide future development.

Estimates of landslide abundance

Maps that show the estimated abundance of landslides have been made for the conterminous United States (Radbruch-Hall et al., 1982), California (Radbruch and Crowther, 1973), and the San Francisco Bay region (Radbruch and Wentworth, 1971). These estimates show that most of the San Francisco Bay region has very abundant landslides and is more prone to slope-stability problems than most other parts of the United States. Maps showing the locations of landslides that caused damage in the Bay region during the rainy seasons of 1968–69 and 1972–73 indicate that landslides generated during a single season have a wide distribution (Taylor and Brabb, 1972; Taylor et al., 1975; Nilsen, Taylor, and Dean, 1976).

Landslide-inventory maps

Photointerpretive maps showing the distribution of existing landslide deposits for nearly the entire Bay region were prepared at scales of 1 : 24 000, 1 : 62 500, and 1 : 125 000; their location and scale are shown in Nilsen and Brabb (1977) and Nilsen et al. (1979). The photointerpretive techniques permit the recognition of large numbers of landslide deposits in most areas and have shown that landsliding is one of the major erosional processes in the Bay region (Figs. 18.4 and 18.5). The techniques depend upon recognition of scarps, anomalous bulges and lumps, hummocky topography, ridgetop trenches and fissures, terraced slopes, abrupt slope changes, altered stream courses, discontinuous drainage patterns, closed depressions, springs, and anomalous color, texture, shade, vegetation, and bedrock patterns (Nilsen et al., 1979).

Although these landslide-inventory maps do not indicate the type of movement, date of most recent activity, and nature of the materials in the landslide deposits, they did provide a useful basis for preparing derivative slope-stability maps at the regional scale of SFBRS. Because many landslides in the region consist of renewed movement of older landslide deposits (Nilsen, Taylor, and Dean, 1976), the inventory maps are a useful guide to those problem areas where site investigations by consulting engineering geologists may be desirable. They provide a reconnaissance-level investigation. Because construction activities typically alter the stability of marginally stable older landslide deposits, causing renewed movement, the maps are a useful indicator of potentially hazardous areas.

In the southern part of the Bay region, the landslide-inventory maps were generalized and quantified in the form of an isopleth map (Wright and Nilsen, 1974). This map shows the distribution of landslide deposits by contours (isopleths) drawn through points representing equal percentages of landslide deposits within a unit area (Wright et al., 1974). The contour format permits easy combination of landslide inventory mapping with other quantified geological, geophysical, and land-use data for land-use decisionmaking.

Slope maps

Slope maps of the bay region were produced by a photomechanical process from contour negatives of 1 : 24 000-scale USGS topographic maps. Intervals

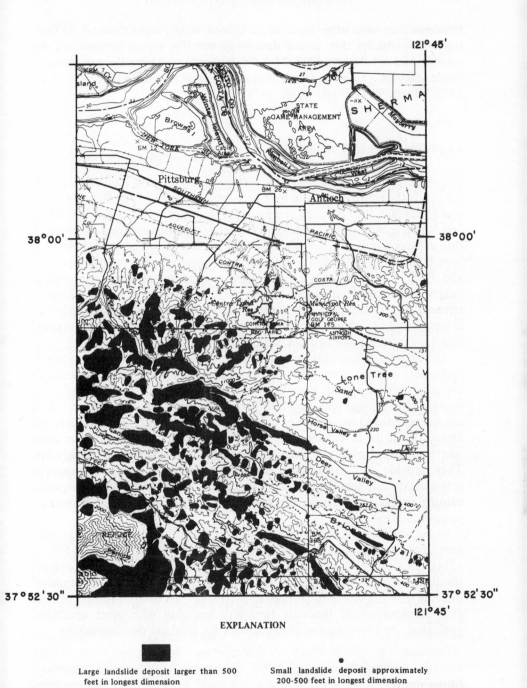

EXPLANATION

Large landslide deposit larger than 500
feet in longest dimension

Small landslide deposit approximately
200-500 feet in longest dimension

Figure 18.5 Landslide inventory map of part of northeastern Contra Costa County, California (from Nilsen et al., 1979, Fig. 44).

shown on the maps are 0–5 percent, 5–15 percent, 15–30 percent, 30–50 percent, 50–70 percent, and greater than 70 percent; these slope intervals are also used by city, county, and regional planners and engineers (U.S. Geological Survey, 1972).

Slope-stability maps

Early maps that depicted the relative slope stability of the Bay region in a cartographic format readily usable by urban planners covered very small areas. Brabb et al. (1972) prepared the first map of a larger area, a landslide-susceptibility map of San Mateo County. This map was prepared from a geologic map, a largely photointerpretive landslide inventory map, and a slope map, each at the scale of 1 : 62 500. (See Nilsen and Brabb (1977) for a discussion of the methodology.)

The map divides the land area of the county into six categories, ranging from a very low to a very high susceptibility to landsliding. The mapped landslide deposits themselves were shown in a separate category. Brabb et al. (1972) determined that slope and strength of the rock unit seem to be the principal factors controlling slope stability in San Mateo County. Other factors, such as structural control, soil type, rainfall, climate, and vegetation, appear to be averaged through time and space so that their relative influence, at least on a regional basis, is minimal. Their map permitted regional and county planners to evaluate areas between landslide deposits on the basis of the average number of slope failures on each geologic unit. The method compensates for stable formations on steep slopes and unstable formations on low slopes. The principal disadvantages are (1) that the landslide susceptibility of flat areas adjacent to unstable slopes is underestimated, and (2) that the map does not express in absolute terms the likelihood that any given slope will fail in any given period of time.

Slope-stability maps can be and have been prepared in many ways and from diverse types of information; no formula or technique has yet been developed that covers all situations, areas, and scales. Nilsen et al. (1979) prepared a regional slope-stability map for SFBRS at a scale of 1 : 125 000 based on analyses of the nature of the underlying bedrock material, the angle of slope, and the presence or absence of older landslide deposits in the area (Fig. 18.6). These three factors were used (1) because they had previously been determined to be significant factors that controlled slope stability in the region, (2) because information was available for these three factors throughout the region, and (3) because they could be effectively incorporated into a regional slope-stability analysis. Although Nilsen et al. (1979) did not use computers for analysis of their data, they suggested that computers could be readily and easily applied to their analysis, a conclusion supported by Newman et al. (1978).

Nilsen et al. (1979) divided the land area of the San Francisco Bay region into five categories and one subcategory of relative slope stability: (1) areas of

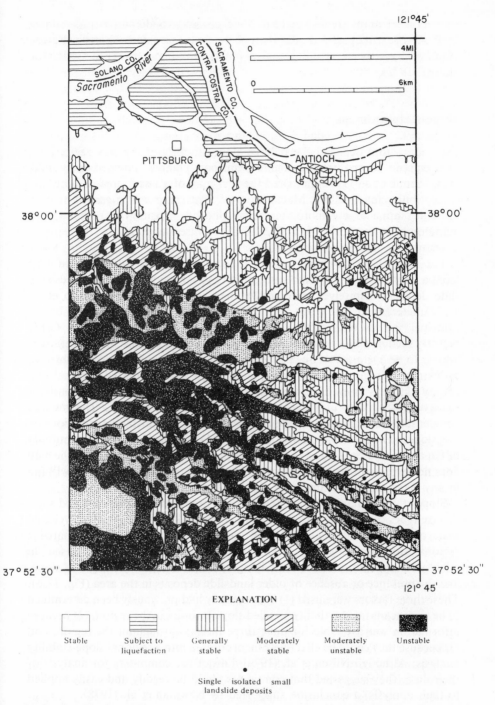

Figure 18.6 Slope-stability map of part of northeastern Contra Costa County, California (from Nilsen et al., 1979, Fig. 50).

0–5 percent slope, (1A) areas of 0–5 percent slope underlain by moist unconsolidated bay-margin muds, (2) areas of 5–15 percent slope, (3) areas of greater than 15 percent slope, (4) areas of greater than 15 percent slope that are underlain by bedrock geologic units considered to be especially susceptible to slope failure, and (5) areas underlain by individual or closely spaced landslide deposits. These categories effectively subdivided the land area of the Bay region into areas ranging from relatively stable to relatively unstable. Nilsen et al. (1979) recommended that the land areas in moderate to high categories or subcategories of slope instability warrant detailed study by engineering geologists, soils engineers, and civil engineers to determine their suitability for development and to estimate the level of risk involved with their development. They concluded, as did Laird et al. (1979), that it may be safer and less costly to leave some of the more vulnerable slopes in open space or low-density development than to subject them to intense development.

Other research into the nature, processes, and varieties of landsliding has led to the publication of other types of slope-stability maps. The effects of earthquakes on landsliding have been of particular concern in the San Francisco Bay region (Nilsen and Brabb, 1975) because earthflows and other types of landslides that formed during the 1906 (magnitude 8.3) earthquake resulted in several fatalities and major damage to man-made structures. Keefer et al. (1979) analyzed the types of landslides generated during 15 historic earthquakes and identified those types that occurred in the greatest numbers and provided the greatest hazard potential. Ground failures associated with liquefaction, commonly in areas of very low slopes along the margins of San Francisco Bay, were particularly destructive locally in 1906 (Youd and Hoose, 1978). The liquefaction potential in the downtown area of San Francisco, where tens of people were killed in 1906 from liquefaction-induced foundation deformation, was mapped by Roth and Kavazanjian (1984). Borcherdt (1975) predicted the geologic effects in part of the Bay region, including landsliding and liquefaction, for a magnitude 6.5 earthquake along the San Andreas Fault.

Radbruch-Hall (1976) determined that the relative slope stability of a small area in the Coast Ranges north of the San Francisco Bay region was controlled significantly by the amount of fracturing and shearing of the bedrock units. CDMG mapped various parts of the San Francisco Bay region and prepared slope-stability maps at varying scales using different types of data. Cotton and Associates (1977) mapped a small area of Santa Clara County, showing relative slope stability in a number of categories based on detailed mapping (Fig. 18.7), and also recommended a local land-use policy. Wieczorek (1982) mapped landslides in part of San Mateo County. He classified them as recently active or dormant and divided them into five different types, permitting a more sophisticated slope-stability map to be prepared of the area. Roth (1983) analyzed the effects of slope angle, rock type, rainfall, soil type, vegetation type, and the distribution of landslide deposits on the slope stability of San Mateo County, concluding that the methodology originally developed by

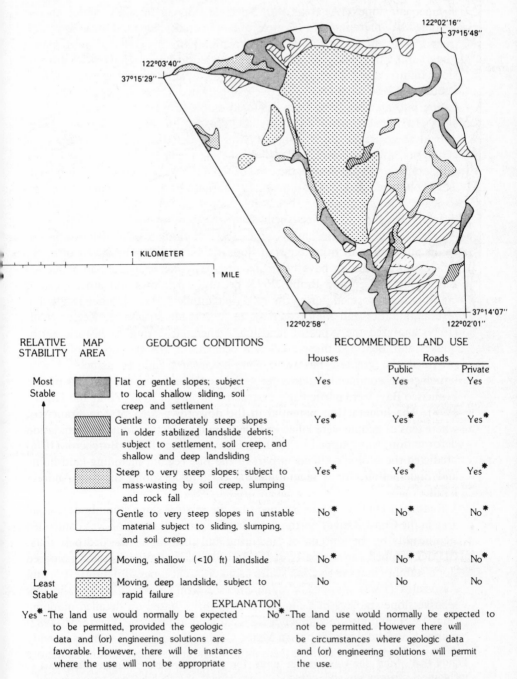

Figure 18.7 Map showing potential ground movement and recommended land use for a small area near Congress Springs, Santa Clara County, California (from Cotton and Associates, 1977, Plate 3).

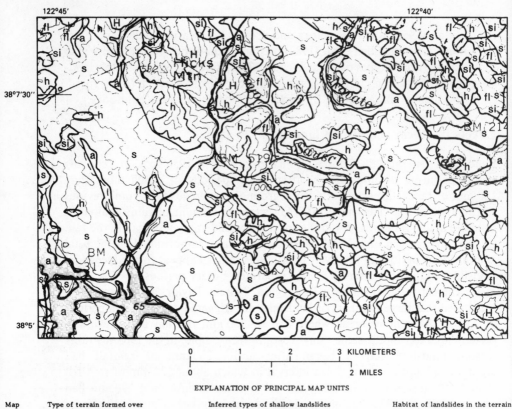

EXPLANATION OF PRINCIPAL MAP UNITS

Map unit	Type of terrain formed over Franciscan bedrock	Inferred types of shallow landslides and hazards they pose	Habitat of landslides in the terrain
H	Very hard terrain: regularly spaced straight ribs between sharply incised flutes; sharp crests, steep slopes.	Debris-avalanche and debris-flow failures in granular material that are characterized by sudden, rapid movement during heavy rainfall. Hazard is impact by rapidly moving debris as thick as several meters.	Debris and avalanche is likely at heads of flutes and along lower slopes of ribs; possibly on upper slopes of ribs. Debris flow is likely from heads of flutes down drainages and out on to slopes below mouths of drainages; possible from lower slopes of ribs.
h	Hard terrain: ribs between sharply incised flutes are somewhat irregular in form or spacing; crests may be rounded, slopes steep.	Chiefly debris avalanche and debris flow, as characterized above for unit H. Local earth flow and slump-earth flow, as characterized below for unit s.	Debris avalanche and debris flow are likely in habitats described above for unit H. Earth flow and slump-earth flow are possible in places, particularly on aprons at the foot of fluted hillslopes.
s	Soft terrain: lacks flutes, although includes irregular and poorly incised drainages; crests broadly rounded, gentle slopes.	Chiefly earth flow, earth-flow complex, and slump-earth flow, failures in clayey material that are characterized by slow movement lasting days to months during rainy season. Hazard is distortion of structures by slow movement of underlying or adjacent material as thick as several meters. Some debris avalanche and debris flow, as characterized above for unit H.	Earth flow, earth-flow complex, and slump-earth flow are likely in concave portions of terrain, possible throughout terrain. Debris avalanche and debris flow are possible in steep portions of terrain.

Other terrain units: a, alluvium; si, soft-intermediate; fl, fluted-intermediate.

Figure 18.8 Map showing terrain units for part of Marin County, California, including descriptions of styles of landsliding, habitat of landslides, and hazards posed by map units H, h, and s (from Ellen et al., 1981). Descriptions of other map units are not shown.

Brabb et al. (1972) was satisfactory and useful. For Marin County and parts of Sonoma County, Ellen et al. (1982) prepared a landslide-susceptibility map showing the various types of hazards caused by different types of shallow landslides (Fig. 18.8). They differentiated a number of different "terrain types," each of which is characterized by particular types of shallow landslides.

Effects of storm activity on landslides

Landslides generated in Contra Costa County from 1950 to 1971 generally occurred during or immediately after storm periods in which more than 180 mm of rain fell, particularly if the ground was already wet from previous storms (Nilsen and Turner, 1975). However, the amount of rain required to generate abundant landslides in the spring is considerably less than in the fall. In fact, the pattern of rainfall appears to be more important than the total amount; long periods of relatively continuous rainfall appear to produce more landslides than short discrete storms separated by dry periods.

For the entire San Francisco Bay region, Nilsen, Taylor, and Dean (1976) concluded that large numbers of landslides are triggered during storm periods of 150 to 200 mm in areas where 250 to 380 mm of rain had previously accumulated during the same rainy season. For example, in the 1968–69 rainy season, landslide activity began after about 280 mm of rainfall had accumulated and during a storm interval in which about 230 mm of rain fell (Fig. 18.9); however, landsliding did not become intense until 790 mm had accumulated. In 1972–73, intense landslide activity began after 330 mm of rain had accumulated and during a storm interval in which 180 mm fell. These comparative data permit the establishment in the San Francisco Bay region of a landslide warning system based on analyses of previous cumulative rainfall and predicted rainfall during storm periods.

In 1982 a catastrophic storm that resulted locally in the precipitation of more than 600 mm of rain over a 24-hour interval triggered a very large number of landslides as well as floods throughout San Francisco Bay region (Brown, 1984). Twelve of these landslides caused 24 fatalities. Large amounts of damage were inflicted on houses, roads, and other man-made structures. Many of the landslides were debris avalanches that moved suddenly and rapidly (Smith et al., 1982). However, the largest landslide, which caused 10 deaths in Santa Cruz County, was a gigantic block glide that had associated with it smaller debris avalanches (Cotton and Cochrane, 1982); it was a dip-slope failure on a slope underlain by Miocene sandstone and mudstone. Although many of these landslides were activated in areas previously shown by Nilsen et al. (1979) to be relatively unstable, others, particularly shallow, rapidly moving debris avalanches and flows, formed in areas of 25–45° slope that had been shown to be relatively stable. The 1982 storm thus signified the need within the San Francisco Bay region for newer, better, and more sophisticated maps of relative slope stability that account for landslides generated during infrequent, major storms.

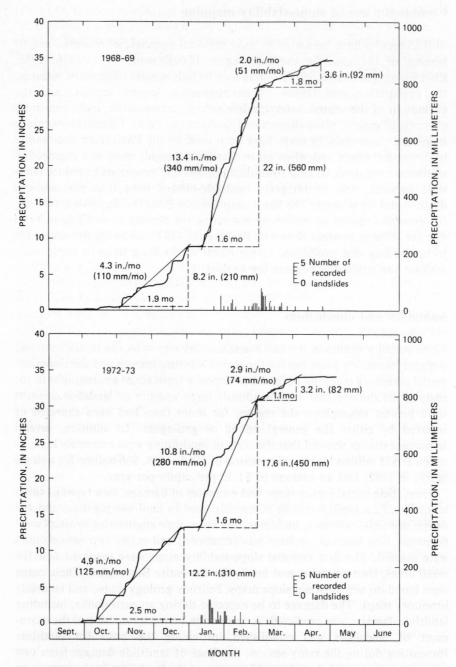

Figure 18.9 Seasonal rainfall accumulation for 1968–69 and 1972–73 at the Burton Ranch Precipitation Recording Station in Contra Costa County and concurrent landslide activity (from Nilsen et al., 1976*b*, Fig. 8).

Community use of slope-stability mapping

SFBRS reports have been used by most regional agencies, all counties, and 90 percent of the cities in the Bay region (Kockelman, 1975, 1976, 1979; Kockelman and Brabb, 1979). These uses include studies of geologic hazards, the preparation and review of environmental impact reports, and the evaluation of the seismic safety, public safety, conservation, and open-space elements of general plans (Brown and Kockelman, 1983). Landslide-inventory maps and slope-stability maps have been used by the Bay region community to determine where and when more detailed geologic, soils, and engineering studies are required, to help formulate policies and programs to reduce landslide hazards, and to integrate landslide-hazard data into the land-use decisionmaking process. The San Mateo County Board of Supervisors (1973), for example, passed an ordinance restricting the density of dwelling units to one per 40 acres in areas shown by Brabb et al. (1972) to be highly susceptible to landsliding and prohibited construction of dwelling units in those areas without site studies by engineering geologists.

Summary and conclusions

Slope-stability studies in the San Francisco Bay region by the USGS have had a major impact on planning for the region's future growth and development. Initial inventory mapping of landslide deposits from aerial photographs in the early 1970s showed that a tremendously large number of landslide deposits were present throughout the region, far more than had been expected or inferred by either the general public or geologists. In addition, several important studies showed that the costs of landsliding were extremely high, as much as $25 million for the rainy season of 1968–1969, $66 million for a single storm in 1982, and an average of $1.30 per capita per year.

From these initial survey maps and estimates of damage, new types of slope maps based on slope intervals commonly used in land-use planning and the design of roads, railroads, buildings, and large-scale engineering projects were prepared. The bedrock geology was remapped, and many new active faults were located. The first regional slope-stability maps were prepared first for small areas, then counties, and finally for the entire Bay region. These maps were based on new regional slope maps, bedrock geology maps, and landslide inventory maps. The damage to be expected during an earthquake, including landslide damage, was predicted. A correlation between rainfall and the movement of landslides was established, providing the basis for landslide forecasting during the rainy season. Analyses of landslide damage from two rainy seasons showed that about 63 percent of the landslides took place on or near underlying and pre-existing landslide deposits, that about 78 percent of the landslides took place on slopes steeper than 15 percent, and that more than 60 percent of the landslides took place in certain geologic formations or the soils overlying them that had previously been mapped as being especially prone to slope failure (Nilsen, Taylor, and Dean, 1976). Finally, the slope-stability maps and related land-use hazard data were used by the Bay region community to varying degrees for land-use planning. Techniques for reducing landslide hazards have been outlined by Kockelman (1985); these techniques are useful to land-use planners for the implementation of hazard-reduction programs and to engineers who serve as advisors to local or state governments.

The narrow time framework of the SFBRS and its regional scale forced geologists, engineers, and land-use planners to take many short cuts in order to provide the community with a usable product. Much detail had to be left out, and many interesting, useful, and sometimes necessary studies were not completed or even started because of time, personnel, and financial constraints. In the case of the slope-stability maps, no other suitable regional-scale maps were previously available to the community. Thus, completion of the USGS maps provided the Bay region with the first suitable regional maps of slope stability that would be incorporated into the land-use planning process. At the same time, SFBRS permitted geoscientists to make significant progress in understanding the processes, scale, timing, and nature of landsliding in the region.

It is now about 15 years since SFBRS began, and many of its important conclusions have been supplanted by subsequent, more detailed studies by the USGS and others. In the field of slope stability, geologists have examined smaller areas in great detail, permitting the preparation of slope-stability maps that consider the types of landslide deposits, their rates of movement, their depths, and the detailed makeup of their rock, soil, and colluvial materials. Landslides have been dated using radiometric techniques, leading to a better understanding of the history of landsliding in the region. These studies have yielded greater understanding of the processes and dangers of landslides and the development of slope-stability maps of greater use to both the geological and general community. The objectives and framework for future studies of slope stability in other parts of the United States (U.S. Geological Survey, 1982), as well as for general studies of geologic and hydrologic hazards (Hays, 1981), have been greatly advanced by SFBRS.

Acknowledgments

I thank Earl E. Brabb, Robert D. Brown, Jr, William J. Kockelman, and an anonymous reviewer for very helpful reviews of this manuscript, and Athol Abrahams, volume editor, for his encouragement and help in various ways.

References

Alfors, J. T., J. L. Burnett, and T. E. Gay, Jr, Urban geology master plan for California, *California Division of Mines and Geology Bulletin 198,* 1973.

Atwater, B. F., C. F. Hedel, and E. J. Helley, Late Quaternary depositional history, Holocene sea-level changes, and vertical crustal movement, southern San Francisco Bay, California, *U.S. Geological Survey Professional Paper 1014,* 1977.

Bailey, E. H., and D. R. Harden, Map showing mineral resources of the San Francisco Bay region, California—present availability and planning for the future, scale 1 : 250 000, *Miscellaneous Investigations Series Map I-909,* U.S. Geological Survey, Washington, D.C., 1975.

Bonilla, M. G., Landslides in the San Francisco South quadrangle, California, *Open-File Report,* 44 pp., U.S Geological Survey, Washington, D.C., 1960*a.*

Bonilla, M. G., A sample of California Coast Range landslides, *U.S. Geological Survey Professional Paper 400-B,* p. B149, 1960*b.*

Bonilla, M. G., Preliminary geologic map of the San Francisco South and part of the Hunters Point quadrangles, scale 1 : 24 000, *Miscellaneous Field Studies Map MF-311,* U.S. Geological Survey, Washington, D. C., 1971.

Borcherdt, R. D. (Ed.), Studies for seismic zonation of the San Francisco Bay region, *U.S. Geological Survey Professional Paper 941-A,* 1975.

Brabb, E. E. (Ed.), Progress on seismic zonation in the San Francisco Bay region, *U.S. Geological Survey Circular 807,* 1979.

Brabb, E. E., Minimum landslide damage in the United States, 1973–1983, *Open-File Report. 84-486,* 3 pp., U.S Geological Survey, Washington, D.C., 1984.

Brabb, E. E., E. H. Pampeyan, and M. G. Bonilla, Landslide susceptibility in San Mateo County, California, scale 1 : 62 500, *Miscellaneous Field Studies Map MF-360,* U.S. Geological Survey, Washington, D.C., 1972.

Brown, R. D., Jr, Faults that are historically active or that show evidence of geologically young surface displacements, San Francisco Bay region; a progress report, October 1970, scale 1 : 250 000, *Miscellaneous Field Studies MF-331,* U.S. Geological Survey, Washington, D.C., 1970.

Brown, R. D., Jr, and W. J. Kockelman, Geologic principles for prudent land use—a decision-maker's guide for the San Francisco Bay region, *U.S. Geological Survey Professional Paper 946,* 1983.

Brown, W. M., III, Summary of the conference, in *Debris Flows, Landslides, and Floods in the San Francisco Bay Region, January 1982,* National Academy Press, Washington, D.C., pp. 1–66, 1984.

Brown, W. M., III, and L. E. Jackson, Jr, Erosional and depositional provinces and sediment transport in the south and central part of the San Francisco Bay region, California, scale 1 : 125 000, *Miscellaneous Field Studies MF-515,* 21 pp., U.S. Geological Survey, Washington, D.C., 1973.

Clague, J. J., Landslides of southern Point Reyes National Seashore, *California Division of Mines and Geology, Mineral Information Service, 22,* pp. 7, 107–110, 116–118, 1969.

Committee on Ground Failure Hazards, Commission on Engineering and Technical Systems, National Research Council, *Reducing Losses from Landsliding in the United States,* 41pp., National Academy Press, Washington D.C., 1985.

Conomos, T. J. (Ed.), *San Francisco Bay: The Urbanized Estuary,* 493 pp., Pacific Division of the American Association for the Advancement of Science, San Francisco, Calif., 1979.

Cotton, W. R., and Associates, Analysis of the geologic hazards of the Congress Springs study area, Unpublished report for Santa Clara County Planning Office, San Jose, Calif., 47 pp., 1977.

Cotton, W. R., and D. A. Cochrane, Love Creek landslide disaster, January 5, 1982, *California Geology, 35,* 153–157, 1982.

Ellen, S., D. M. Peterson, and G. O. Reid, Map showing areas susceptible to different hazards from shallow landsliding, Marin County and adjacent parts of Sonoma County, California, scale 1 : 62 500, *Miscellaneous Field Studies MF-1406,* U.S. Geological Survey, Washington, D.C., 1982.

Fleming, R. W., and F. A. Taylor, Estimating the costs of landslide damage in the United States, *U.S. Geological Survey Circular 832,* 1980.

Harding, R. C., Landslides—a continuing problem for Bay area development, in *Urban Environmental Geology in San Francisco Bay Region,* edited by E. A. Danehy, pp. 65–74, Association of Engineering Geologists Special Publication, San Francisco section, 1969.

Hays, W. W. (Ed.), Facing geologic and hydrologic hazards, earth-science considerations, *U.S. Geological Survey Professional Paper 1240-B,* 1981.

Helley, E. J., K. R. Lajoie, W. E. Spangle, and M. L. Blair, Flatland deposits of the San Francisco Bay region, California—their geology and engineering properties, and their importance to comprehensive planning, *U.S. Geological Survey Professional Paper 943,* 1979.

Hines, W. G., Evaluating pollution potential of land-based waste disposal, Santa Clara County, California: an application of earth-science data for planning, *U.S. Geological Survey Water Resources Investigations 31-73,* 1973.

Kachadoorian, R., Engineering geology of the Warford Mesa Subdivision, Orinda, California, *Open-File Report,* 13pp., U.S. Geological Survey, Washington, D.C., 1956.

Keefer, D. K., G. F. Wieczorek, E. L. Harp, and D. H. Tuel, Preliminary assessment of seismically induced landslide susceptibility, in Progress on Seismic Zonation in the San Francisco Bay Region, edited by E. E. Brabb, pp. 49–60, *U.S. Geological Survey Circular 807,* 1979.

Kockelman, W. J., Use of USGS earth-science products by city planning agencies in the San Francisco Bay region, California, *Open-File Report 75-276,* 110pp., U.S. Geological Survey, Washington, D.C., 1975.

Kockelman, W. J., Use of USGS earth-science products by county planning agencies in the San Francisco Bay region, California, *Open-File Report 76-547,* 186pp., U.S. Geological Survey, Washington, D.C., 1976.

Kockelman, W. J., Use of USGS earth-science products by selected regional agencies in the San Francisco Bay region, California, *Open-File Report 79-221,* 173 pp., U.S. Geological Survey, Washington, D.C., 1979.

Kockelman, W. J., Some techniques for reducing landslide hazards, *Bulletin of the Association of Engineering Geologists,* in press, 1985.

Kockelman, W. J., and E. E. Brabb, Examples of seismic zonation in the San Francisco Bay region, in Progress on Seismic Zonation in the San Francisco Bay Region, edited by E. E. Brabb, pp. 73–84, *U.S. Geological Survey Circular 807,* 1979.

Kockelman, W. J., T. J. Conomos, and A. E. Leviton (Eds.), *San Francisco Bay: Use and Protection,* 310 pp., Pacific Division of the American Association for the Advancement of Science, San Francisco, Calif., 1982.

Kojan, E., Mechanics and rates of natural soil creep, *Proceedings 5th Engineering Geology and Soil Engineering Symposium,* pp. 233–253, 1968.

Laird, R. T., J. B. Perkins, D. A. Bainbridge, J. B. Baker, R. T. Boyd, D. Huntsman, P. E. Staub, and M. B. Zucker, Quantitative land-capability analysis, *U.S. Geological Survey Professional Paper 945,* 1979.

Limerinos, J. T., K. W. Lee, and P. E. Lugo, Flood-prone areas in the San Francisco Bay region, scale 1 : 125 000, *Water Resources Investigations, 3-73,* U.S. Geological Survey, Washington, D.C., 1973.

McCulloch, D. S., D. H. Peterson, P. R. Carlson, and T. J. Conomos, A preliminary study of the effects of water circulation in the San Francisco Bay estuary, *U.S. Geological Survey Circular 637,* pp. A1–26 and B1–8, 1970.

Newman, E. B., A. R. Paradis, and E. E. Brabb, Feasibility and cost of using a computer to prepare landslide susceptibility maps of the San Francisco Bay region, California, *U.S. Geological Survey Bulletin 1443*, 1978.

Nilsen, T. H., and E. E. Brabb, Preliminary photointerpretation and damage maps of landslide and other surficial deposits in northeastern San Jose, California, scales 1 : 12 000 and 1 : 24 000, *Miscellaneous Field Studies MF-361*, U.S. Geological Survey, Washington, D.C., 1972.

Nilsen, T. H., and E. E. Brabb, Landslides, in Studies for Seismic Zonation of the San Francisco Bay Region, edited by R. D. Borcherdt, pp. A75–87, *U.S. Geological Survey Professional Paper 941-A*, 1975.

Nilsen, T. H., and E. E. Brabb, Slope-stability studies in the San Francisco Bay region, California, in Landslides, edited by D. R. Coates, pp. 235–244, *Reviews in Engineering Geology, 3*, Geological Society of America, Boulder, Colo., 1977.

Nilsen, T. H., F. A. Taylor, and E. E. Brabb, Recent landslides in Alameda County, California (1940–71); an estimate of economic losses and correlations with slope, rainfall and ancient landslide deposits, *U.S. Geological Survey Bulletin 1398*, 1976.

Nilsen, T. H., F. A. Taylor, and R. M. Dean, Natural conditions that control landsliding in the San Francisco Bay region—an analysis based on data from 1968–69 and 1972–73 rainy seasons, *U.S. Geological Survey Bulletin 1424*, 1976.

Nilsen, T. H., and B. L. Turner, Influence of rainfall and ancient landslide deposits on recent landslides (1950–1971) in urban areas of Contra Costa County, California, *U.S. Geological Survey Bulletin 1388*, 1975.

Nilsen, T. H., R. H. Wright, T. C. Vlasic, and W. E. Spangle, Relative slope stability and land-use planning in the San Francisco Bay region, *U.S. Geological Survey Professional Paper 944*, 1979.

Pampeyan, E. H., 1970, Geologic map of the Palo Alto $7\frac{1}{2}$ minute quadrangle, San Mateo and Santa Clara Counties, California, scale 1 : 24 000, *Open-File Map*, U.S. Geological Survey, Washington, D.C., 1970.

Perkins, J. B., Y. C. San Jule, and P. Y. Chiu, *Environmentally dangerous waste in the San Francisco Bay; Land disposal and its alternatives, 92 pp.*, Association of Bay Area Governments, Final Project Report for California Solid Waste Management Board, 1977.

Radbruch, D. H., Areal and engineering geology of the Oakland West quadrangle, California, scale 1 : 24 000, *Miscellaneous Geologic Investigations Map I-239*, U.S. Geological Survey, Washington, D.C., 1957.

Radbruch, D. H., Areal and engineering geology of the Oakland East quadrangle, California, scale 1 : 24 000, *Geologic Quadrangle Map GQ-769*, U.S. Geological Survey, Washington, D.C., 1969.

Radbruch, D. H., and K. C. Crowther, Map showing areas of estimated relative amounts of landslides in California, scale 1 : 1000 000, *Miscellaneous Geologic Investigations Map I-747*, U.S. Geological Survey, Washington, D.C., 1973.

Radbruch, D. H., and L. M. Weiler, Preliminary report on landslides in a part of the Orinda Formation, Contra Costa County, California, *Open-File Report*, 35pp., U.S. Geological Survey, Washington, D. C. 1963.

Radbruch, D. H., and C. M. Wentworth, Estimated relative abundance of landslides in the San Francisco Bay region, California, scale 1 : 500 000, *Open-File Map*, U.S. Geological Survey, Washington, D.C., 1971.

Radbruch-Hall, D. H., Maps showing areal slope stability in part of the northern Coast Ranges, California, scale 1 : 62 500, *Miscellaneous Investigations Series Map I-982*, U.S. Geological Survey, Washington, D.C., 1976.

Radbruch-Hall, D. H., R. B. Colton, W. E. Davies, I. Lucchitta, B. A. Skipp, and D. J. Varnes, Landslide overview map of the conterminous United States, *U.S. Geological Survey Professional Paper 1183*, 1982.

Rogers, T. H., Environmental geologic analysis of the Santa Cruz Mountains study area, Santa Clara County, California, *California Division of Mines and Geology,* 64 pp., Sacramento, Calif., 1971.

Roth, R. A., Factors affecting landslide-susceptibility in San Mateo County, California, *Bulletin of the Association of Engineering Geologists 20,* 353–372, 1983.

Roth, R. A., and E. Kavazanjian, Jr, Liquefaction susceptibility mapping, San Francisco, California, *Bulletin of the Association of Engineering Geologists, 21,* 459–478, 1984.

San Mateo County Board of Supervisors, Adding a resource-management district and regulations to the county zoning ordinance, *Ordinance No. 2229,* 24pp., Redwood City, Calif., 1973.

Schlocker, J., The geology of the San Francisco Bay area and its significance in land use planning, *Association of Bay Area Governments Supplemental Report IS-3,* 47pp., Berkeley, Calif., 1968.

Schlocker, J., Generalized geologic map of the San Francisco Bay region, California, scale 1 : 500 000, *Open-File Map,* U.S. Geological Survey, Washington, D.C., 1971.

Schlocker, J., M. G. Bonilla, and D. H. Radbruch, Geology of the San Francisco North quadrangle, California, scale 1 : 24 000, *Miscellaneous Geologic Investigations Map I-272,* U.S. Geological Survey, Washington, D.C., 1958.

Schuster, R. L., Introduction, in Landslides—Analysis and Control, edited by R. L. Schuster and R. J. Krizek, pp. 1–10, *National Academy of Science, Transportation Research Board Special Report 176,* Washington, D.C., 1978.

Smith, T. C., E. W. Hart, J. E. Baldwin, and R. J. Rodrigues, Landslides and related storm damage, January 1982, San Francisco Bay region, *California Geology, 35,* 139–152, 1982.

Taylor, F. A., and E. E. Brabb, Map showing distribution and cost by counties of structurally damaging landslides in the San Francisco Bay region, California, winter of 1968–69, scale 1 : 1000 000, *Miscellaneous Field Studies MF-327,* U.S. Geological Survey, Washington, D.C., 1972.

Taylor, F. A., T. H. Nilsen, and R. M. Dean, Distribution and costs of landslides that have damaged manmade structures during the rainy season of 1972–1973 in the San Francisco Bay region, California, scale 1 : 1000 000, *Miscellaneous Field Studies MF-327,* U.S. Geological Survey, Washington, D.C., 1975.

U.S. Geological Survey, *Slope map of the San Francisco Bay region,* scale 1 : 125 000, three sheets, 1972.

U.S. Geological Survey, Goals and tasks of the landslide part of a ground-failure hazards reduction program, *U.S. Geological Survey Circular 880,* 1982.

U.S. Geological Survey and U.S. Department of Housing and Urban Development, *Program Design 1971, San Francisco Bay Region Environment and Resources Planning Study,* 123pp., Report PB2-06826, National Technical Information Service, Springfield, Va., 1971.

Waananen, A. O., J. T. Limerinos, W. J. Kockelman, W. E. Spangle, and M. L. Blair, Flood-prone areas and land-use planning—selected examples from the San Francisco Bay region, California, *U.S. Geological Survey Professional Paper 942,* 1977.

Waltz, J. P., An analysis of selected landslides in Alameda and Contra Costa Counties, California, *Bulletin of the Association of Engineering Geologists, 8,* 153–163, 1971.

Wieczorek, G. F., Map showing recently active and dormant landslides near La Honda, central Santa Cruz Mountains, California, scale 1 : 4800, *Miscellaneous Field Studies MF-1422,* U.S. Geological Survey, Washington, D.C., 1982.

Wright, R. H., R. H. Campbell, and T. H. Nilsen, Preparation and use of isopleth maps of landslide deposits, *Geology, 2,* 483–485, 1974.

Wright, R. H., and T. H. Nilsen, Isopleth map of landslide deposits, southern San Francisco Bay region, California, scale 1 : 125 000, *Miscellaneous Field Studies MF-550,* U.S. Geological Survey, Washington, D.C., 1974.

Youd, T. L., Liquefaction, flow, and associated ground failure, *U.S. Geological Survey Circular 688*, 1973.

Youd, T. L., and S. N. Hoose, Historic ground failures in northern California triggered by earthquakes, *U.S. Geological Survey Professional Paper 993*, 1978.

Youd, T. L., D. R. Nichols, E. J. Helley, and K. R. Lajoie, Liquefaction potential, in Studies for Seismic Zonation of the San Francisco Bay Region, edited by R. D. Borcherdt, pp. A68–74, *U.S. Geological Survey Professional Paper 941-A, 1975*.

Index

References to figures are printed in italics.